AF552948

Climate Resilient Agriculture
Adaptation and Mitigation Strategies

About the Editors

Dr. Manish Bhan did his PhD programme from University of Florida Gainesville, USA with specialization in pest suppression in organic cropping systems. He joined J.N. Krishi Vishwa Vidyalaya, Jabalpur in 2011 after completing post- doctoral research at the University of Florida. At present, he is working as Assistant Professor with work on climate change and agriculture, crop growth modeling, weather based agro- advisories services and other agrometeorological aspects of major crops and associated weeds under All India Coordinated Research Project on Agrometeorology at Jabalpur. He has guided many students, published research papers in referred journals and printed technical bulletins related to climate assessment of the state. He is also a Principal Investigator of research projects funded by ICAR and IMD related to agrometeorology, climate resilient agriculture, crop-weather modeling, and enhancing area under direct seeded rice production in Madhya Pradesh.

Dr. Vijay Singh Tomar received his PhD degree in Soil Science with specialization in Soil Physics from G.B. Pant University of Agriculture and Technology, Pantnagar with an experience of nearly forty five years, served largely at GBPUAT and J.N. Krishi Vishwa Vidyalaya, Jabalpur as Head, Department of Soil Science and Agriculture Chemistry; Dean, RAK College of Agriculture, Sehore; Director Extension Services, Director Research Services, and Dean Faculty of Agriculture at JNKVV, Jabalpur. He served as a founder Vice Chancellor of R. V. S. Krishi Vishwa Vidyalaya, Gwalior from 2008-12, and presently serving as Vice Chancellor, JNKVV, Jabalpur since 2012. He is the recipient of Jawaharlal Nehru Award-1974 (ICAR), Best KVK Award 2000-01 (ICAR), Best Agricultural University Vice Chancellor Award - 2015(AIASA); Lifetime Achievement Award-2015 as most impactful mentor (Pantnagar Clan-GBPUA&T), and others. He has the honor of being recognized as Fellows of "Indian Society of Soil Science", "NAAS", and "Range Management Society of India". He has guided several M.Sc and PhD students, published more than 110 research papers in National and International peer reviewed journals, 100 technical popular articles, 9 books/ manuals/ bulletins, chapters in books on various aspects of Soil Science, Natural Resource Management, Integrated Nutrient Management, and Agri-policy analysis. He established Centre for Advanced Studies in Soil Science (ICAR) at JNKVV, Jabalpur, and developed agroclimatic zone plan for increasing agricultural production and doubling the income of the farmers.

Dr. Kamal Kishore Agrawal received his Ph.D degree from Rani Durgawati Vishwavidyalaya, Jabalpur. He joined J.N. Krishi Vishwa Vidyalaya, Jabalpur in 1986 with work in Agronomy and Agrometeorological aspects in agriculture. Currently, he is working as a Professor in Agronomy with work on organic farming, and crop-weather relationship of major crops of Madhya Pradesh. In addition, he also involves in teaching and extension activities. He has guided more than 15 MSc and 06 PhD students in Agronomy. During his carrier, he has undergone many training programmes related to agriculture and agrometeorology. He has published more than 100 research papers in reviewed journals, review articles, book chapters, technical bulletins, teaching manuals and popular articles. He is recognized by several universities as an external examiner. Dr Agrawal visited Israel for advance training in agrometeorology. He has organized- workshop/ seminar/ICAR winter school in JNKVV, Jabalpur. He has handled many research projects funded by ICAR and DST. His significant contributing in the field of agrometorology , include crop- weather relationship, development of weather data bank, climate characterization and preparing contingent crop plan of Madhya Pradesh.

Dr. Sahib Singh Tomar has been a recipient of ICAR Fellowships during his B.Sc. (Ag), M.Sc. (Ag) and Ph.D. programmes. In the four decades of his career in J. N. Krishi Vishwa Vidyalaya, Jabalpur, he served almost all the administrative positions and superannuated as Director Research. During last six years of service, he established three agriculture colleges at Ganjbasoda, Tikamgarh and Waraseoni and one Dryland Horticulture Research and Training Center at Sagar in the state of Madhya Pradesh. His significant contribution towards research, teaching and extension being on administrative positions and experience of addressing challenging administrative and financial issues effectively. He also served as member of many Institute Management Committee, Research Advisory Committee, Member of Professional Societies, and organized National and International scientific seminars and conferences.

Dr. Sunil Dutta Upadhyaya did his PhD. in Botany with specialization in Ecology, and currently is the Head of Department of Forestry, JNKVV, Jabalpur. He made a significant contribution in wasteland development through agroforestry research, carbon sequestration, tree-crop interaction, and climate change. He has more than 30 years of experience in teaching, research, and extension in the field of agricultural botany, agroforestry and medicinal plants, and supervised several MS and PhD. students in this area.

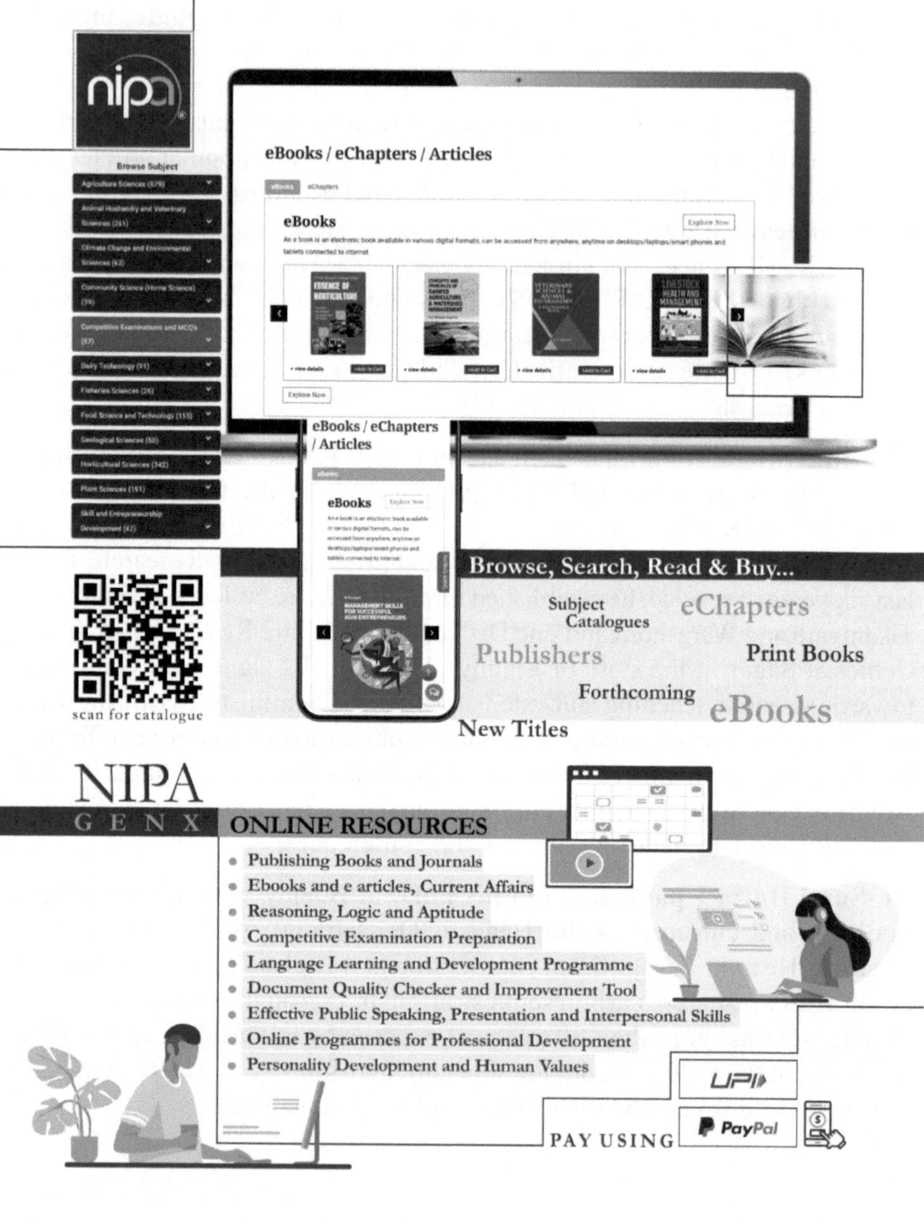

nipa
eBooks / eChapters / Articles
eBooks
eBooks / eChapters / Articles
eBooks
scan for catalogue
Browse, Search, Read & Buy...
Subject Catalogues
eChapters
Publishers
Print Books
Forthcoming
eBooks
New Titles
NIPA
GENX
ONLINE RESOURCES
Publishing Books and Journals
Ebooks and e articles, Current Affairs
Reasoning, Logic and Aptitude
Competitive Examination Preparation
Language Learning and Development Programme
Document Quality Checker and Improvement Tool
Effective Public Speaking, Presentation and Interpersonal Skills
Online Programmes for Professional Development
Personality Development and Human Values
UPI
PayPal
PAY USING

Climate Resilient Agriculture
Adaptation and Mitigation Strategies

Editors
Manish Bhan
Vijay Singh Tomar
Kamal Kishore Agrawal
Sahib Singh Tomar
Sunil Dutta Upadhyaya

NEW INDIA PUBLISHING AGENCY
New Delhi – 110 034

NEW INDIA PUBLISHING AGENCY
101, Vikas Surya Plaza, CU Block, LSC Market
Pitam Pura, New Delhi 110 034, India
Phone: + 91 (11) 27 34 17 17 Fax: + 91 (11) 27 34 16 16
Email: info@nipabooks.com
Web: www.nipabooks.com

Feedback at feedbacks@nipabooks.com

ISBN: 978-93-87973-07-7

Composed and Designed by NIPA

त्रिलोचन महापात्र, पी.एच.डी.
एफ.एन.ए., एफ.एन.ए.एस.सी., एफ.एन.ए.ए.एस
सचिव एवं महानिदेशक
TRILOCHAN MOHAPATRA, Ph.D.
FNA, FNASc, FNAAS
SECRETARY & DIRECTOR-GENERAL

भारत सरकार
कृषि अनुसंधान और शिक्षा विभाग एवं
भारतीय कृषि अनुसंधान परिषद
कृषि एवं किसान कल्याण मंत्रालय, कृषि भवन, नई दिल्ली, 110 001

GOVERNMENT OF INDIA
DEPARTMENT OF AGRICULTURAL RESEARCH & EDUCATION
AND
INDIAN COUNCIL OF AGRICULTURAL RESEARCH
MINISTRY OF AGRICULTURE AND FARMERS WELFARE
KRISHI BHAVAN, NEW DELHI 110 001
TEL. : 23382629, FAX : 91-11-23384773
E-mail : dg.icar@nic.in

Foreword

Climate is a vital factor that influences land use, crop quality, its productivity as well as all the other of agricultural systems. The significant impact of climate change is visible on human societies and natural ecosystems around the world. This impact will be more severe on agriculture if global warming continues. As per estimates of IPCC (2014), the agriculture, forestry and other land use contributes 24% of global greenhouse gas emissions. These emissions need to be reduced to avoid the serious impact of climate change using mitigation measures and adaptation strategies.

Currently, unreliable and seasonal variations in weather have emerged as a serious challenge for sustainability influencing vegetation, biodiversity, livestock, soil, water, and other natural resources. In the last decade, more occurrence of extreme weather events affected farming community directly in their agricultural growth. The matter is of great concern to country like India, which require more produce from rainfed fields and shrinking crop land.

To understand the problems occurring due to climate change, concerted efforts are required for mitigation and adaptation to reduce the vulnerability of rainfed agriculture and making it resilient. Agricultural output as well as the livelihood of people who depend on it, are particularly vulnerable to climate change, and it is important that we assess adaptation mechanisms to reduce these vulnerabilities. These practices should play a vital role to reduce GHG emissions by improving efficiency of farm inputs and others like agroforestry interventions for green agricultural technologies. Similarly, adoption of conservation agriculture, suitable cultivars, changing sowing dates, irrigation scheduling, and recycling waste water and solid waste in agriculture are some of the options for developing climate resilient agriculture.

The book will serve as a platform for basic, fundamental, and advance learning for researchers and students from SAUs and ICAR institutes.

Dated: 28th August 2017 New Delhi

(T. Mohapatra)

Preface

Climate change and globalization are the two main processes of global change, and it is assumed that both have major impact on India's agriculture. The current agriculture is still dependent on climate despite the impressive advances in agricultural technology over the last half a century. Any adverse change in climate can cost food production thereby affecting farmers' income. In the wake of climate change, many organizations are creating awareness and provide funding for research programmes to check its impact on crop production, human life and their survival; more in the last two decades. The result outcome showed that climate variability has the potential to influence crop productivity, insect populations, irrigation use, etc in agriculture. In Madhya Pradesh, agriculture will face major challenges pertaining to decline in soil fertility and ground water levels. The climate related factors like droughts, excess rainfall, frost and hailstorm will cause significant year to year variation in production and productivity. Large areas in the state, particularly in western part are prone to droughts due to uncertain rainfall, while rice based cropping system in eastern part records lower productivity due to poor water management despite receiving relatively high rainfall. Additionally, black soil in the state suffer from poor drainage and heavy rainfall in short period results in water logging and yield reduction in soybean and pulses. The impact of climate change are complex and no single strategy will address the issue adequately.

A lot of research work is undergoing at National and International levels to overcome this problem. In India, substantial data has been generated to mitigate the effect of climate change along with identification of climate vulnerable districts. In addition, technologies for sustainable production through climate resilient agriculture have been identified and tested that need to be rationalized. This would require assessment and impact of climate change along with vulnerabilities, mitigation measures, capacity building, changes in policies, etc.

JNKVV (Jabalpur) **Editors**

Contents

Greenhouse Gas Emissions and Indian Agriculture

New Technologies in Relation to Climate Change

List of Contributors

A. Kashyapi, Director, Mausam Kendra, India Meteorological Department, Bhopal, Madhya Pradesh

A.R. Sharma, Director, ICAR-Directorate of Weed Science Research, Jabalpur, Madhya Pradesh

Anil Kumar Singh, Ex-Vice Chancellor, Rajmata Vijyaraje Scindia Krishi Vishwavidayalay, Gwalior Madhya Pradesh

B. Venkateswarlu, Hon'ble Vice Chancellor, Vasantrao Naik Marathwada Krishi Vidyapeeth Parbhani, MAH.

Himanshu Pathak, Principal Scientist and Head, Environmental Sciences Division ICAR-Indian Agricultural Research Institute, New Delhi

K. A. Gopinath, Principal Scientist, Central Research Institute for Dryland Agriculture Hyderabad

K.K. Agrawal, Principal Scientist (Agronomy), Jawaharlal Nehru Krishi Vishwa Vidyalaya Jabalpur, Madhya Pradesh

K.K. Singh, Head, Agromet Division, India Meteorological Department, New Delhi

M.P. Jain, Princiapl Scientist, Dryland Agriculture, College of Agriculture, Rajmata Vijayaraje Scindia Krishi Vishwavidyalaya, Indore, Madhya Pradesh

Manish Bhan, Assistant Professor-Agrometeorology, Jawaharlal Nehru Krishi Vishwa Vidyalaya, Jabalpur, Madhya Pradesh

Navin Sharma, Global Programme Director-Biofuel, World Agroforestry Centre, Nairobi, Kenya

P. Vijaya Kumar, All India Coordinated Research Project on Agrometeorology, Central Research Institute for Dryland Agriculture, Hyderabad

Pradeep Dey, Project Coordinator (STCR), ICAR-Indian Institute of Soil Science, Bhopal Madhya Pradesh

Pratap Bhattacharyya, Senior Scientist-Crop Production Division, ICAR- National Rice Research Institute, Cuttack, Odisha

Raj K. Gupta, Head, Borlaugh Institute for South Asia, Jabalpur, Madhya Pradesh

S.K. Bal, Principal Scientist- Agrometeorology, ICAR-National Institute for Abiotic Stress Management, Pune, Maharashtra

S.S. Tomar, Director Research, Jawaharlal Nehru KrishiVishwaVidyalaya, Jabalpur, Madhya Pradesh

Ram Ratan Simaiya, State Knowledge Management Centre on Climate Change, EPCO, Bhopal Madhya Pradesh

Surabhi Purohit, India Meteorological Department, Arera Hills, Jail Road, Bhopal, Madhya Pradesh

V.U.M. Rao, Project Coordinator, All India Coordinated Research Project on Agrometeorology Central Research Institute for Dryland Agriculture, Hyderabad

CLIMATE CHANGE AND INDIAN AGRICULTURE

1

Impact of Climate Change on Indian Agriculture

B. Venkateswarlu

Evidence over the past few decades has conclusively established that significant changes in climate are taking place worldwide as a result of enhanced anthropogenic activities. The fast pace of development and industrialization and indiscriminate destruction of natural environment, more so in the last century, have altered the concentration of atmospheric gases that lead to global warming. The major cause for climate change and global warming has been ascribed to the increased levels of greenhouse gases (GHGs) like carbon dioxide (CO_2), methane (CH_4), nitrous oxide (N_2O), chlorofluorocarbons (CFCs) etc. beyond their natural levels due to the uncontrolled activities such as burning of fossil fuels, increased use of refrigerants, and enhanced agricultural related practices.

The IPCC (2007) report has projected that by 2100, earth's mean temperature will rise by 1.4 to 5.8 °C, precipitation will decrease in the sub-tropical areas, and frequency of extreme events will increase significantly. As of now, in reality, in the past 100 years, the global mean temperature has increased by 0.74 °C. IPCC reported that eleven of the last twelve years (between 1995 to 2009) rank among the twelve warmest years in the instrumental record of global surface temperature (since 1850). The temperature increase is widespread over the globe and is greater at higher northern latitudes. Land regions have warmed faster than the oceans which has already started affecting the climatic phenomenon in different parts of world. Melting of glaciers, rising sea level are some of the most important manifestations of it. IPCC (2007) have further reported change in their frequency and /or intensity of extreme events over the last 50 years. The report has made the following significant observations:

- It is very likely that cold days, cold nights and frosts have become less frequent over most land areas, while hot days and hot nights have become more frequent

- It is likely that heat waves have become more frequent over most land areas.
- It is likely that the frequency of heavy precipitation events has increased over most areas.
- It is likely that the incidence of extreme high sea level has increased at a broad range of sites worldwide since 1975.

Since climate is central to many critical agricultural decisions ranging from farm to policy level (Venkateswarlu and Shanker, 2009), it is also imperative to realize that most of these decisions have to be made well in advance to counter the impacts of climate change at the earliest. Therefore, the Indian Government has taken a stance to address climate issues by fully participating in international conventions such as the United Nation Framework Convention on Climate Change (UNFCCC) and, secondly, developing National Climate Change Response Policies. At the domestic level, it is essential to develop programmes and policies that increase farm productivity and incomes; make agriculture climate resilient and promote stability and security for the farmers; ensure that the agriculture sector becomes a part of the solution to the climate change problems rather than the cause and there is a need to make climate change adaptation and mitigation measures as an integral part of overall planning and development strategy of the country on long term.

A National Action Plan on Climate Change (NAPCC) was unveiled by the Hon'ble Prime Minister of India in June 2008. It outlined a national strategy which would enable the country to adapt to climate change and ensure that development and ecological sustainability go hand in hand. This National Action Plan consists of eight missions *viz.* National Solar Mission; National Energy Efficiency Mission; National Sustainable Habitat Mission; National Water Mission; National Sustaining Himalayas Mission;A Green India Mission; National Sustainable Agriculture Mission; and National Strategic Knowledge Mission. At least three of them have a bearing on Indian Agriculture *viz.*, National Water Mission, National Mission on Green India, National Mission for Sustaining the Himalayan Ecosystem, in addition to the National Mission for Sustainable Agriculture.

Building state of the art infrastructure for research and training of scientists in frontier areas and tools, increasing climate change literacy to different levels of stakeholders, mainly farmers; enhancement of national capacity on decision support systems developing best weather insurance products for vulnerable areas and farmers and carbon trading in agriculture; and international collaboration are some other key areas through which challenges of climate change and global warming can be tackled. The Indian Council of Agricultural

Research (ICAR) has recently launched a mega project entitled "National Initiative on Climate Resilient Agriculture" which includes all these components of strategic research, strengthening R&D infrastructure, capacity building and technology demonstration on farmer's fields.

Since the beginning of 21st Century, India has experienced droughts in quick succession, of which the 2009 one was the most recent one that significantly affected *kharif* crop production. It was the 2nd largest all India monsoon rainfall deficit since 1972 (23% below normal). Incidentally, 2009 also achieved the distinction of being the warmest year in past several centuries across the world. However, after 2009, 2010 proved to the warmest year on record since 1850. 2011 is now the 11th warmest year on record since 1850. Apart from that, 1998 was one of the warmest years; 2003 experienced unprecedented heat and cold waves across the globe; occurrence of high temperature in March 2004 adversely affected crops like wheat, apple, potato etc. across northern India; 2005 witnessed destructive hurricanes/cyclones across the globe and again, 2007 was as warm as 1998 in the entire northern hemisphere and unusual summer rains and floods were experienced in many parts of India. Besides that, the amount and distribution of rainfall is becoming more and more erratic which is causing greater incidences of droughts and floods globally. The increase in frequency of heavy rainfall events in last 50 years over Central India points towards a significant change in climate pattern in India (Goswami, 2006).The data clears indicates increase in heavy (>10 cm) and very heavy (>15cm) rainfall events and decrease in light to moderate rainfall events. The projected change in seasonal temperature and rainfall for the 2030 scenario has been shown in fig.

A change in ocean acidity is likely to reduce the ocean's capacity to absorb CO_2 from the atmosphere, thus compounding the effects of climate change, and that will affect the entire marine food chain. Large-scale, irreversible system disruption and the destabilization of the Antarctic ice sheets are serious risks: changes to polar ice, glaciers and rainfall regimes have already occurred.

The rainfall analysis of India (Rao *et al.*, 2008) showed that significant negative trends of rainfall were observed in the eastern parts of Madhya Pradesh, Chhattisgarh and parts of Jharkhand, U.P., and northeast India. Lal (2001) reported that annual mean area-averaged surface warming over thc Indian sub-continent is to likely to range between 3.5 and 5.5 °C by 2080s. These projections showed more warming in winter season over summer. The spatial distribution of surface warming suggests a mean annual rise in surface temperatures in North India by 3 °C or more by 2050. The study also indicated that during winter, the surface mean air temperature could rise by 3 °C in Northern and Central parts, while it would rise by 2 ^{0}C in Southern parts by the same period.

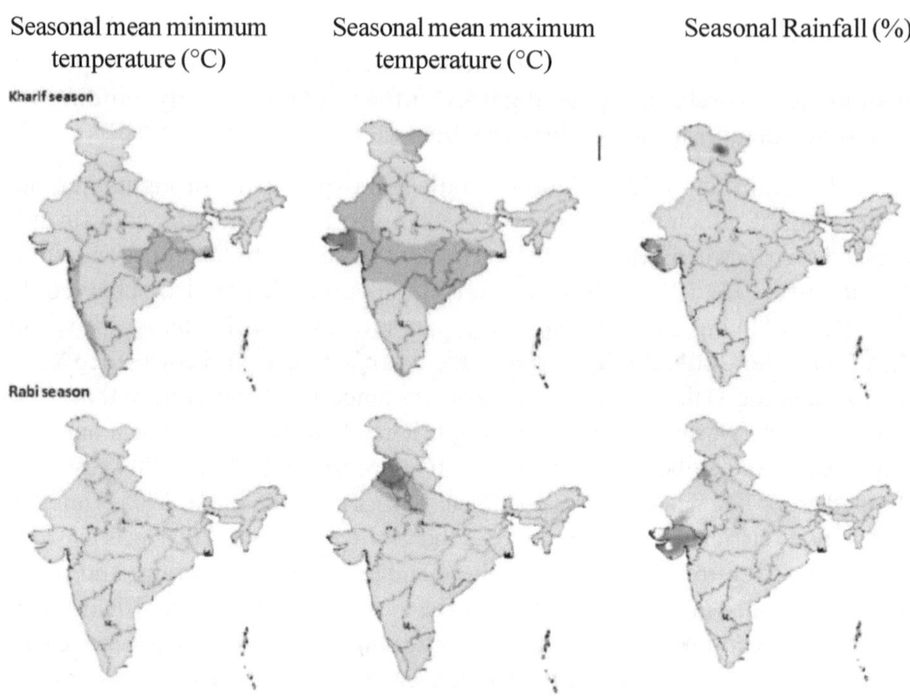

Fig. 1: Projected change in seasonal maximum and minimum temperatures and rainfall in India in A1B 2030 scenario

In case of rainfall, a marginal increase of 7-10% in annual rainfall is projected over the sub-continent by 2080. Nevertheless, the study suggests a fall in rainfall by 5-25% in winter, while it would increase by 10-15% in summer, Marked variability is seen even in the onset and withdrawal of monsoon over the period. The glaciers and the snowfields in the Himalayas are on the decline as a result of climate variability. The rate of retreat of the snow of Gangotri glacier demonstrated a sharp rise in the first half of the 20th century. Glacial melt would lead to increased summer river flow and floods over the next few decades, followed by a serious reduction in flows thereafter.

Role of greenhouse gases

The increasing levels of greenhouse gases (GHG's) in the atmosphere have been attributed as one of the major driving force behind the rapid climate change phenomenon. The main GHGs contributing to this phenomenon are CO_2, CH_4 and N_2O. Apart from fossil fuel burning, the frequent volcanic eruptions are also contributing to this increase in concentration, in the atmosphere. Though the increase in the level of CO_2 is expected to produce some beneficial effects on crop dry matter production, it may soon be nullified by associated water and thermal stresses leading to overall deterioration of agro-climatic conditions for food production systems. At the global scale, the historical temperature-yield relationships indicate that warming from 1981 to 2002 is very likely to offset some of the yield gains from technological advance, rising CO_2 and other non-climatic factors (Lobell and Field, 2007). The recent release of greenhouse gas inventory by Indian government revealed that the Net Greenhouse Gas (GHG) emissions from India in 2007, including LULUCF, were 1727.71 million tons of CO_2 equivalent (eq) of which CO_2 emissions were 1221.76 million tons, CH4 emissions were 20.56 million tons; and N_2O emissions were 0.24 million tons. GHG emissions from Energy, Industry, Agriculture and Waste sectors constituted 58%, 22%, 17% and 3% of the net CO_2 eq emissions, respectively. The energy sector emitted 1100.06 million tons of CO_2 eq, of which 719.31 million tons of CO_2 eq were emitted from electricity generation and 142.04 million tons of CO_2 eq from the transport sector. The industrial sector emitted 412.55 million tons of CO_2 eq. The concentration of CO_2 in the atmosphere has been increasing continuously and had reached a concentration of 392 ppm in 2011. If the CO_2 concentration continues to increase at this rate, the "point of no return" may be reached within a very short time span.

Role of carbondioxide

A growing body of research suggests that atmospheric carbon dioxide levels may affect water availability through its influence on vegetation. Controlled experiments indicate that elevated CO_2 concentrations increase the resistance of plant "pores," that is stomata, to water vapor transport. Experiments suggest that a doubling of CO_2 would increase stomatal resistance and reduce the rate of transpiration — the passage of water vapor from plants by about 50 percent on average. The resulting decrease in transpiration would tend to increase runoff. On the other hand, CO_2 also has been demonstrated to increase plant growth, leading to a larger area of transpiring tissue and a corresponding increase in transpiration. Other factors that might offset increases in plant water-use efficiency associated with a CO_2 -enriched atmosphere are a potential increase in leaf temperatures caused by reduced transpiration rates and species changes in vegetation communities. The net effect of opposing influences on water supplies would depend on the type of vegetation and other interacting factors, such as soil type and climate.

The management and labour costs of farm production will rise to a great level owing to increased incidences of pests and diseases as well as weeds. Additionally, extreme events like hails and frosts will also negatively impact crop production. As of now, most of the developing countries, including India, are not fully prepared to deal with the adverse impacts expected as a consequence of climate change and are therefore relatively vulnerable.

Impact on water resources

A warmer climate will modify the hydrologic cycle, altering rainfall, magnitude and timing of run-off. Warm air holds more moisture and increases evaporation of surface moisture. With more moisture in the atmosphere, rainfall and snowfall events tend to be more intense, increasing the potential for floods. However, if there is little or no moisture in the soil to evaporate, the incident solar radiation will lead to increase in the temperature, contributing to longer and more severe droughts (Trenberth, 1999). Therefore, change in climate will affect the soil moisture, groundwater recharge and frequency of flood or drought episodes and finally groundwater level in different areas.

At present, available statistics on water demand show that the agriculture sector is the largest consumer of water in India using more than 80% of the available water. The quantity of water used for agriculture has increased progressively through the years as more and more areas were brought under irrigation. Since 1950-51, the net irrigated area in India rose from 28.85 to 62.29 million ha up to 2007-08. Contribution of surface water and groundwater resources for irrigation has played a significant role in India attaining self-sufficiency in food production during the past three decades, but water is likely to become more scarce and critical input in future. By judicious utilization, the demand for water from farm sector can be pegged at 68% by the year 2050, but agriculture will still remain the largest consumer. In order to meet this demand, augmentation of the existing water resources by development of additional sources of water or conservation of the existing resources through impounding more water in the existing water bodies and their conjunctive use will be needed (Mall *et al.*, 2006). Table 1 (adopted from Mall *et al.*, 2006) summarizes the possible impacts of climate change on water resources during the next century over India. The enhanced surface warming over the Indian subcontinent by the end of the next century would result in an increase in pre-monsoonal and monsoonal rainfall and no substantial change in winter rainfall over the central plains. This would result in an increase in the monsoonal and annual run-off in the central plains, with no substantial change in winter run-off and increase in evaporation and soil wetness during the monsoon on an annual basis.

Table 1: Impact of climate change on water resources during the next century over India (adopted from Mall *et al.*, 2006)

Region/location	Impact
Indian subcontinent	• Increase in monsoonal and annual run-off in the central plains • No substantial change in winter run-off • Increase in evaporation and soil wetness during monsoon and on an annual basis
Orissa and West Bengal	One metre sea-level rise would inundate 1700 km^2 of prime agricultural land
Indian coastline	One metre sea-level rise on the Indian coastline is likely to affect a total area of 5763 km^2 and put 7.1 million people at risk
All-India	Increases in potential evaporation across India
Central India	Basin located in a comparatively drier region is more sensitive to climatic changes
Kosi Basin	Decrease in discharge on the Kosi River Decrease in run-off by 2–8%
Southern and Central India	Soil moisture increases marginally by 15–20% during monsoon months
Chenab River	Increase in discharge in the Chenab River
River basins of India	General reduction in the quantity of the available run-off, increase in Mahanadi and Brahmini basins
Damodar Basin	Decreased river flow
Rajasthan	Increase in evapotranspiration

In another study made by IITM, Pune in collaboration with DEFRA, UK reported that there will be an increase in annual rainfall and annual flow in Godavari and Ganga basins by 2071-2100 period.

Impact on agriculture, livestock and fisheries agriculture Sector

The impact of climate change on agriculture will be one of the major deciding factors influencing future food security of the world including India. A major part of the agriculture in India is rainfed (~80 million ha out of 142 million ha net cultivated area) and will remain so for at least for a foreseeable future. Rainfed agriculture in India contributes to 44% of total foodgrain production. Around 66% of livestock population is also dependent on rainfed areas. The crop losses due to climate variability will vary from region to region depending on regional climate, crop and cropping systems, soils and management practices. Rainfed crops are likely to be worst hit by climate change because of the limited options for coping with variability of rainfall and temperature. The major crops like wheat and rice are expected to undergo all the weather aberrations and their sustainability will be determined by crop's capacity of natural adaptability as well as appropriate mitigation measures adopted.

The impact of climate change on agriculture may accentuate at regional level creating more vulnerability in food security rather than the global level as a whole. The potential impact will be felt in shifts of sowing time and length of growing seasons, which may necessitate effective adjustment in sowing and harvesting dates, change in genetic traits of cultivars and sometimes total adjustment of cropping system itself. With warmer environment associated with erratic rainfall distribution, the rate of evapotranspiration will increase and quick depletion of soil nutrient reservoir would call for much greater efficiency in use of water and nutrients to sustain crop productivity. Apart from these, tackling frequent and more intense extreme events like heat and cold waves, droughts and floods may become a "norm" for the common farming community (IPCC, 2001). Such phenomena will impact agriculture considerably through their direct and indirect effects on crops, livestock, and incidences of pest-disease-weeds, increasing deterioration of soil health in totality, and, thereby, threatening the food security like never before.

The Indian Council of Agricultural Research indicated an All India Network Project on Climate Change in 2004 to study the impact of climate change on major crops, livestock, fisheries, soils and other biotic factors as well as to understand different natural adaptation capabilities of both flora and fauna. The possible interventions to increase the adaptability of crop-livestock systems and mitigation measures to minimize the adverse impacts were studied across length and breadth of different agro-ecosystems of India (Naresh Kumar *et al.*, 2012).

Assessment of climate change impact over Indian agriculture (Aggarwal, 2009) so far has indicated that a marginal 1 °C increase in atmospheric temperature along with increase in CO_2 concentration would cause very minimal reduction in wheat production of India if simple adaptation strategies like adjustment of planting date, increased fertilizer use, irrigation water availability and varieties are adopted uniformly. But in absence of any adaptive mechanism, the yield loss in wheat may cross 4-5 million tons.

The availability of viable pollen, sufficient numbers of germinating pollen grains and successful growth of pollen tube to the ovule are of fundamental importance in grain formation. The Network study on wheat and rice suggested that high temperature around flowering reduced fertility of pollen grains as well as pollen germination on stigma. These effects were more pronounced in *Basmati* rice as well as *Durum* wheat cultivars. A positive finding of the study was that the *Aestivum* wheat cultivars were more or less tolerant to such adverse affects. But differential impact of increasing temperatures were observed with respect to grain quality of wheat where it was found that *Aestivum* wheat cultivars are more prone to reduced grain quality due to increasing temperature during the fruit setting stage than *Durum* cultivars. Field experiments, using advanced

'Temperature gradient tunnels' with different dates of sowing to study the impact of rising temperature on growth and development of different crops, revealed that an increase of temperature from 1 to 4 °C reduced the grain yield of rice (0-49%), potato (5-40%), green gram (13-30%) and soybean (11-36%). However, one of the important pulses, chickpea, registered 7-25% increase in grain yield by an increase in temperature up to 3 °C, but was reduced by 13% with a further 1 °C rise in temperature. The increase in crop productivity due to higher CO_2 in the atmosphere is likely to be negated by rising temperature.

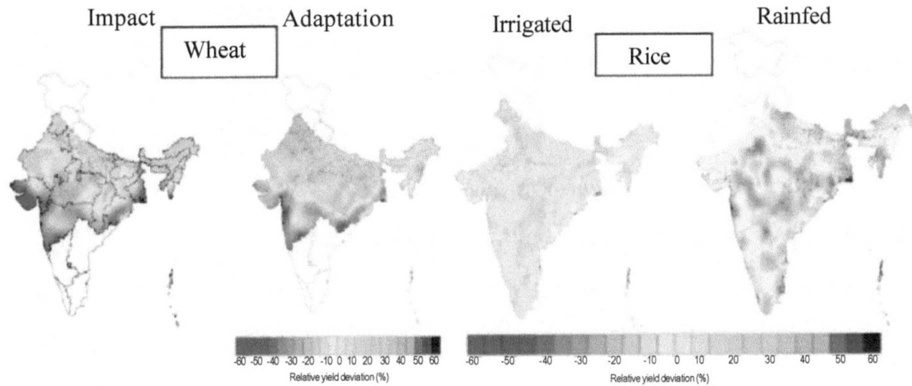

Fig. 2: Impacts on Wheat and Rice-2020 scenario

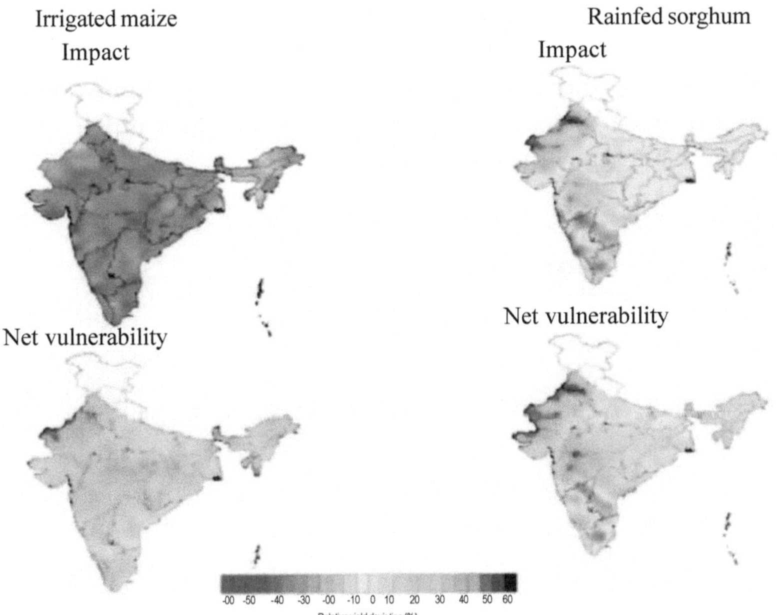

Fig. 3: Climate change impacts and net vulnerability of Maize and Sorghum

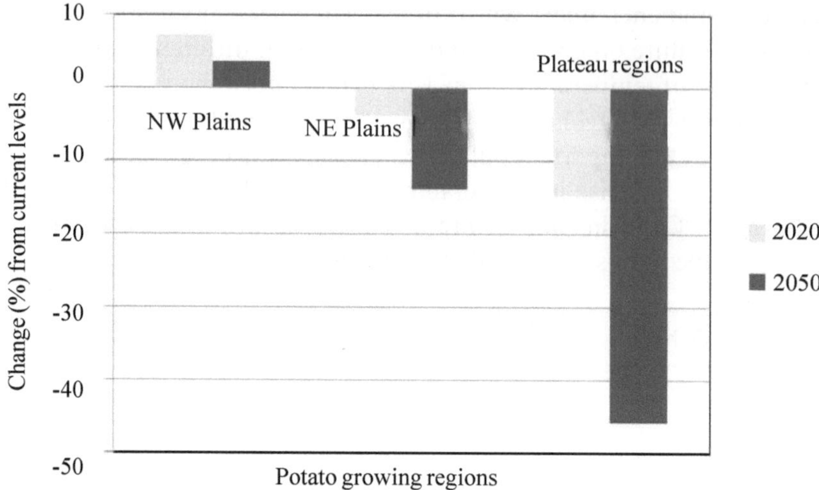

Fig. 4: Impact of climate change on Potato production in different regions of India

A significant decrease in average productivity of apples in Kullu and Simla districts of Himachal Pradesh have been reported which is attributed mainly to inadequate chilling required for fruit setting and development. Reduction in cumulative chill units of coldest months may have caused shift of apple belt to higher elevations of Lahaul-Spitti and upper reaches of Kinnaur districts of Himachal Pradesh. However, results from simulation models suggest that climate change could benefit coconut crop. Coconut yields are likely to increase by 4, 10, and 20% by 2020, 2050 and 2080, respectively, in the western coastal areas of Kerala, Maharastra, Tamil Nadu and Karnataka. But the impact may be negative in east coast areas as they are already facing a much warmer atmospheric thermal regime than western coast.

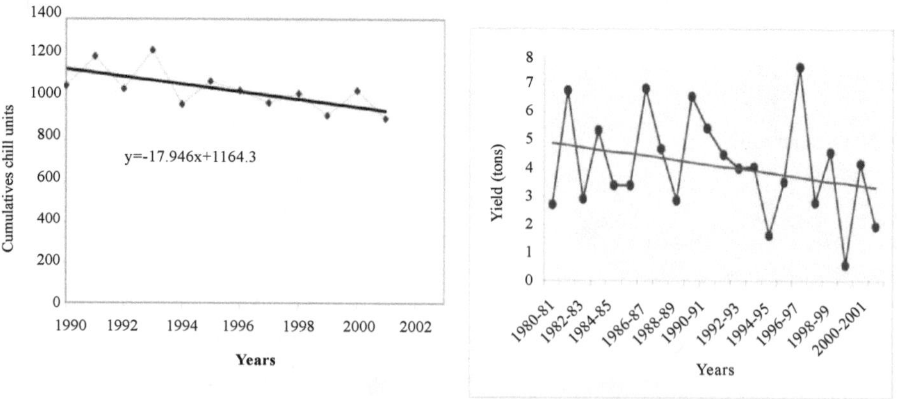

Fig. 5: Shift in apple cultivation to higher elevations due to non fulfillment of chilling requirement

Besides, the nutrient loss from soil through high rate of mineralization, CO_2 emissions from soil could be accelerated as a result of increase in temperature. Low carbon soils of mainly dryland areas of India are likely to emit more CO_2 compared to high or medium carbon temperate region soils. Simulation of water balance using Global and Regional Climate Models revealed likely increase in annual as well as seasonal stream-flows of many Indian river basins pointing to the need for adoption of more effective runoff and soil loss control measures to sustain crop production across the country.

Agroforestry can play a significant role in mitigation through carbon sequestration. It can been observed Carbon sequestration potential of agro-forestry in Bundelkhand region is currently of the order of 22.42 Mg C ha^{-1} and is likely to increase up to 35.78 Mg C ha^{-1} at end of rotation period. In the middle Gangetic plains (Jaunpur, Pratapgarh, Azamgarh and Basti districts), it ranged from 18.5 to 21.80 Mg C ha^{-1} and after 24 years it is projected to increase in the range of 28-38 Mg C ha^{-1}. The annual C sequestration in coconut above ground biomass varied from 15 Mg CO_2/ha/year to 35 Mg CO_2/ha/year out of which 2-3 Mg CO2/ha/year stocked into stem. The carbon sequestration potential of arecanut plantations ranged from 5.14-10.94 Mg CO2/ha/year while that of cocoa plantation ranged from 2.02-3.89 Mg CO_2/ha/year.

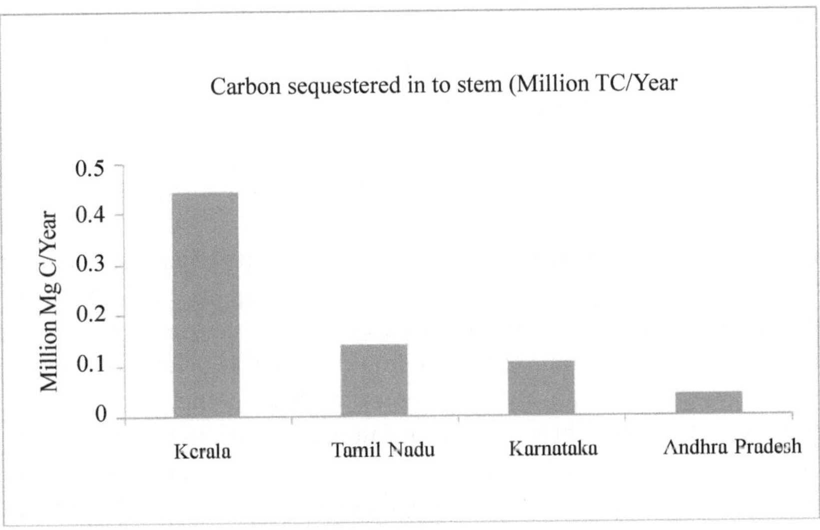

Fig. 6: Mitigation options - Carbon sequestration potential of plantation crops and agroforestry

Crop-Pest Interactions

The change in climate may bring about changes in population dynamics, growth and distribution of insects and pests. Changes in rainfall, temperature and wind speed pattern may influence the migratory behaviour of the locusts. Most crops have C_3 photosynthesis (responsive to CO_2), while many weeds are C_4 plants (non-responsive to CO_2). The climate change characterized by higher CO_2 concentration will favour crop growth over weeds. The impact of rising temperature and CO_2 are also likely to change insect pest dynamics. Dilution of critical nutrients in crop foliage may result in increased herbivory of insects. For example, Tobacco caterpillar (*Spodoptera litura*) consumed 39% more castor foliage under elevated CO_2 conditions than controlled treatments (Srinivasa Rao *et al.,* 2009).

Dairy sector

Global warming is likely to lead to a loss of 1.6 million tons in milk production by 2020 and 15 million tons by 2050. High producing crossbred cows and buffaloes will be affected more by climate change. Based on temerature-humdity index (THI), the estimated annual loss in milk production at the all-India level by 2020 is valued at Rs. 2661.62 crores at current prices. The economic losses were highest in UP followed by Tamil Nadu, Rajasthan and West Bengal. Stressful THI with 20 h or more daily THI-hrs (THI >84) for several weeks affects animal responses. Under climate change scenario, increased number of stressful days with a change in Tmax and Tmin and decline in availability of water will further impact animal productivity and health in Punjab, Rajasthan and Tamil Nadu (Upadhyay *et al*., 2009a; Annual Reports 2004 to 2010).

Increased heat stress associated with rising temperature may, however, caused distress to dairy animals and possibly impact milk production. A rise of 2 to 6 °C in temperature is expected to negatively impact growth, puberty and maturation of crossbred cattle and buffaloes. As of now, India loses 1.8 million tons of milk production annually due to climatic stresses in different parts of the country. The low producing indigenous cattle are found to have high level of tolerance to these adverse impacts than high yielding crossbred cattle. This needs to be explored by a systematic breeding programme emphasizing on increasing the productivity of indigenous breeds. Lactating cows and buffaloes have higher body temperature and are unable to maintain thermal balance. Body temperature of buffaloes and cows producing milk is 1.5- 2 °C higher than their normal temperature, therefore, more efficient cooling devices are required to reduce thermal load of lactating animals as current measures are becoming ineffective (Upadhyay *et al*., 2009b; Annual Reports 2004 to 2010).

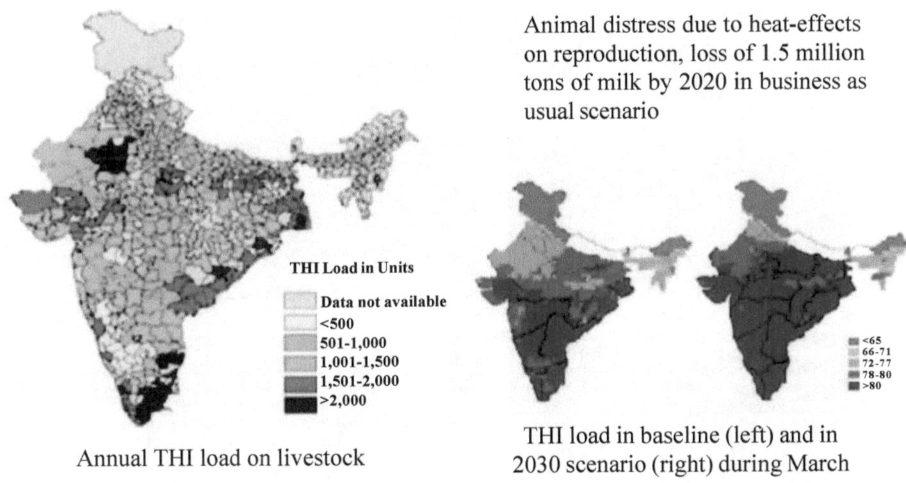

Fig. 7: Annual Temperature Humidity Index load on cattle in India

Fishery Sector

Marine fisheries

A rise in temperature as small as 1°C could have important and rapid effects on the mortality of fish and their geographical distributions. Oil sardine fishery did not exist before 1976 in the northern latitudes and along the east coast of India as the resource was not available/and sea surface temperature (SST) were not congenial. With warming of sea surface, the oil sardine was able to find temperature of its preference especially in the northern latitudes and eastern longitudes, thereby, extending the distributional boundaries and establishing fisheries in larger coastal areas. (Vivekanandan *et al.*, NPCC Annual Reports 2004 to 2010)

The dominant demersal fish, the threadfin breams have responded to increase in SST by shifting the spawning season off Chennai. During past 30 years period, the spawning activity of *Nemipterus japonicus* reduced in summer months and shifted towards cooler months. A similar trend was observed in *Nemipterus mesoprion* too. Analysis of historical data showed that the Indian mackerel is able to adapt to rise in sea surface temperature by extending distribution towards northern latitudes, and by descending to deeper depths (Vivekanandan *et al.*, 2009; Annual Reports 2004 to 2010)

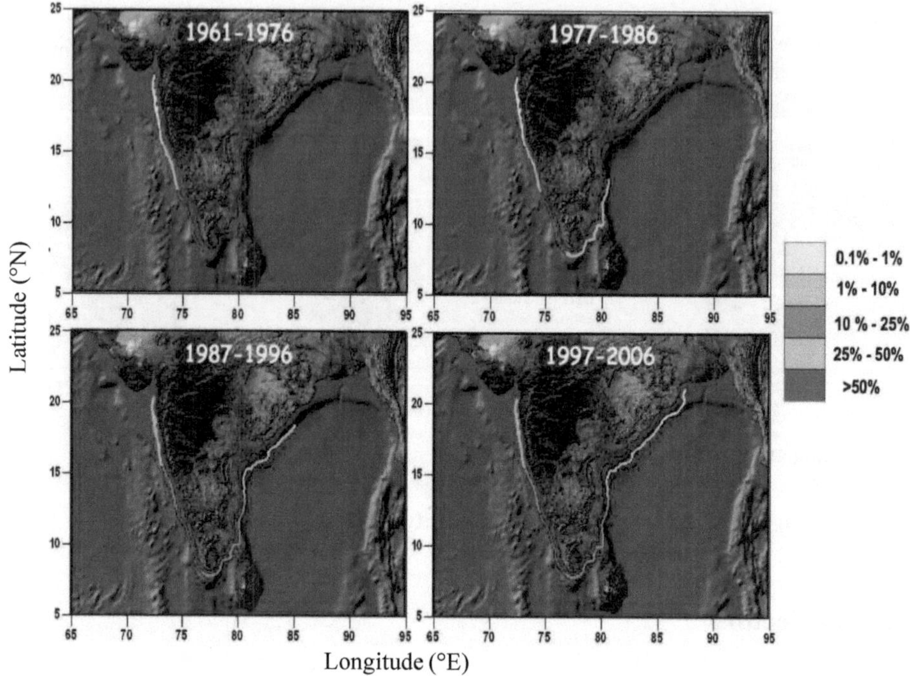

Fig. 8: Extension of abundance of Oil sardine along Indian coast due to rise in sea surface temperature

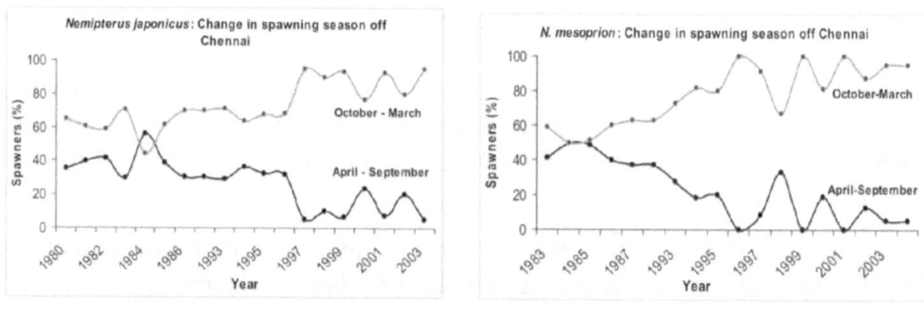

Fig. 9: Change in spawning season of threadfin breams (N*emipterus Japonicas* and N. *Mesoprion)* off Chennai

Inland fisheries

In recent years, the phenomenon of Indian Major Carps maturing and spawning as early as March was observed in West Bengal with its breeding season extending from 110-120 days (Pre1980-85) to 160-170 days (2000-2005). As a result, it is now possible to breed them twice in a year at an interval ranging from 30-60 days. A prime factor influencing this trend is elevated temperature, which stimulates the endocrine glands and helps in the maturation of the gonads of Indian major carp. The average minimum and maximum temperature throughout the state has increased in the range of 0.1 to 0.9°C (Das *et al.*, 2009; Annual Reports 2004 to 2010).

Recent climatic patterns have brought about hydrological changes in the flow pattern of river Ganga. This has been one major factor resulting in erratic breeding and decline in fish spawn availability. As a result the total average fish landing in the Ganga river system declined from 85.21 tons during 1959 to 62.48 tons during 2004. In the middle and lower Ganga 60 genera of phytoplankton were recorded during 1959 and declined to 44 numbers by 1996. In case of Zooplankton during the same period, the number diminished from 38 to 26. A number of fish species, which were predominantly only available in the lower and middle Ganga in 1950s, are now recorded from the upper cold-water stretch up to Tehri (Das *et al.*, 2009; Annual Reports 2004 to 2010)

Poultry Sector

In poultry sector, mortality due to heat stress, occurrs at about 34°C. It was significantly high in heavy meat type chickens (8.4%) as compared to light layer type (0.84%) and native type (0.32%) chickens. Increase in temperature from 31.6°C to 37.9°C decreased feed consumption by about 36% and egg production up to 7.5% in broiler breeders and up to 6.4% in commercial layers as compared to their standard egg production percentages. The critical body temperature at which the birds succumb to death was 45°C which was observed at the shed temperature of 42°C. The naked neck birds performed significantly better than the normal birds with respect to thermo-tolerance, growth, feed efficiency and immunity in high temperatures as compared to normal broilers. The sperm viability and fertilizing ability, and live sperm counts were significantly reduced during high ambient temperature (Annual Reports 2004 to 2010)

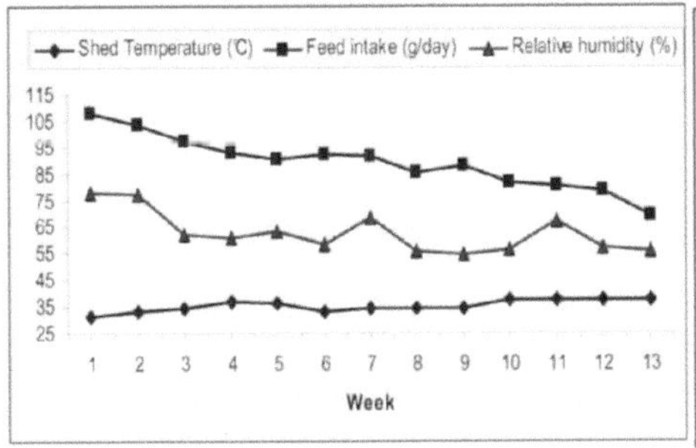

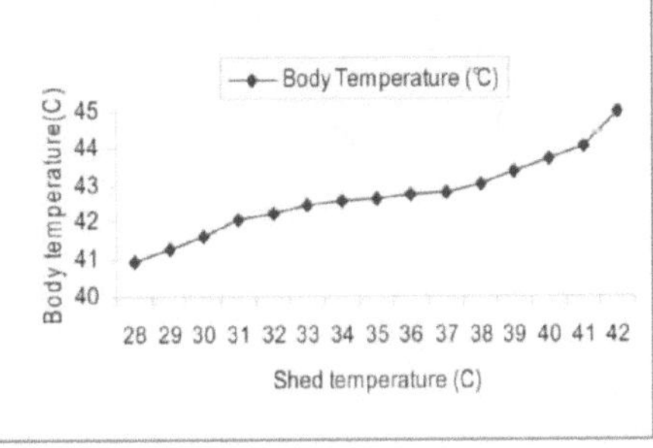

Fig. 10: Influence of elevated ambient temperature on feed consumption

Technology dissemination for empowering farmers

The technology demonstration component of National Initiative on Climate Resilient Agriculture (NICRA) deals with demonstrating an integrated package of proven technologies for adaptation of the crop and livestock production systems to climate variability. This component is being implemented in selected vulnerable districts of the country by implementing location specific interventions through Krishi Vigyan Kendras in a participatory mode. The project is implemented in 130 districts (see map) involving over one lakh farm families across the country.

The selection of districts was based on analysis of climatic constraints of villages based on long term data, assessment of the natural resources status of the village, identification of major production systems, and study of existing institutional structures and identification of gaps through focused group discussion with the community to finalize the interventions.

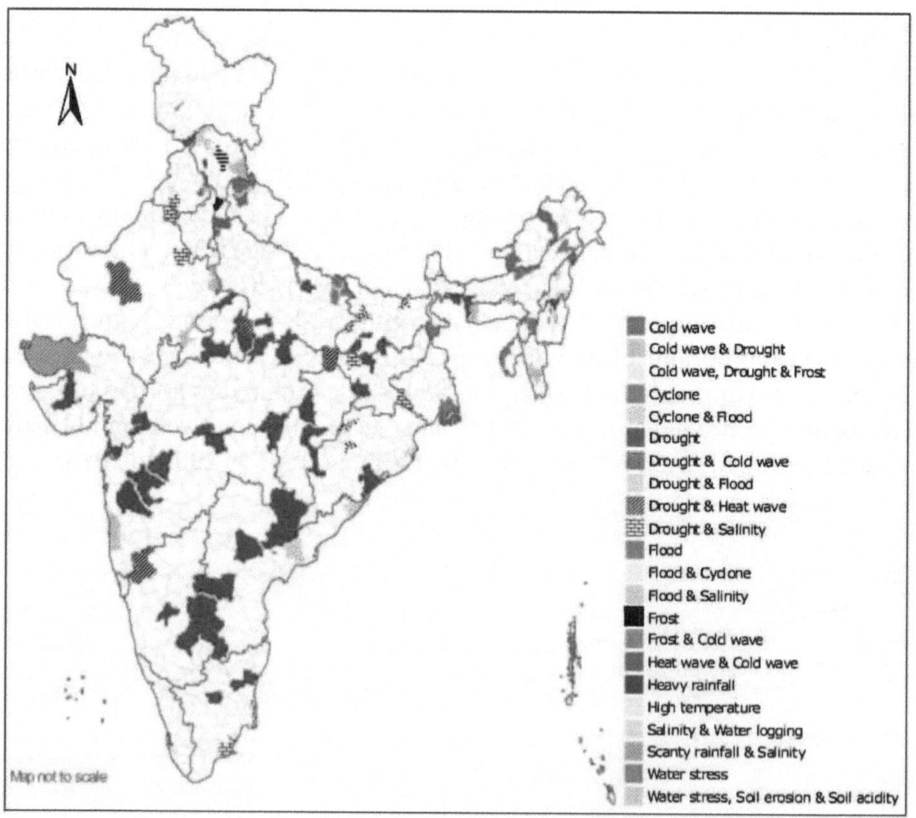

Fig. 11: The location of 130 districts being covered in NICRA

Rationale for Technology Demonstration Component (TDC)

Over the years, a plethora of technologies tailor made to suit different situations has been developed. Though these technologies cannot be termed as climate resilient they were applied in situations challenged by climate variability in different agro climatic environments. Therefore, the TDC took a fresh look at these technologies to demonstrate them from the climate resilience perspective, as current efforts underway for developing climate resilient technologies may take some time before getting ready for implementation. The following points captured the rationale for implementing technology demonstration component under NICRA.

- Availability of technologies
- Availability of indigenous practices
- Long experience of NARS in evolving drought/flood resilient technologies
- Inherent resilience with the community for coping with disasters With the following objectives:
- To enhance the resilience of Indian agriculture (including crops, livestock and fisheries) to climatic variability and climate change through strategic research on adaptation and mitigation
- To demonstrate site specific technology packages on farmers' fields to cope with current climatic variability
- To enhance the capacity of scientists and other stakeholders in climate resilient agricultural research and awareness of impacts

The interventions covered under this component are broadly classified into four modules, *viz*: Natural resources, Crop Production, Livestock and Fisheries, and Institutional Interventions.

The Process

The KVK team for each district carried out a detailed exercise on the needs of the village, the climatic vulnerability (drought/floods/heat wave/frost/cyclone) and the available technology options from the concerned Zonal Agricultural Research Station of the SAUs. After a careful study of the gaps, specific interventions from each of the modules were selected and an integrated package from all modules was formulated. The majority of the farmers were covered with one or more of the interventions in order to demonstrate a discernable effect. The project was launched in each village after giving wide publicity and by involving all the line departments under the leadership of the district administration. The launch event was used to generate wide spread awareness within the community and across line departments so as to prepare a platform for exploiting synergy through convergence of other government projects.

The unique features of the project are that a baseline for all the project sites were established through a systematic benchmark survey, emphasis was on natural resource management interventions to build the communities' capacity to cope with climate variability through need based investment and establishment of a network of automatic weather stations across 100 KVKs as well as custom hiring centers in each project village for promoting mechanization on small farms and friendly setting up of Village Climate Risk Management Committees (VCRMC) – grassroots peoples' institutions to take need based decisions at the village level.

The project has already made a visible impact in helping the farmers in the vulnerable districts to cope up with the vagaries of weather.

Small weather station in project village to raise weather

VCRMC meeting in progress

Improved implements at village custom hiring

Adaptation/mitigation strategies

A comprehensive strategy of utilization of existing knowledge, strengthening R&D in key areas and evolving a policy frame work that builds on risk management and providing incentives to sustainable use of natural resources will be required for successful adaptation to climate by the farm sector. The goal of this strategy is to minimize as risks associated with farming and enable farms to cope with these risks (Singh *et al.*, 2009).

The main adaptation strategies include development of new genotypes; intensifying search for genes for stress tolerance across plant and animal kingdom; intensifying research efforts on marker aided selection and transgenics development for biotic and abiotic stress management; development of heat and drought tolerant genotypes; attempt conversion of C3 plants to C4 plants; development of new land use systems; evolving new agronomy for climate change scenarios; explore opportunities for restoration of soil health; use multipurpose adapted livestock species and breeds; development of spatially differentiated operational contingency plans for weather related risks, supply management through market and non-market interventions in the event of adverse supply changes; enhancement research on applications of short, medium and long range weather forecasts for reducing production risks; development of knowledge based decision support systems for translating weather information into operational weather management sources; development of pest and disease forewarning systems covering range of parameters for contingency planning; conducting an integrated study of 'climate change triangle' and 'disease triangle', especially in relation to viruses and their vectors. Development of a compendium of indigenous traditional knowledge and explore opportunities for its utilization forms an important part of this strategy.

While adaptation measures are important, it also important to focus simultaneously on mitigation measures so that we contribute to a reduction in the pace of global climate change (Venkateswarlu and Arun Shanker, 2009). The important mitigation options include efficient water and nutrient management options to enhance use efficiency; evaluate carbon sequestration potential of different land use systems including opportunities offered by conservation agriculture and agro-forestry; identify cost effective opportunities for reducing methane emission in ruminants by modification of diet, and in rice paddies by water and nutrient management. Renewed focus is needed on nitrogen fertilizer use efficiency with added dimension of nitrous oxide mitigation. However, there is a need to assess the socio-economic implications of proposed mitigating options before developing a policy frame work.

References

Aggarwal, P. K. (2009). Global Climate Change and Indian Agriculture: Case studies from the ICAR Network Project. ICAR, New Delhi. p. 148.

Annual Reports 2004-2010. ICAR Network Project on Climate Change

Das M.K. (2009). Impact of recent changes in weather on inland fisheries in India. In Global Climate Change and Indian Agriculture Case Studies from the ICAR Network Project (PK Aggarwal ed). ICAR Pub. Pp 101-103.

DEFRA (2005). http://www.defra.gov.uk/environment/climatechange/internat/dangerous-cc.htm Goswami, B. N. (2006). Increasing trend of extreme rain events and possibility of extremes of seasonal mean Indian monsoon in a warming world (http://saarc-sdmc.nic.in/pdf/ workshops/ kathmandu/pres16.pdf)

IPCC (2001). Intergovernmental Panel on Climate Change, Climate Change 2001: Impacts, adaptation and vulnerability-Summery for policymakers. Cambridge University Press, pp. 1032.

IPCC (2007). Intergovernmental Panel on Climate Change, 4th Assessment Report-Synthesis Report.

Lal, M. (2001). Future climate change: Implications for Indian summer monsoon and its variability. Current Science, 81 (9), p. 1205.

Lobell, D. B. and Field, C. B. (2007). Global scale climate-crop yield relationships and the impacts of recent past. Environmental Research Letters, 2, p. 7.

Mall, R. K., Akhilesh Gupta, Ranjeet Singh, R. S. Singh and L. S. Rathore (2006). Water resources and climate change: An Indian perspective. Current Science, 90 (12): 1610-1626.

Naresh Kumar, S., Singh, A.K., Aggarwal, Rao, V.U.M. and Venkateswarlu, B. 2012. Climate Change and Indian Agriculture: Impact, Adaptation and Vulnerability. Indian Agriculture Research Institute, New Delhi, p.1-26

Singh, A. K., Aggarwal, P. K., Gogoi, A. K., Rao, G. G. S. N. and Ramakrishna, Y. S. (2009). Global Climate Change and Indian Agriculture: Future priorities. In: Global Climate Change and Indian Agriculture: Case studies from the ICAR Network Project (Ed. Aggarwal, P. K.), ICAR, New Delhi. 146-148.

Srinivasa Rao, M., Srinivas, K., Vanaja, M., Rao, GGSN, Venkateswarlu, B. and Ramakrishna, Y.S. (2009). Host plant (Ricinus communis Linn.) mediated effects of elevated CO2 on growth performance of two insect folivores. Current Science, 97 (7) 1047-1054.

Trenberth, K. E. (1999). Conceptual framework for changes of extremes of the hydrological cycle with climate change. Climatic Change, 42, 327–339.

Upadhyay, R.C., Sirohi, S., Ashutosh, Singh, S.V., A. Kumar, and Gupta, S.K. (2009a). Impact of climate change on milk production of dairy animals in India. In Global Climate Change and Indian Agriculture Case Studies from the ICAR Network Project (PK Aggarwal ed). ICAR Pub. 104-106.

Upadhyay R.C., Ashutosh, Raina, V.S., and Singh, S.V. (2009b). Impact of climate change on reproductive functions of cattle and buffaloes. In Global Climate Change and Indian Agriculture Case Studies from the ICAR Network Project (PK Aggarwal ed). ICAR Pub. 107-110.

Venkateswarlu, B. and Shanker, A.K. (2009). Climate change and agriculture: Adaptation and mitigation strategies. Indian Journal of Agronomy. 54(2): 226-230.

Vivekanandan, E., Rajagopalan, M. and Pillai, N.G.K. (2009). Recent trends in sea surface temperature and its impact on oil sardine In Global Climate Change and Indian Agriculture Case Studies from the ICAR Network Project (PK Aggarwal ed). ICAR Pub. 89-92.

Vivekanandan, E., Hussain Ali, M. and Rajagopalan, M. (2009). Impact of rise in seawater temperature on the pawning of threadfin beams. In Global Climate Change and Indian Agriculture Case Studies from the ICAR Network Project (PK Aggarwal ed). ICAR Pub. 93-96.

2

Climate Change, El Niño, Monsoon and Impact on Crops in India

A. Kashyapi and Surabhi Purohit

Introduction

Increase in atmospheric carbondioxide (CO_2) and other green house gases *viz.* methane (CH_4), nitrous oxide (N_2O) and CFC (Chloro-Flouro- Carbon) due to fossil fuel burning, rapid industrialization and deforestation, which reduces the amount of earth's radiation, that escapes to space. Role of these gases to change in climate over a longer period of time is the key issue of climate change. Climate change is a change in climate over time, whether due to natural variability or as a result of human activity (IPCC 2007).

IPCC : The UNEP (United Nations Environment Programme) and WMO (World Meteorological Organization) jointly established IPCC (Inter-governmental Panel on Climate Change) to assess available scientific information on climate change as well as to study its impact during 1988. Since establishment, IPCC has produced five major assessment reports. According to IPCC, climate change is defined as a movement in climate system as a result of internal changes within climate or interaction among its components. Frequency of extreme events (*viz.* heat waves, intense rain, floods, droughts etc.) are reported to follow increasing trend. Fig.1 is showing natural climatological cycle in the earth throughout different seasons as various components of climate system (IPCC 1996).

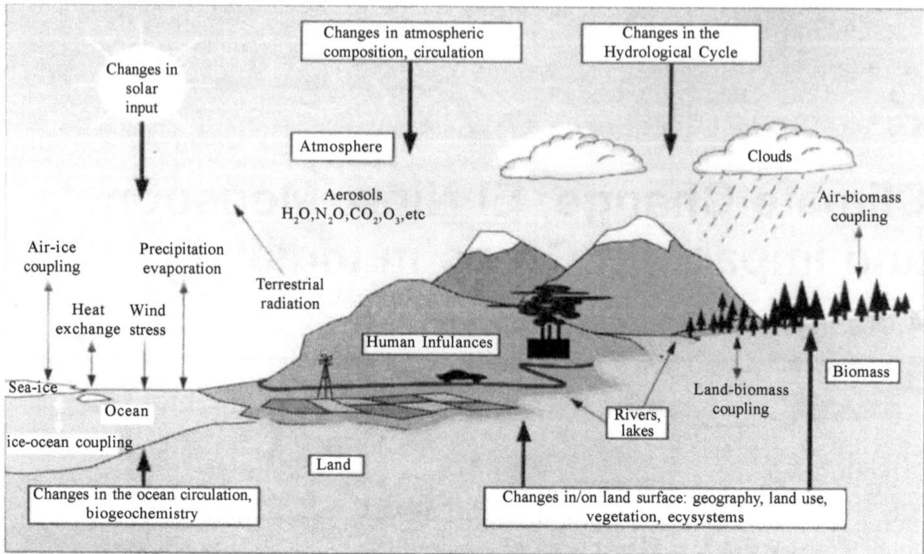

Fig. 1: Climatological cycle of earth's climate system

Agriculture and climate change

Agriculture is one of the most vulnerable sector (Fig. 2), which is likely to be affected adversely by climate change. Around 70% people in rural India are dependent on agriculture. Agriculture and allied sectors (animal husbandry, poultry and fisheries) are inherently sensitive to climatic conditions and are vulnerable to weather and climate risks (Fig. 3). Although, increase in carbon dioxide is likely to be beneficial to several crops, associated increase in temperatures and increased variability of rainfall would considerably impact food production.

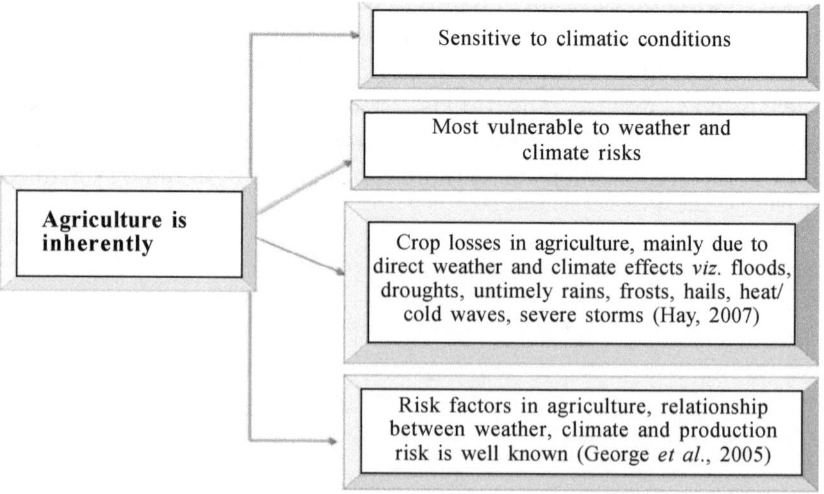

Fig. 2: Agriculture is inherently sensitive to climatic conditions

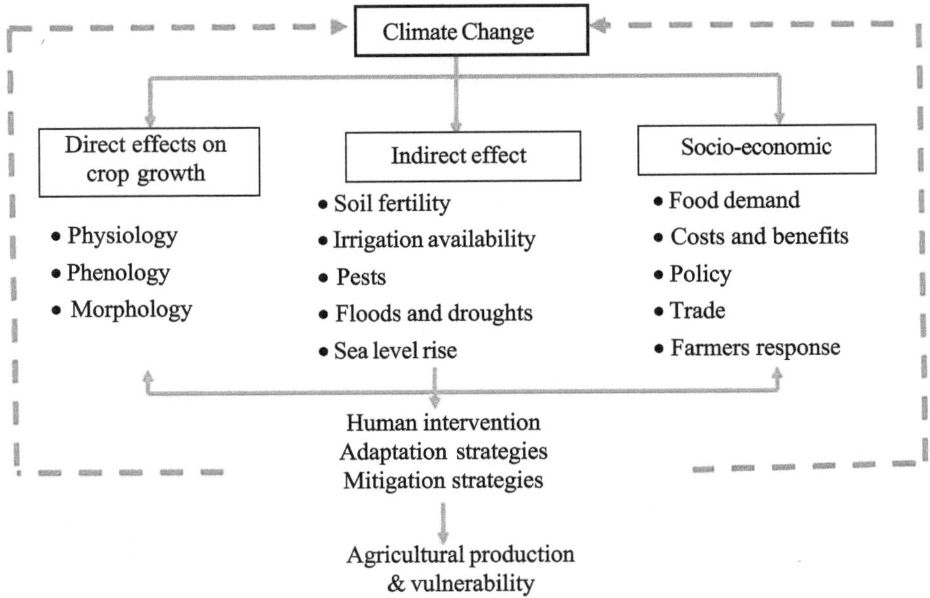

Fig. 3: Showing vulnerability of agriculture to climate change

Of the total annual crop losses in agriculture, a major share is directly due to weather and climate effects *viz*. floods, droughts, untimely rains, frosts, hails, heat, cold waves and severe storms (Hay 2007). In India, 28% land is vulnerable to droughts, 12% to floods and 8% to cyclones. *Kharif* crops *viz*. rice, maize, cotton, oilseeds, soybean and coarse grains are heavily dependent on monsoon rains (Kashyapi *et al*., 2012). It will lead to a lower yield of *kharif* crops. A high temperature during *rabi* season will affect yield of wheat, potato and vegetables. Agriculture is likely to be worst affected in the coastal regions of Gujarat and Maharashtra, where agriculturally fertile areas are vulnerable to inundation and salinization. Agriculture will be adversely affected not only by an increase or decrease in the overall amount of rainfall, but also by shifts in the timing of the rainfall. Chhattisgarh region has received less than its share of pre-monsoon showers in May and June, which are important to ensure adequate moisture in fields being prepared for rice crops. In Rajasthan, a 2°C rise in temperature was estimated to reduce production of pearl millet by 10-15 per cent. For couping up with impact of climate change on agriculture, modern era IAAS (Integrated Agromet Advisory Services) with district level forecast for generating location and crop specific agromet advisories on real time basis can serve as useful tool towards sustainable development and food security. For IAAS bulletins, extended range district level forecast with five days validity (useful for farming community) are issued from IMD New Delhi. Value addition by respective MCs (for their States) based on their analysis, helps in further enrichment of the forecast, which serves the need for ideal MRF (medium range forecast). Now-a-days, Agromet Advisories are issued through a network 130 AMFUs (Agrometeorological Field Units) located in 127 agroclimatic zones, in SAUs (State Agricultural Universities) throughout the country, on every Tuesday and Friday). The State of M.P. is divided into two meteorological sub-divisions *viz.* East M.P. and West M.P. The State is divided into 9 Agro-climatic zones covering all 51 districts.

Under current five year plan IAAS has been modified as GKMS (Gramin Krishi Mausam Seva). GKMS is supposed to provide block level agrometeorological advisories based on block level forecast. The real time dissemination of block level advisories will fulfil the location and crop specific demand of advisories by the farmers, which will help them to coup up with negative impact of climate change. The paper also discusses climate variability and adaptive measures with reference to M.P.

Hence, aim of the study are 1) to review and summarize key effects of climate change and climate variability in various parts of India with special reference to M.P., 2) to familiarize with terminologies *viz.* El Niño, La Niño, ENSO etc., 3) to study impact of climate change on Indian summer monsoon, 4) to study climate variability with respect to Central India, 5) to summarize impact of climate variability on crops and to provide adaptive measures as sustainable solution and 6) to identify GKMS project as a platform for adaptive measures as tool against climate change with reference to the country as well as the State of Madhya Pradesh in Central India.

Extreme events: It has been observed that there is increasing frequency of extreme events nowadays. Summary of observed changes in extreme events and severe climate anomalies in India and neighbourhood are presented in Table 1.

Table 1: Summary of observed changes in extreme events and severe climate anomalies in India and adjacent countries

Country/region	Key trend
A) Heatwaves	
India	Frequency of hot days and multiple-day heat waves have increased in past century.
B) Intense rains and floods	
South Asia	Serious and recurrent floods in Bangladesh, Nepal and N-E States of India during 2002 to 2004; a record 944 mm of r/f in Mumbai, India on 26 to 27 July 2005; floods in Surat, Barmer and in Srinagar during 2006.
C) Droughts	
South Asia	Consecutive droughts in 1999 and 2000 in Pakistan and N-W India lead to sharp decline in water tables; consecutive droughts between 2000 and 2002 caused crops failure in Orissa; droughts in N-E India during 2006.
D) Cyclones and typhoons	
South Asia	Frequency of monsoon depressions and cyclones formation in Bay of Bengal and Arabian sea on decline since 1970, but intensity increased causing severe floods.

The frequency of extreme weather events in India are depicted in Fig.4. Heat wave in 2003 over Andhra Pradesh resulted in the loss of 27 crores of poultry as a result the industry in the State suffered a lot (Fig.5). Floods in A.P. during 2005 to 2007 resulted in huge damage to public as well as crops (Fig. 6). Cold wave in Hissar resulted in loss of papaya, mustard and Jatropha crops (Fig.7)

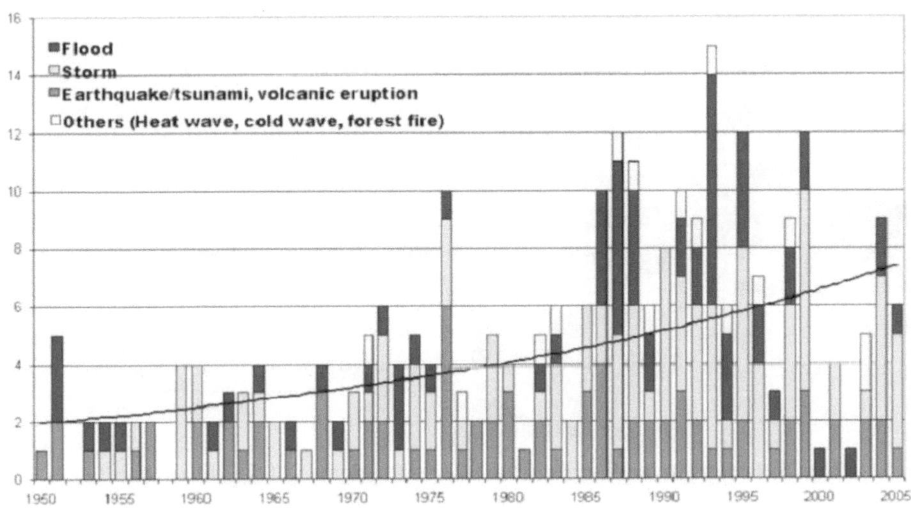

Fig. 4: Extreme weather events in India

Fig. 5: Heat Wave damage to poultry in Andhra Pradesh in 2003 (20 lakh birds died in May and June 2003 in East and West Godavari Districts with a loss of 27 crores rupees)

Fig. 6: Floods in Andhra Pradesh during the years 2005, 2006 and 2007

Fig. 7: Frost damage to different crops in Hisar, 2005-06

Monsoon, IMD data and climate change

Heavy south-west monsoon rains in central India were more intense between 1981 to 2000 than in 1950s, 1960s. Severe rains doubled over the same period. This was linked to rising global temperatures, as opined by IITM (Indian Institute of Tropical Meteorology, Pune) scientists. Study shows that changing monsoon (SMW) would have fewer rainy days. More than half the world's population depends on Asian monsoon to bring much needed water for agriculture. The monsoon rainfall in India, which accounts for 25% of GDP and 70% of the population may be delayed by 5-15 days in next hundred years due to climate change, which can affect *kharif* agriculture in parts of India. Rising temperatures may weaken the monsoon causing less or intense precipitation, longer breaks, floods, droughts, cyclones. Future projections predicted that by end of 21st century, rainfall will increase by 15-31% and mean annual temperature by 3°C to 6°C. The warming will be more pronounced over land areas, over northern India. Climate change can bring, as opined by IPCC in 1994, changes in parts of Asia *viz.* increase in rainfall, strengthening monsoon and increase in monsoon variability.

A huge IMD data from 1901 to 2006 were analyzed. The data revealed that annual mean temperature anomalies (Fig.8) based on the data set showed increasing trend especially from 1991 onwards. The annual maximum (Fig.9) and minimum (Fig.10) temperature anomalies based on data set from 1901 to 2006 also showed increasing trend from 1991 onwards. Though the trend was prominent for annual maximum temperature anomalies, while for annual minimum temperature anomaly the trend was meagre.

Table 2: Summary of observed past and present changes in precipitation

Region	Country	Change in precipitation
South Asia	India	Increased rains in N-W during summer monsoon in recent decades; lower no. of rainy days along east coast.
	Pakistan	10 to 15% decrease in coastal belt increase in summer and winter precipitation over last 40 years in N-Pakistan.
	Bangladesh	Decadal rain anomalies above long term averages since 1960s.
S-E Asia	Philippines	Increase in annual mean R/F since 1980s and number of rainy days since 1990s; increase in inter — annual variability of onset of R/F.

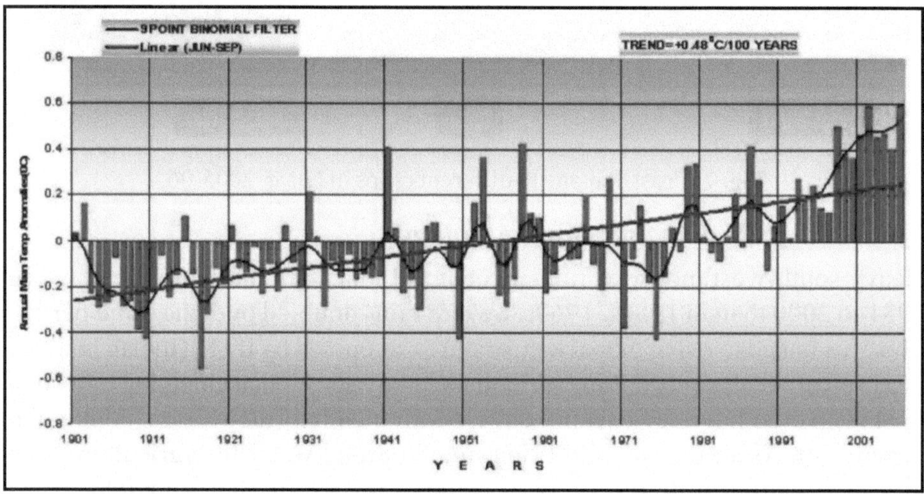

Fig. 8: Annual mean temperature anomalies (1901-2006)

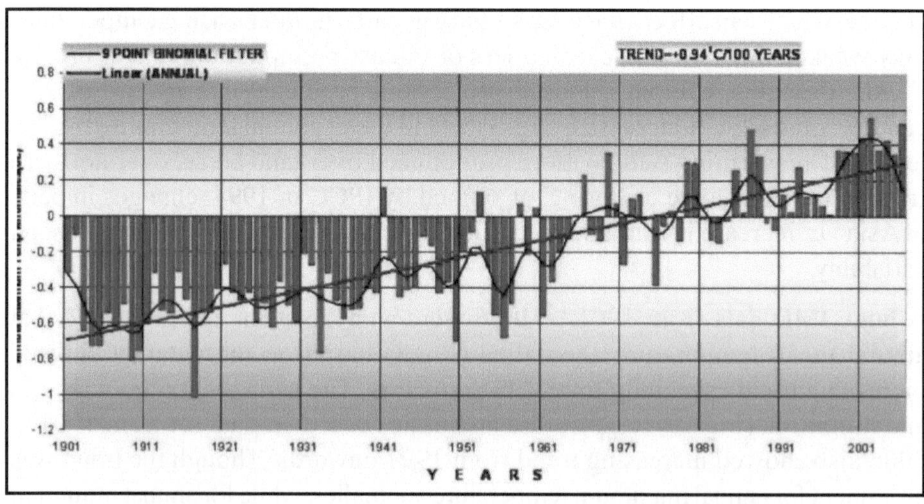

Fig. 9: Annual maximum temperature anomalies (1901-2006)

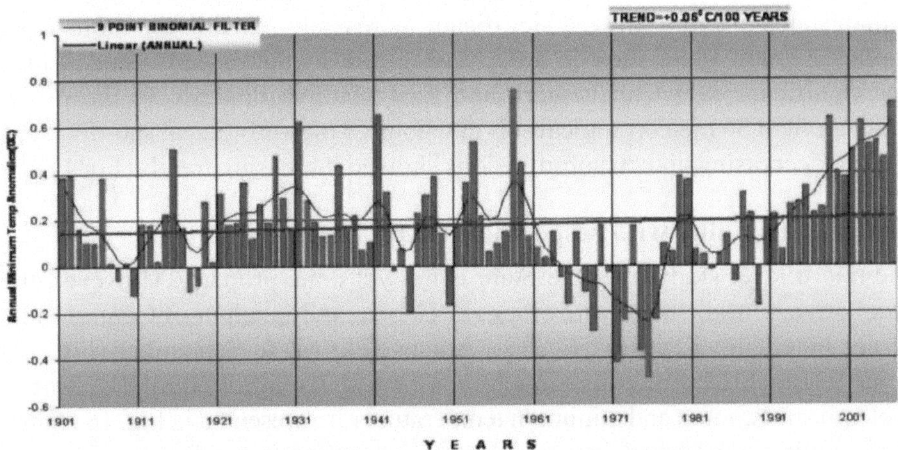

Fig. 10: Annual minimum temperature anomalies (1901-2006)

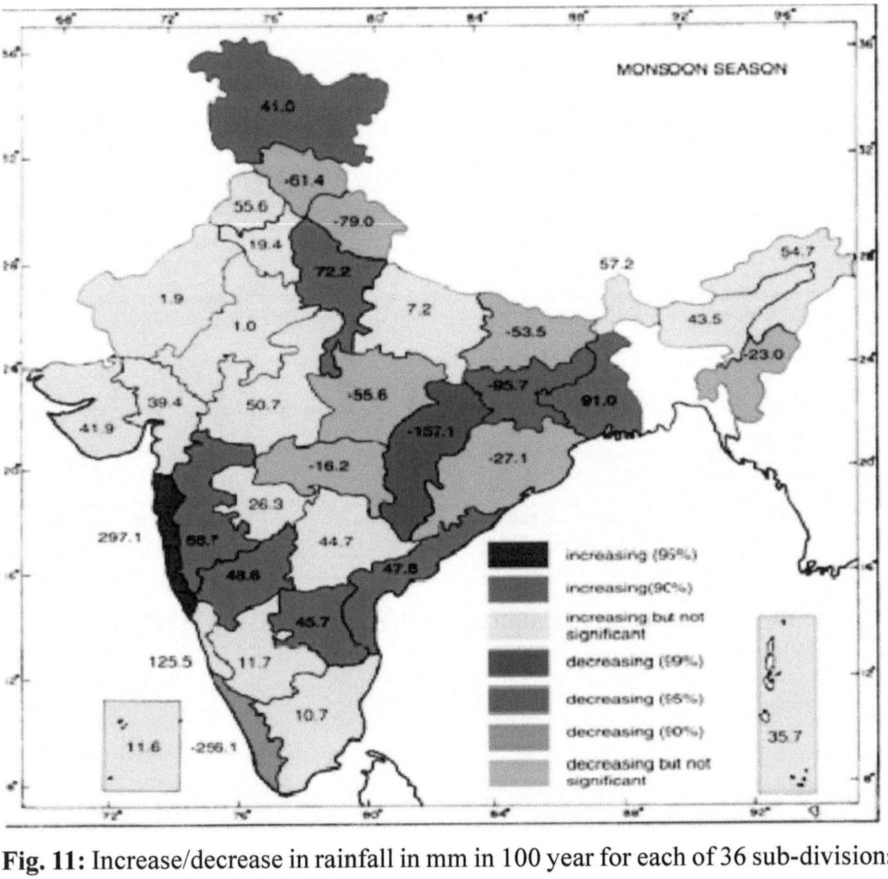

Fig. 11: Increase/decrease in rainfall in mm in 100 year for each of 36 sub-divisions for the south-west monsoon season

In the other studies, it was observed that Coastal Andhra, West Bengal and Punjab states of India showed significant increasing trends in the monsoonal rainfall. Central India showed a decreasing monsoonal rainfall trend, which was significant over Chhattisgarh and East Madhya Pradesh. In the recent decades out of 36 meteorological sub-divisions 16 meteorological sub-divisions showed decreasing and 20 sub-divisions showed increasing trends (Fig.11).

Climate variability with respect to Central India

The micro level study over the State is not yet done properly. However, the cumulative rainfall study for 2 cities *viz.* Bhopal and Jabalpur of Central India during June (Fig. 12), July (Fig.13), August (Fig.14) to September (Fig. 15) from 1991 to 2014 showed not much/minor trend in rainfall. Similarly, June to September maximum and minimum temperatures are presented in Fig. 16 (June), Fig.17 (July), Fig.18 (August) to Fig.19 (September). Basically, 24 years data over Bhopal and Jabalpur showed almost no trend or minor trend in temperatures over the cities.

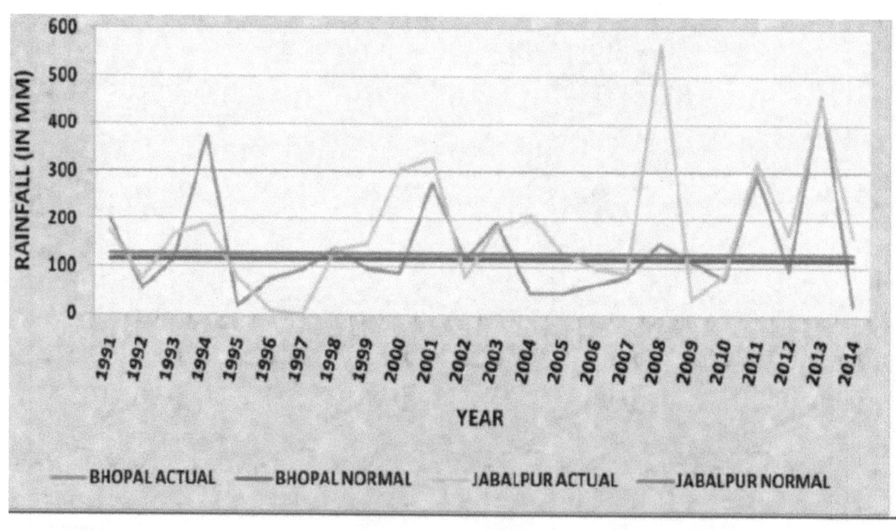

Fig. 12: Cumulative rainfall during June at Bhopal and Jabalpur

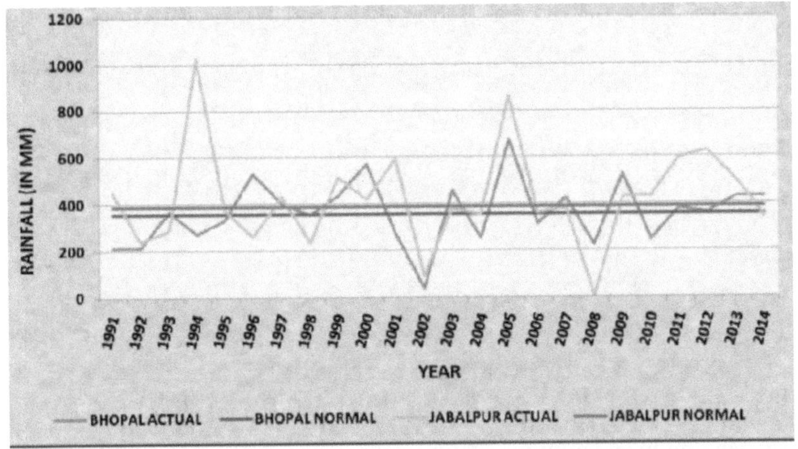

Fig. 13: Cumulative rainfall during July at Bhopal and Jabalpur

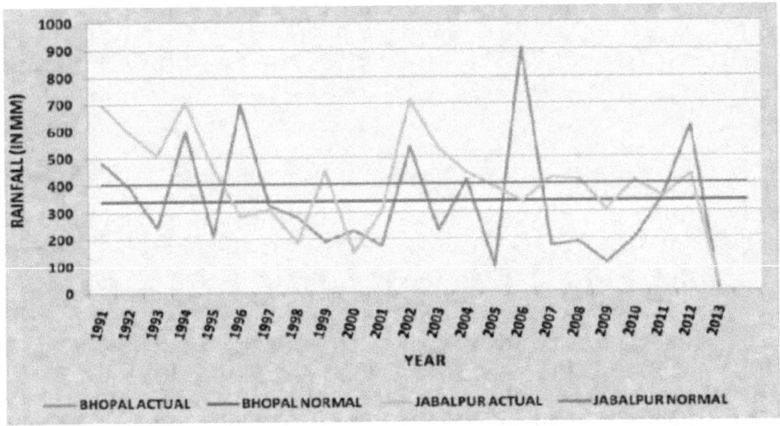

Fig.14: Cumulative rainfall during August at Bhopal and Jabalpur

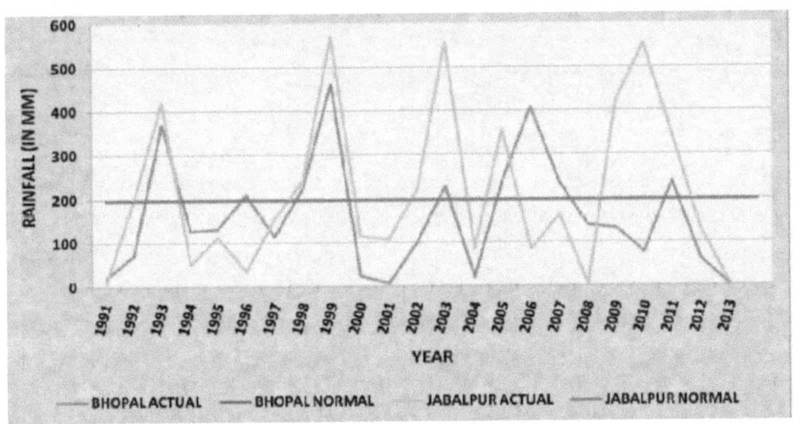

Fig. 15: Cumulative rainfall during September at Bhopal and Jabalpur

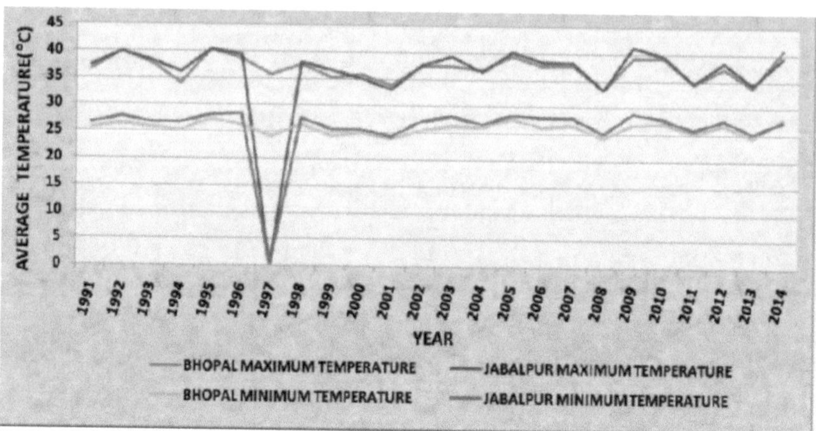

Fig. 16: Maximum and minimum temperatures during June at Bhopal and Jabalpur

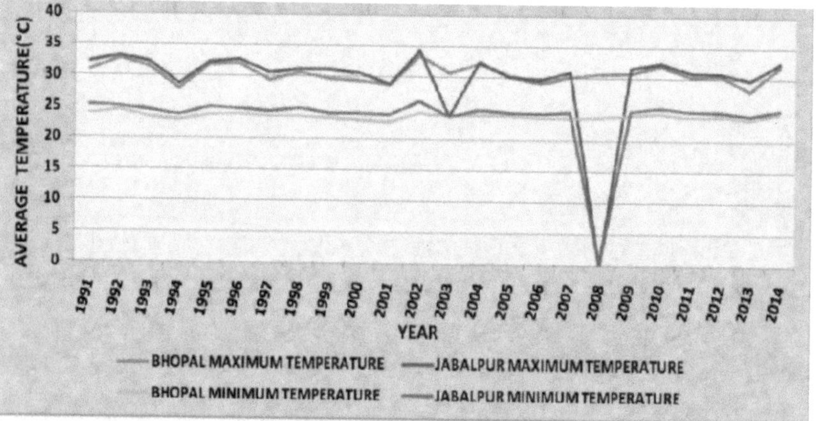

Fig. 17: Maximum and minimum temperatures during July at Bhopal and Jabalpur

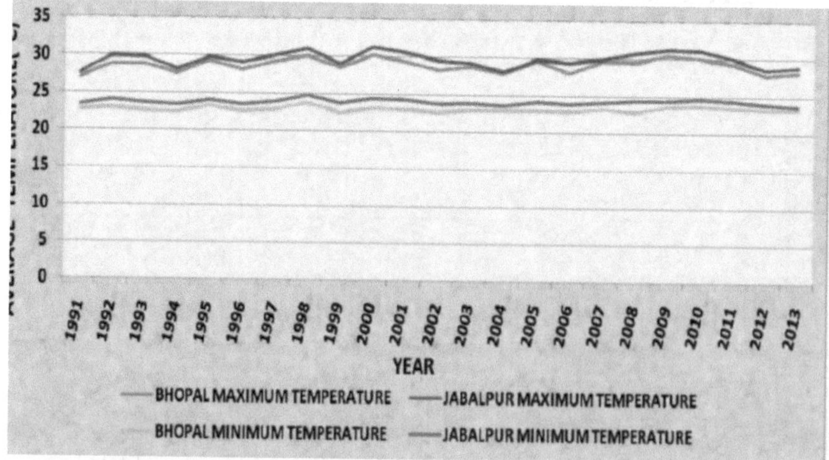

Fig. 18: Maximum and minimum temperatures during August at Bhopal and Jabalpur

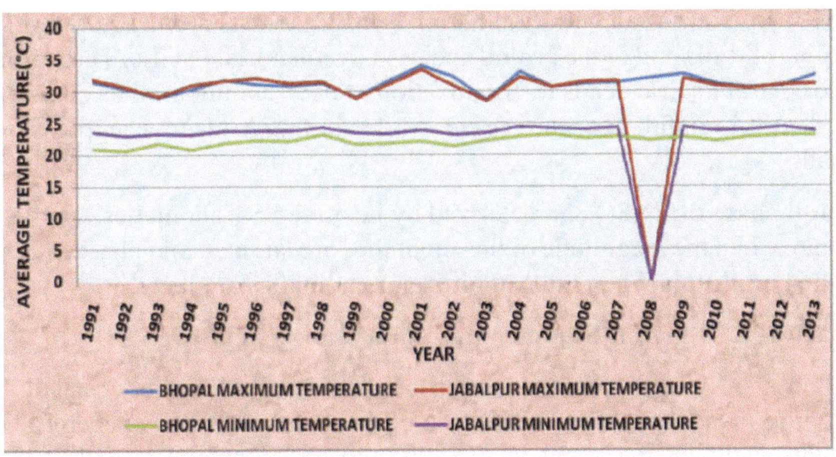

Fig. 19: Maximum and minimum temperatures during September at Bhopal and Jabalpur

El Niño, La Niña and ENSO

Increasing frequency and intensity of droughts in many parts of Asia is due to rise in temperature during summer ENSO events (Lal 2003). El Niño causes above normal rains and droughts in many areas, while ENSO provides influences on weather and climate extremes. The expected occurrence of an undesirable phenomenon- the El Nino over the Pacific Ocean, is considered to be a major factor in determining the fate of the monsoon. El Niño years directly have impact on India's agrarian economy as their effect tends to lower the production of summer crops such as rice, soybean, sugarcane and oilseeds.

El Niño: El Niño (in Spanish) means "little boy" or "Child Jesus" in Spanish, is a weather system which re-emerges after a gap of about 3 to 5 years in the Pacific Ocean and lasts for about 12 months on an average. During this time the warming of sea surface temperature takes place, affecting wind pattern. The warm sea- surface temperature (SST) anomaly is associated with reduced upwelling- reduced nutrients in the surface layer and hence, the disappearance of fishes and birds on which the economy of Ecuador and Peru are dependent. El Niño now refers to as a warm event, is a much larger scale phenomenon characterized by warm SST anomalies stretching from eastern to central equatorial Pacific. This phenomenon affects rainfall in India during Monsoon.

La Niña: The cold events are known as, La Niña or a girl. During La Niña which is just the opposite of El Niño, less heating leads to colder sea water off western South America coast, thus making it a high pressure zone, which pushes the moist sea winds towards the Indian Ocean. This increases chances of normal or excessive rainfall in the Indian sub-continent.

ENSO : El Ninño-Southern Oscillation (ENSO) events are irregular cycle of warming and cooling of the central and eastern equatorial Pacific. The warming of the central Pacific, leads to organization of deep cumulonimbus convection and plentiful rainfall over the region, while cooling leads to a suppression of rainfall.

A warm phase of ENSO characterized by positive SST anomalies is called an El Niño, when the magnitude of the anomalies are large. A cold phase of ENSO is called a La Niña when the magnitude of anomalies are large.

Impact of El Niño on Summer monsoon and agriculture

El Niño and the Indian summer monsoon are mostly inversely related. The most prominent droughts in India—six of them since 1871 have been El Niño droughts, including the recent ones in 2002 and 2009 (Table 3). Usually, El Niño years affect the weather in India in terms of monsoon rain. During this time, the rainfall is generally below normal, which has its bearing on crop production. El Niño, La Niña years during last decade is presented in Table 3. The drought years in India since last 120 years, decade-wise are presented in Table 4.

Table 3: Occurrence of El NiÆo, La NiÆo in current decade

Year	Occurrence	State	Precipitation
2004	El Nino	Drought	-12%
2005	Neutral	Normal	+1%
2006	Neutral	Normal	+3%
2007	La Nina	Excess	+ 10%
2008	La Nina	Above Normal	+5%
2009	El Nino	Severe Drought	-21%
2010	La Nina	Normal	0%
2011	La Nina	Normal	+4%
2012	Mild El Nino	Below Normal	-8%
2013	Neutral	Above Normal	+6%

Table 4: Drought years in India since 1801

Period	Drought Years	Number of Droughts
1801-1830	1801,1804, 1806, 1812, 1819, 1825	6
1831-1860	1832,1833,1837, 1853, 1860	5
1861-1890	1862, 1866, 1868, 1873, 1877, 1883	6
1891-1920	1891, 1897, 1899, 1901, 1904, 1905, 1907, 1911,1918, 1920	10
1921-1950	1939, 1941	2
1951-1980	1951. 1965, 1966, 1968, 1971, 1972, 1974, 1979	8
1981-2010	1982, 1986, 1987, 2002, 2004, 2009	6

Climate change has direct effect on crops' growth; indirect effect and socio-economic effect on weather and environment. Impact of climate change on rice (Fig. 20) and wheat (Fig. 21) production as projected during 2000, 2020, 2050 and 2080 showed differences. There will be a loss in rice production without adaptation from 2020 to 2080 (Fig. 20) marginal reduction in wheat yield with projection from 2020 to 2080. (Fig. 21)

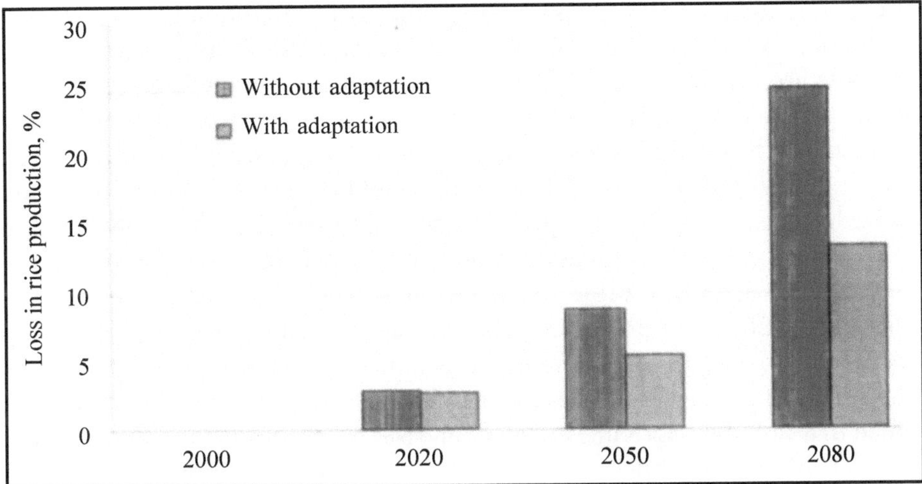

Fig. 20: Impact of climate change on rice production

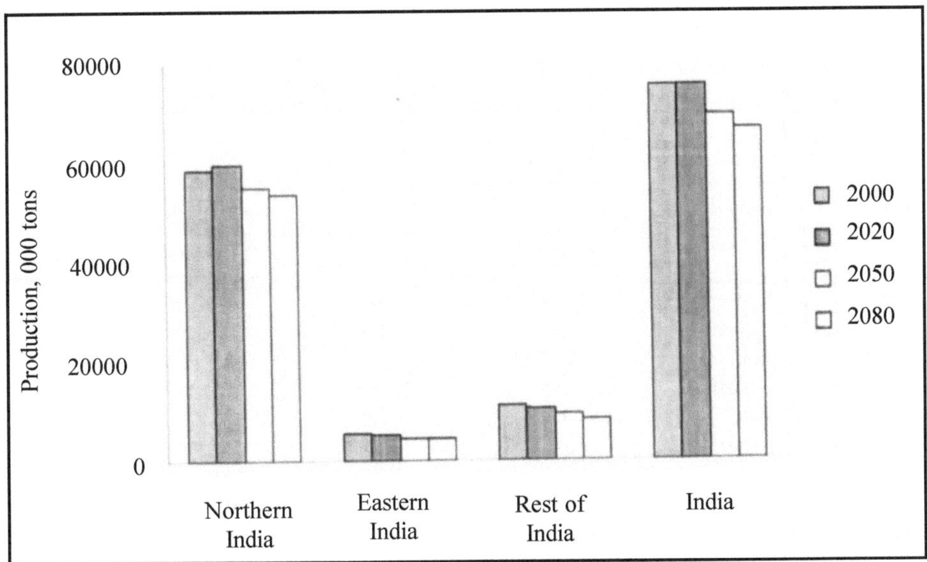

Fig. 21: Impact of climate change on wheat production

Quantified response of climate change on crops

Quantified responses of climate change are basically products after simulation of models, which provides projection after certain years. Simulated climate change scenario indicated that a 0.5°C rise in temperature will reduce wheat yield by 0.45 tonnes per hectare in India (Kalra *et al.* 2003). 10-40% loss in crop production in India with increase in temperature by 2080-2100. In this direction, IARI indicated the possibility of loss of 4-5 million tonnes in wheat production in future with every rise of 1°C temperature. Temperature increase of 2-4°C are predicted to reduce rice yields. Eastern regions are predicted to be most impacted by increased temperatures and decreased radiation, resulting in relatively fewer grains and shorter grain filling durations. Study showed temperature rise of 0.5 to 1.5°C in India would decrease yield of wheat and maize by 2 to 5% (Aggarwal 2003). A rise of 2°C temperature could decrease rice yield by 0.75 tonnes/ hectare in the country. The net cereal production in South Asian countries is projected to decline at least by 4 to 10% by end of this century (Lal 2007). At IRRI, rice yield is observed to decline by 10% for every 1°C rise in growing season minimum temperature. More intense rain and frequent flash floods during monsoon results in runoff and loss of *kharif* rice. Projection stated that in Indian agriculture, quantum rise in temperature would be higher during *rabi* season, while rise in rainfall would be higher during *kharif* season (Table 5). In India, *kharif* rice yield is expected to be adversely affected by heat stress. Recent IPCC reports and other studies indicated 10-40% loss in crop production in India due to rise in temperature by 2080 to 2100. Production of rice, maize and wheat in past few decades have declined in many parts of Asia due to increasing water stress due to rise in temperature (Tao *et al.* 2004).

Table 5: Climate change impact in India for different crop seasons

Year	Season	Change in Rainfall %		Increase in temperature °C	
		Lowest	Highest	Lowest	Highest
2020s	Rabi	- 1.95	4.36	1.08	1.54
	Kharif	1.81	5.10	0.87	1.12
2050s	Rabi	-9.22	3.82	2.54	3.18
	Kharif	7.18	10.52	1.81	2.37
2080s	Rabi	-24.83	-4.50	4.14	6.31
	Kharif	10.10	15.18	2.91	4.62

Semi-arid regions of western India are expected to receive higher rainfall, while central regions likely to experience 10-20% decrease in winter rainfall by 2050. In semi-arid agro-ecological regions, water storages have already impact on agriculture. Large areas in Rajasthan, Andhra Pradesh, Gujarat and Maharashtra and some areas of Karnataka, Orissa, Madhya Pradesh, Tamil Nadu, Bihar,

West Bengal and Uttar Pradesh are already experiencing recurrent drought with several regions currently experiencing water deficits due to temperature increase. Study showed that the climate change expected to reduce crop yield in tropical areas. Increased temperatures have multiple impacts on crop productivity depending on the biological characteristics of the specific crop and the time of the heat stress in relation to its development. Higher daytime temperature accelerated plant maturity and resulted in reduced grain filling, while higher night temperatures increased yield losses due to higher rate of respiration. In dry farming regions even moderate warming of 1°C for wheat, maize and 2°C for rice will reduce yields significantly. *Kharif* rice is found to be more vulnerable to climate change. High temperatures around flowering reduced fertility of pollen grains as well as pollen germination on stigma in rice crop (IPCC 2014).

High temperatures reduced 1000-grain weight and amylase content and adversely affected important quality traits *viz.*, grain elongation and aroma in rice cultivars. A rise in temperature from 1° to 4°C reduces green gram and soybean yield. Variation in rainfall, Tmax, Tmin during monsoon will negatively impact maize crop. Hence, a pattern of resource use to be thought, which aims at meeting current need with sustainable development. Adaptive strategies are better suited for local climatic variability due to climate change.

Adaptive measures and role of IAAS/ GKMS Project

Impact of climate change on some agricultural crops in some areas is sure to occur. Integrated Agromet Advisory Services (IAAS) with crop and location specific district level advisories followed by current Gramin Krishi Mausam Seva (GKMS) with block level advisories is a strong mechanism, which can help farmers to coup up against day to day impact of climate change. The different tiers in IAAS/ GKMS is presented in Fig. 22. Under the National level project, advisories are issued for the major crops on every Tuesday and Friday, which are disseminated on real time basis through Doordarshan, All India Radio for the benefit of farming community. Constant feedback from end users' helped in enriching quality of the bulletins.

In M.P. 9 Agrometeorological Field Units (AMFUs) are there, which are issuing district level advisories throughout the State. The major agronomic crops under different agroclimatic zones along with vegetables are presented in Table 5. For these crops, timely agrometeorological advisories and agronomic practices can help farmers to coup up with negative impact of climate change. Currently on pilot mode block forecast as well as agrometeorological advisories have started in 39 blocks of the State. The verification of the same are carried out at Meteorological Centre, Bhopal. Soon, the same will be implemented at all the block levels of the State.

Table 5: Agroclimatic zones of the State of M.P. and he major crops/ Vegetables (*kharif*)

Agrodimatic Tones/ (AMFU)	Districts(% Departure of Rainfall)	CropsReconun ended	Vegetables
Kyuiorc plateau & Sutpurai Jills (Jabalpur)	Jabalpur, Satna, Panna, Seoni, Kewa, Sidlii, Kaliii, Annuppnr, Singrauli, Balaghat, 11 maria, Mandla, Dlndnri, Shahdol	Rice, maize, Jawar, Soybean(E), Moong, Tur	Tomato, Okra, Brinjal
Vindliyan Plateau (Schorc)	Sagar, Damoh, Bhopal, Srhore, Raisen, Vidisha	Rice(E), Maize, Jawar, Soybean (E)	Tomato, Chilli, Gourd, Pumpkin
Central Narmada (Powarklicda)	Narshinghpur, Hoshangabad	Soybean (K), Rice (E)	Tomato, Chilli, Gourd, Pumpkin
Gird (Morena)	Morena, Bhind, Gwalior, Shivpuri, Guna, Slicopnr, Ashnlcnagar	Bajra, Moong, Urad, Soybean (K), Tur	Tomato, Chilli, Gourd, Pumpkin
Jhabua hills (Jhabua)	Jhabua,Alirajpur, Dhar	Urad, Moong, Til, Tur	Tomato, Okra, Brinjal
Bundclkliand(Tikamgarh)	Datia, Tikamgarh, Chhatarpur	Soybean (E), Urad, Til, Moong	Tomato, Okra, Brinjal
Salpura Plateau (Clihindwan)	Belul, Chliindwara	Soybean (E), Maize (E)	Tomato, Okra, Brinjal
Malwa Plateau (Indore)	Ujjain, Indore, Mandsaur, Ratlaui, Shajapur, Rajgarh, Dewas, Nimacli	Moong, Urad, Soybean(E),H1	Tomato, Okra, Brinjal
Nimar Valley (Khargonc)	Kliargone, Khandwa, Barwani, Bhurhanpur, Uarda	Maize (K), Soybean (E), Urad, Moong	Tomato, Okra, Brinjal

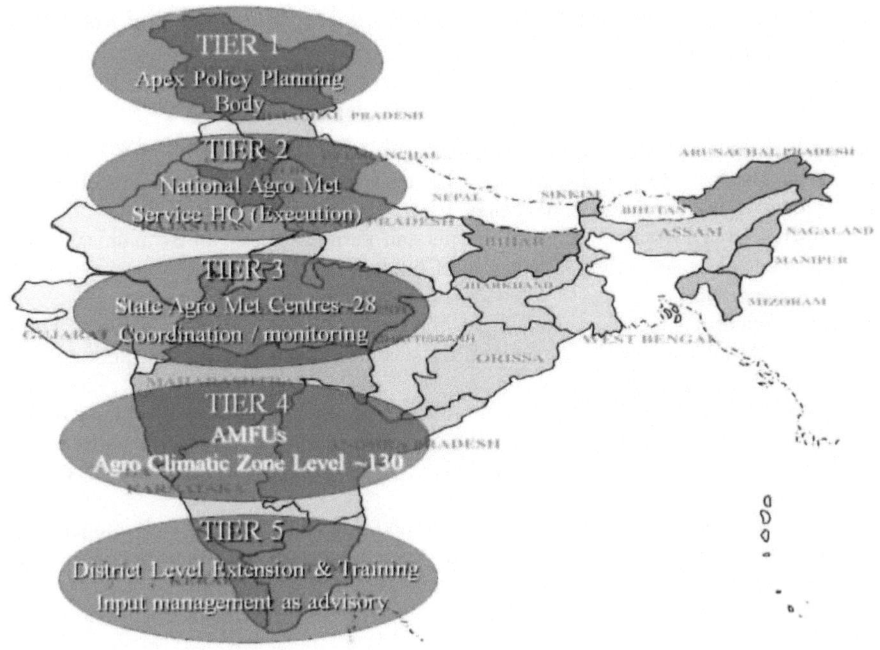

Fig. 22: Different tiers in integrated AAS in India

Conclusions

Frequency of extreme events have increased in the recent years. El Niño, La Niña, ENSO events are partially governing Indian summer monsoon. Climate change is modifying Indian monsoon by reducing number of rainy days and by delaying monsoon by 5 to 15 days. Climate change brings some changes *viz.* increase in rainfall, increase in monsoon variability. Over the State also some changes are conspicuous. Agricultural crops are most vulnerable to climate change. The crops are not only affected by an increase or decrease in overall amount of rainfall but by shift in timing of rainfall. Modern era IAAS with district level advisories or future plan of GKMS with block level advisories will help agricultural sector to coup up with the impact of climate change. In Madhya Pradesh 9 agroclimatic zones have 9 AMFUs, which are issuing AAS bulletins on every Tuesday and Friday. The real time dissemination of advisories on major crops based on weather and ongoing agricultural operations will help farmers to save their crops against negative impact of weather and climate change.

Acknowledgements

The authors are thankful to Mrs. D.A. Kulkarni, Mrs. A.S. Bahot and to Mrs. A.P. Hage, Agricultural Meteorology Division, IMD, Pune and to Ms. A.

Singhroul, M.C. Bhopal for their assistance in preparation of the manuscript (review materials).

References

Aggarwal, P. K. 2003. Impact of climate on Indian agriculture, in J. of Plant Biology, 30: 189-198.

George, D. A., Birch, C, Buckley, D., Partridge, I.J. and Clewett, J.F. 2005. Assessing climate risk to improve farm business management. Extension Farming System J. Vol 1: 1.

Hay, J. 2007. Extreme Weather, Climate Events and Farming Risks. In Sivakumar, M.V.K., Motha, R., Eds. Managing Weather and Climate Risks in Agriculture, Springer, Berlin, Heidelberg. pp.1-19.

IPCC, 1996. Scientific- technical analysis of impacts, adaptations and mitigation of climate change, Vol.2. Climate change 1995. IPCC second Assessment Report, Cambridge Univ. Press, Cambridge, U.K.

IPCC, Climate change. 2007. Impacts, Adaptation and Vulnerability. Working Group II Contribution to Fourth Assessment Report of the Intergovernmental Panel on Climate Change, Parry, M., Canziani, O., Palutikof, J., Linden, P.V. D. and Hanson, C., Eds., Cambridge Univ. Press, Cambridge, 1976.

IPCC, Climate change. 2014. Impacts, Adaptations and Vulnerability, Part- B, Regional aspects of contribution of Working Gr. II to 5th Assessment Report of the Intergovernmental Panel on Climate change Field, C.B., Barros, V.R., Dokken, D.J., Mach, K.J., Mastrandrea, T.E., Bilir, T.E., Chatterjee, M., Ebi, K.L., Estrada, Y.O., Genova, R.C., Girmma, B., Kissel, E.S., Levy, A.N., MacCracken, S., Mastrandrea, P.R. and White, L.L., Eds. Cambridge Univ. Press, U.K. and New York. pp. 688.

Kalra, N., Aggarwal, P.K., Chander, S., Pathak, H., Choudhary, R., Choudhary, A., Mukesh, S., Rai, H.K., Soni, U.A., Anil, S., Jolly, M., Singh, U.K., Owrs, A. and Hussain, M.Z. 2003. Impacts of climate change on agriculture. Climate change and India: Vulnerability Assessment and Adaptation, Shukla, P.R., Sharma, S.K., Ravindranath, N.H., Garg, A. and Bhattacharya, S., Eds, Orient Longman Pvt., Hyderabad, pp. 193-226.

Kashyapi, A., Kulkarni, D.A., Bahot, A.S. and Hage, A.P., 2012. Asian monsoon in a changing climate—Does it have impact on kharif season agriculture? Int. Conf. On opportunities and Challenges in Monsoon Prediction in a Changing Climate, OCHAMP-2012, Pune, 21-25 Feb.

Lal, M. 2003. Global climate change: India's monsoon and its variability in J. of Environmental studies and policy. 6:1-34.

Lal, M. 2007. Implications of climate change on agricultural productivity and food security in South Asia in key vulnerable regions and climate change, indentifying thresholds for impacts and adaptation in relation to Article 2 of UNFCCC, Springer, Dordrecht.

Tao, F., Yokozawa, M., Zhang, Z., Hayashi, Y., Geressl, H. and Fu, C. 2004. Variability in climatology and agricultural production in China in association with the East Asia Summer monsoon and El Nino Southern Oscillation. Climate Res. 28: 23-30.

3

Current Status and Future Climate Projections of Madhya Pradesh

Lokendra Thakkar and Ram Ratan Simaiya

Introduction

Agriculture is the mainstay of livelihood activity of about 70% of the state's population. Madhya Pradesh is an agrarian state and the state's economy is highly dependent on agriculture. Considering the sector's sensitivity to climatic parameters it can be easily understood that the sector would be highly impacted by any change in these parameters.

Madhya Pradesh has a subtropical climate with three distinct seasons' viz. winter from December to February, followed by summer season from March to May and rainy season extending from June to October. During winters the mean temperature remains around 10°C and mean maximum temperature remains 25°C and the minimum temperature goes down to 1°C in some regions. During summers, the mean minimum temperature is 22°C and mean maximum is 38°C. The maximum temperature during summer can go up to 48°C, especially in May and June which are the hottest months.

The average annual rainfall for the state is 1160 mm. Rainfall is heaviest in the south-eastern parts of the state and it decreases as one move towards the north-west. Balaghat in the south receives the maximum rains; where average rainfall is above 1600 mm. The other southern districts of Mandla, Dindori and Anuppur receive between 1200-1400 mm of rain (Goswami *et.al.*2006). The rainfall drops to the 1000-1200 mm as one move further north and west. Western districts in MP including most of those in the Malwa plateau and Sheopur and Shivpuri in the north receives 800-1000 mm of rain. Average annual rainfall is below 800 mm in the south-western tip of the state, including southern half of Jhabua and western half of Barwani district. The rainfall is below 800 mm in the districts lying to the northern most part of the state including Morena, Datia, Gwalior and Bhind. Most of the rainfall in the state is received from the south-west monsoon during June to September. (MP Resource atlas 2007, MPCST)

Impact of climate change on agriculture sector, including horticulture, animal husbandry and fisheries, is considered to be significant. These activities are not only highly sensitive to weather parameters but are also extremely sensitive to availability of water, rate of evapo-transpiration, rate and intensity of precipitation and humidity etc.

Vulnerability of the Madhya Pradesh with respect to agriculture is very high. As per the Atlas on Vulnerability of Indian Agriculture to climate change prepared by CRIDA 30 districts come under high to very high vulnerable (Rao *et al.*, 2013)., This atlas quoted that change in June rainfall in entire MP may change >5 % in 2021-2050 over 1961-1990 except Gird zone. Study also suggest that maximum changes (>5%) of rainy days comes under Nimar velly and Jhabua hills. Changes in incidence of dry spell of more than 14 days may increase in entire MP except Jabua and Nimar Zones.

Climate change projections made for India indicate an overall increase in temperature by 1 to 4°C and precipitation by 9-16% towards 2050s (Krishna Kumar et al., 2011). However, different regions are expected to experience differential change in the amount of rainfall that is likely to be received in the coming decades. Another significant aspect of climate change is the increase in the frequency of occurrence of extreme events such as droughts, floods and cyclones. All of these expected changes will have adverse impacts on climate sensitive sectors such as agriculture, forest and costal ecosystems and also on availability of water for different uses and on human health. Historical trends also show a noticeable increase in mean temperature in the country though there is no discernible trend in the rainfall during the last several decades. However, regional variation in behaviour of monsoon rainfall was observed over the years.

Materials and Methods

Climate projections for 2030s (2021-2050) and 2080s (2071-2098) have been derived from PRECIS (Providing Regional Climate for Impact Studies), which is a desktop version of the HadRM3 model with a grid resolution of 0.44° x 0.44°. PRECIS simulation data set is provided by the Indian Institute of Tropical Meteorology, Pune. The Climate Change scenarios are driven by the GHG emission scenarios - A1B which assumes a future world of very rapid economic growth, a global population that peaks in mid-century and declines thereafter, and assumes rapid introduction of new and more efficient technologies. Change in precipitation and temperature in 2030s and end of the century 2080s have been derived with respect to 1970s (1961-1990) as per the MP SAPCC.

A very high resolution (0.50 x 0.50 degree) daily rainfall and temperature data set for the period 1971-2005 prepared by IMD for the Indian region has been used for this assessment. The data set was developed using quality controlled data from more than 6000 rain-gauge stations over India. A well tested interpolation method (Shepard's method) was used to interpolate the station data into regular grids of 0.5 X 0.5 Lat x Long (*Rajeevan and Bhate, 2009)*.

Result and Discussion

Temperature: The average surface daily maximum temperatures, in the period 2030s is projected to rise by 1.8-2.0°C throughout Madhya Pradesh and the daily minimum temperature is projected to rise between 2.0°C to 2.4°C during the same period; the eastern half of the state experiencing more warming than the western half depicted in fig. 2. By 2080s, the maximum temperature is projected to rise between 3.4°C to 4.4°C with northern region experiencing warmer temperatures. The minimum temperatures are likely to rise by more than 4.4°C all over Madhya Pradesh (NATCOM Report II 2008).

Rainfall: Projections of rainfall in Madhya Pradesh for the period 2021 to 2050 indicates that there is likely to be decrease in winter rainfall as one move from eastern part of MP to western part of MP. In pre-monsoon period, the rainfall is increasing only in the Southern part of MP, with decrease in rain fall in all other parts. In the Monsoon period, there is a slight increase in rainfall all over MP (the increase being 1.25 times the rainfall observed in the current climate), and with no change in the Morena, Shivpuri, Bhind, Gwalior area.

During post monsoon period, slowly again the western end of MP is likely to face decrease in rain fall, with no change or little increase in rainfall in most other parts of the state. In 2100 there is an Overall increase in rain fall with southern states likely to receive more rain fall in the northern states. The increase in rain fall during the post monsoon and pre monsoon periods are projected to be more than the increase in rain falls projected for the monsoon period mentioned in Fig. 2.

Climate change related concerns

Madhya Pradesh is situated on the upper watersheds of 9 river basins with relatively poor quality soils of low depth and high slopes and some black soils of medium to deep soil depth with flat slopes with underlying impervious hard rock, making the underlying storage capacity low and subsequently low water availability for irrigation. The increasing dependence on the ground water for irrigation is helping to get good productivity but there is felt need for having appropriate ground water recharge strategy for augmenting the fast declining ground water table. Over exploitation of ground water for irrigation is also a

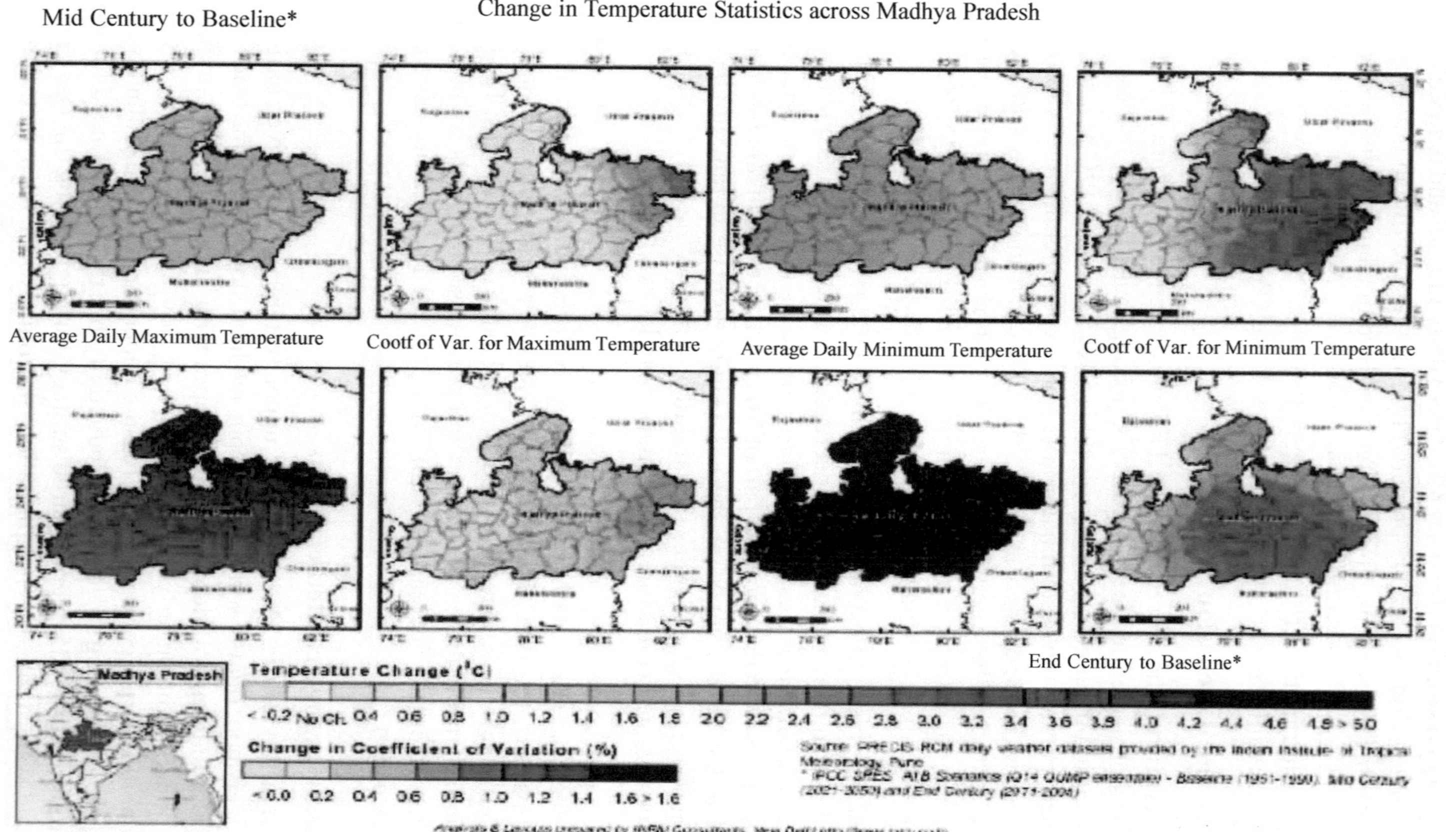

Fig. 1: Average temperature across of Madhya Pradesh

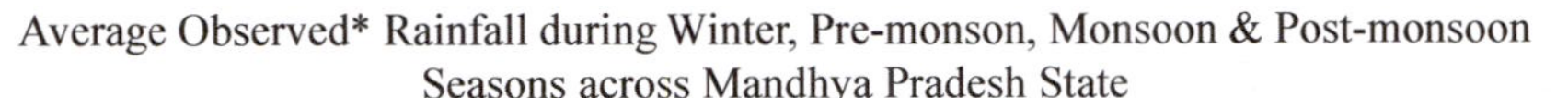

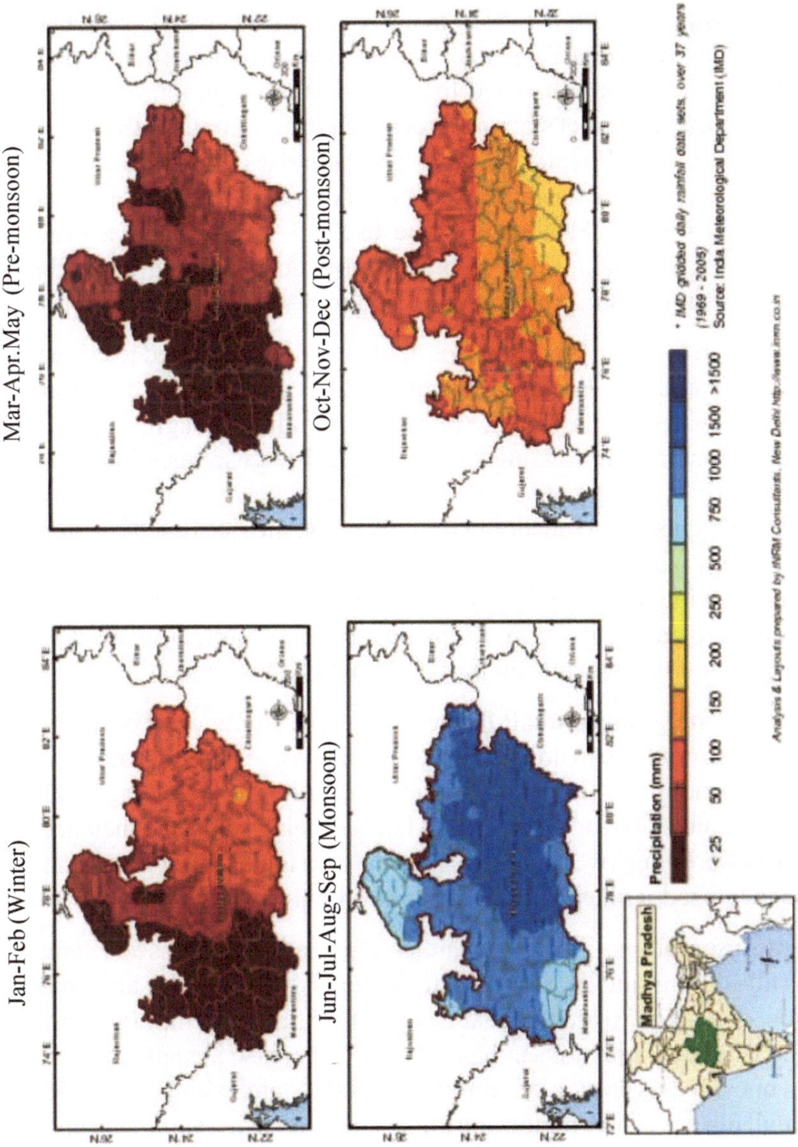

Fig. 2: Average rainfall across Madhya Pradesh

cause of concern. Though about 43% of the cropped area is irrigated, but the facilities that exist function at 50% of their potential and thus the diversion towards use of ground water for irrigation.

Though the first green revolution has bypassed Madhya Pradesh, the state has been witnessing very high growth in agriculture in recent years. During 2013-2014, MP has recorded 25 per cent growth in production. The phenomenal performance has been attributed to expansion in irrigation, farmer favorable policies, excellent market support/ procurement and finally efficient delivery of extension services. Wheat, rice, soybean and pulses are the major crops in the state. Madhya Pradesh has produced 193 lakh tones of wheat, 50 lakh tones of soybean and 69.59 lakh tones of rice in 2013-2014. Madhya Pradesh also produces nearly 50 lakh tones of pulses, of which Bengal gram and red gram are the most important. The state is a leader in soybean area and production.

Agriculture in Madhya Pradesh faces major challenges related to climate change and climate variability. Besides, general problems related to decline in soil fertility and ground water levels, climate related factors like droughts, excess rainfall, frost and hailstorm are causing significant year to year variation in production and productivity. Large areas in the state, particularly in the western part are prone to droughts due to uncertain rainfall, while the rice based cropping system in the eastern part records lower productivity due to poor water management despite receiving relatively high rainfall. The black soils in the state suffer from poor drainage and heavy rainfall in short period results in water logging and yield reduction in soybean and pulses (Vyankateshwarulu *et al*., 2014).

Capacity of marginal & small farmers constituting majority of the population engaged in agriculture to utilise the latest technologies for improving production of crops, horticulture, livestock products and fisheries is also minimal. Further, the land holdings of poor and marginal farmers being small and their accessibility to farm inputs being low, productivity of the area under this type of land holdings can be increased with enhanced agricultural inputs.

Current observations of climate indicate a gradual increase in temperatures across all seasons and decrease in rainfall in all the agro-climatic zones during the monsoon period with erratic and uneven spatial and temporal distribution of rainfall. Climate projections indicate a further increase in temperatures by 2050s and an uneven distribution of rainfall across the state, with perceptible decrease in rainfall during winter period and almost no change in rainfall during monsoon with respect to current climate. However, the frequency and intensity of droughts and heavy rainfall events with respect to the current situation is likely to increase further (MoEF, 2010). These predicted changes could lead to spatial and temporal shift of cropping centres and decline of productivity of crops. Also as thc cvapo-transpiration rate increases with increase in temperature, it will lead to depletion in moisture retention capacity of the different soil types and can pose a threat to agriculture.

On the other hand, increase in intensity of rainfall is likely to lead to faster run off causing higher soil erosion in the ravine areas with little or no scope of ground water recharge in the alluvial plains causing further depletion in the ground water tables in MP. The status of soil health and its fertility is likely to deteriorate further with increase in soil erosion and higher temperatures, causing stored carbon to be released from the soil. MPSAPCC quoted, it is likely that the onset of monsoon may shift from June to first fortnight of July in the state which would likely to affect the cropping sequence and sowing time.

Therefore, even if farm inputs are adequately provided through development of effective delivery mechanism, changes in climate in the future may lead to conditions which may not support sustainable crop production as productivity of crops is an integrated process of favorable climatic conditions as well as farm inputs. MPSAPCC has taken into account the agriculture, horticulture, animal husbandry and fisheries together under the agriculture sector. The projected increase in temperature towards mid and end centuries coupled with reduced availability of water and excessive water loss through evapo-transpiration is projected to aggravate drought like conditions, especially in western MP where sign of xerofication can be seen.

Climate change associated impacts on crops are believed to be shifts in date of sowing and maturity period of crops, negative impacts on grain filling and formation, deterioration in quality of seeds though production may increase, negative impact on the quality of aromatic produce like basmati rice, increase in pest infestation, loss of soil nutrient and soil carbon. Animal husbandry and fisheries too could be adversely impacted by climate change. Negative impacts on livestock health as well as its productivity is projected associated with variation in ambient temperature and humidity.

The non-climatic stresses like monocropping, dwindling crop diversities, near extinction stage of coarse grains or millets, deteriorating soil health, lack of awareness about soil nutrition management and irrigation management among farmers at large, burning crop residues in fields causing CO2 emissions, unavailability of quality feed and fodder for animals, knowledge gap among community about sustainable and best practices of animal rearing and development etc. may aggravate the situations in the times to come.

Conclusion

Growing climate-related risks result in greater needs for effective social protection systems and services shielding the most vulnerable from shocks and stresses. The poorest and most food insecure households and communities face significant barriers to adopting more resilient and productive livelihood strategies that would allow them to enhance their food and nutrition security in the face of climate change. Specific policies and interventions need to be deployed to ensure that these challenges are turned into opportunities for the most vulnerable.

Broad strategies suggested in MPSAPCC for Agriculture sector are:

- Promoting Soil and Water Conservation technologies,
- Planning cropping systems suitable for each agro-climatic zone,
- Management of risks for sustainable productivity,
- Enhancing dissemination of new and appropriate technologies developed by researchers and strengthening research in agriculture sector
- Agriculture Information management and Creation of Rural Business hubs and accessibility to market
- Ensuring availability of adequate feed, fodder and water for livestock
- Ensuring nutrient solvency in Livestock
- Enhanced capacity for disease forecast, monitoring and management
- Promote research on native species breeding and rearing
- Ensure proper housing and dedicated water bodies for livestock to overcome heat stress
- Promote new verities of poultry and native species of small ruminants
- Promote use of livestock and poultry waste for use as organic manure
- Integrate livestock water requirements in watershed management programmes
- Infrastructure for processing, storage and transport of livestock products
- Capacity building for sustainable agriculture.

References

Goswami B. N., Venugopal V., Sengupta D., Madhusoodanan M. S., Xavier K. Prince 2; Increasing Trend of Extreme Rain Events over India in a Warming Environment, Science 1 December 2006: 314(5804):1442–1445.

MP Resource Atlas (2007), Madhya Pradesh Council of Science & Technology.

MP State Action Plan on Climate Change (2014) Climate Change and Risk Assessment Report. NATCOM Report II -2008 Ministry of Environment and Forest, Government of India.

Krishna Kumar, K., Patwardhan, S.K., Kulkarni, A.,Kamala, K., Koteswara Rao, K. and Jones, R. (2011). Simulated projections for summer monsoon climate over India by a high-resolution regional climate model (PRECIS). Current Science, 101(3): 312-326 .

Rama Rao, C.A., Raju, B.M.K., Subba Rao, A.V.M., Rao, K. V., Rao, V. U. M., Ramachandran, Kausalya., Venkateswarlu,B., and Sikka, A.K., (2013). Atlas on Vulnerability of Indian Agriculture to Climate Change. Central Research Institute for Dry land Agriculture, Hyderabad.

Rajeevan, M.; Bhate, Jyoti (2009) A high resolution daily gridded rainfall dataset (1971-2005) for mesoscale meteorological studies Current Science, 96 (4):558-562.

Vyankateshwarulu,B.; Rao, V.U.M. (2014) Agriculture in Madhya Pradesh: Impacts of climate change and adaptation strategies, Policy brief on Agriculture. Madhya Pradesh Climate Change Knowledge Portal: www.skmccc.net.

[illegible]

[illegible] Climate Change [illegible] Report [illegible]

[illegible] (2017). [illegible] climate [illegible]

[illegible]

[illegible]

[illegible]

4

Managing Atmospheric Stressors Forced Challenges in Agriculture under Future Climate

S.K. Bal, P. Suresh Kumar, Y. Singh, Nilesh More and P.S. Minhas

Introduction

Climate is the primary determinant of agricultural productivity. Plant systems, and hence crop yields, are influenced by many environmental factors such as moisture and temperature, may act either synergistically or antagonistically with other factors in determining yields (Waggoner, 1983). In the recent years, climatic change/variability is a biggest challenging factor that affects agriculture in India and elsewhere and concern has been expressed by world community regarding the potential effects of climate change on agricultural productivity. Climate change is expected to influence crop production, hydrologic balances, input supplies and other components of agricultural systems. Agricultural systems are managed ecosystems. The climate change induced atmospheric stress will be the primary cause of crop loss worldwide, reducing average yields for most major crop plants. Environmental stresses such as erratic and insufficient rainfall, extreme temperatures, salinity, alkalinity, acidity, heavy metal toxicity, oxidative stress and others limit yield and productivity of many cultivated crop plants. Drought, extreme temperatures and oxidative stress are interconnected and affect the water relations of a plant on the cellular as well as whole plant level causing specific as well as unspecific reactions (Beck *et al.*, 2007). This leads to a series of morphological, physiological, biochemical and molecular changes that adversely affect plant growth and productivity (Wang *et al.*, 2001). Thus, the combined effects of warming, precipitation change, and CO_2 fertilization on crop yields are expected to vary by crop, location, the magnitude of warming, the direction and magnitude of precipitation change, and the nature of the CO_2 fertilization effect. Increasing temperatures, holding other factors constant, decreases crop yields while increases in precipitation lessen or offset this result.

Inclusion of estimated effects of increasing concentration of CO_2 on crop yields (i.e. a fertilization effect) significantly raises the yields of many crops. There are likely to be regional winners and losers from climate change, given that the potential for net reductions in crop yields is greatest in warmer, low latitude areas and semi-arid areas of the world. This implies that climate change may affect the comparative advantage of agricultural production regions. Changes in comparative advantage can be expected to shift geographically the areas in which specific crops are grown, both within countries and internationally.

Forecasting climate change is imperfect, complex, important, and often controversial. While disputes remain, the consensus foresees accelerating increases in average annual temperature and changes in precipitation coupled with increasingly erratic intra-annual weather patterns. Generally, developing countries are vulnerable to climate change because they depend heavily on agriculture, tend to be relatively warm already, lack infrastructure to respond well to increased variability, and they lack capital to invest in innovative adaptations. Concerns about mitigating and adapting to climate change are now renewing the impetus for investments in agricultural research and are emerging as additional innovation priorities. In the coming decades, the development and effective diffusion of new agricultural technologies will largely shape how well farmers mitigate and adapt to climate change. This adaptation and mitigation potential is nowhere more pronounced than in developing countries where agricultural productivity remains low; poverty, vulnerability and food insecurity remain high; and the direct effects of climate change are expected to be especially harsh. Future farming will have to be better adapted to a range of stresses to cope with the direct and indirect consequences of a climate change (Pathak *et al.,* 2012).

In this chapter, an effort has been made to link the impact of future climate on various atmosphere related abiotic stresses and possible management options to counter their negative impacts in agricultural fields whereas, dealing with the stresses in reality is a highly onerous task owing to their complexity, uncertainty and differential temporal and spatial impacts.

1. Temperature extremes

Susceptibility to high (heat injury inducing) or low (chill injury inducing) temperatures is known to be reduced by prior exposure of the sensitive fruit or vegetable to low ambient temperatures (Wang, 1993). However, if the pre harvest temperature leads to chilling induced injury in the field, then susceptibility to postharvest chilling injury can be increased. Therefore, the level of the pre harvest temperature extreme will be a determinant as to if the exposure will have positive or negative effects on postharvest stress sensitivity. Extreme high temperatures can occur in the field and apple fruit exposed to direct sunlight can reach in excess of 40 °C. High temperatures during the late season (leading up to harvest) can enhance susceptibility of apples to superficial scald which develops in storage. Sun scald of grapes, soft nose of mango, sun burn in papaya and citrus are mostly related to high temperature. Fruit cracking in pomegranate and litchi is associated with temperature aberrations coupled with fluctuation in moisture. Similarly in low temperature when plant tissue temperatures fall below critical values, sensitive perennial crops such as grapevines and tender tree fruits can suffer irreversible cold injury, causing malfunction or death of plant cells.

It has been suggested that higher temperatures reduce net carbon gain by increasing plant respiration more than photosynthesis. In fact, the light-saturated photosynthesis rate of C_3 crops such as wheat and rice is at a maximum for temperatures from about 20–32 °C, whereas total crop respiration shows a steep nonlinear increase for temperatures from 15 to 40 ^{0}C, followed by a rapid and nearly linear decline (Porter & Semenov, 2005). Increased temperature could potentially reduce photosynthetic capacity due to the heat lability of Rubisco on the one hand and the limitation of electron transport in the chloroplast on the other hand (Sage *et al.,* 2008). In addition, by increasing the air evaporative demand, higher temperatures are often implicated with stomatal closure, which further decreases photosynthesis due to smaller CO_2 flux into leaves. The predicted increase in average global temperature will accelerate crop development rates, and the negative effect end to be larger for grain yield than for total biomass. These constraints associated with other on-going climatic changes such as more frequent and severe droughts and more intense precipitation events will probably offset gains in carbon assimilation associated with elevated CO_2 (Sage *et al.,* 2008).

At warmer temperatures, the yield of wheat may decline up to 10% per 1 ^{0}C rise in mean seasonal temperature. The timing of wheat flowering and grain maturity may be considerably earlier at warmer temperatures, thus shortening the time for carbon fixation and biomass accumulation before seed set. Hot temperatures (>32–36 ^{0}C) can also greatly reduce seed set in many annual crops if elevated temperatures coincide with a brief critical period of only 1–3 days around the time of flowering (Craufurd and Wheeler, 2009). In groundnut, Vara Prasad *et al* (2000) noted that, between 32 and 36 ^{0}C and up to 42 ^{0}C, the percentage fruit set fell from 50% of flowers to zero and the decline in rate was linear, illustrating the sharpness of response of crop plants to temperatures during the flowering and fruiting periods. Similar patterns have been identified for other food crops; for example, in maize, pollen viability is reduced at temperatures above 36 ^{0}C, while grain sterility is brought on by temperatures in the mid-30 0C in rice. In fact, the reproductive limits for most crops are narrow, with temperatures in the mid-30 °C representing the threshold for successful grain set (Porter & Semenov, 2005). In Indian Punjab, decrease in productivity of rice-wheat system is expected because of the overpowering yield decreasing effect of increased temperature over yield-enhancing effect of rising CO_2 (Jalota *et al.,* 2013).

Different crop species have different T optima as well as different sensitivities to CO_2 and O_3. One useful distinction is between crops that originated in temperate environments, such as wheat and barley, versus crops from tropical environments, such as cassava (*Manihot esculenta*) and sorghum (*Sorghum bicolor*). A recent synthesis of the literature (Hatfield *et al.,* 2011) identified optimal season average T of 15°C for wheat, 18°C for maize, 22°C for soybean, 23°C for rice and bean (*Phaseolus vulgaris*), and 25°C for cotton (*Gossypium hirsutum*) and sorghum. An important distinction for CO_2 sensitivity is between C4 grains (least responsive), C_3 grains (more responsive), and root and tuber crops (most responsive). For example, a recent field study of cassava showed roughly a doubling of dry mass for a CO2 increase from 385 to 585 μL L"1 (Rosenthal *et al.,* 2012).

Cold wave

The prolonged cold spell causes large-scale damage to agricultural and horticultural crops. The damage to fruit plantations is so severe that many a times farmers up root their well-established orchards. Crop yields were lower by 10 to 40 per cent in wheat, 25 per cent in gram, 50 to 70 per cent in mustard and 60 to 95 per cent in aonla compared to a normal year. The orchards of mango, litchi, guava, ber and kinnow in Punjab were hit resulting in reduced size or poor quality of the fruit. The plant mortality in mango varieties was highest in Dashehari, followed by Amrapali and least in Langra with 18 to 48 per cent flowering damage in parts of Jammu, Punjab, Haryana, Chandigarh and Himachal Pradesh. The mortality rate in papaya ranged from 40 to 83 per cent in the Shiwalik region, plains of Uttar Pradesh, Bihar and the Northeast. The damage to fruit trees was relatively more in low-lying areas where cold air settled and remained for a longer time on ground. But temperate fruits such as apple, peach, plum and cherry gave higher yield due to extended chilling. One of the best ways to reduce the impact of the cold wave is to go in for proper selection of fruit species and varieties as per the site conditions. Apart from that, wind breaks or shelter belts should be provided. There should be frequent irrigation and smoking. Protective cultivation covering young fruit plants with thatches or plastic sheets are one of the best options.

Heat wave

A heat wave is a run of unusually hot days (Three or more consecutive days of unusually high maximum temperatures) but what constitutes a heat wave will differ from region to region given the normal temperatures for that area. Inland areas can expect to be normally hotter than areas closer to the coast. Heat-wave intensity is classified as low, severe or extreme based on the actual temperatures and the previous 30 days' temperatures. Fruit is less likely to set if heat waves occur during blossoming. There are no precise threshold temperatures or periods for reduced pollination because it is a complex interaction between bee activity and the availability of viable pollen and receptive stigmas. Sunburned fruit is caused by combined high air temperature and solar radiation. Sunburn typically occurs in the afternoon on the western side of trees and is often worse where cool, cloudy weather is suddenly followed by a change to hot, sunny conditions. This problem is becoming more common with the increased frequency of heat-waves. An early sign of fruit sunburn is white, tan or yellow skin. Severe sunburn results in dark brown to black sunken patches on the fruit surface. Browning associated with sunburn is likely when the fruit surface reaches 46-49°C. Skin necrosis occurs at 52°C. Air temperature provides a reasonable estimate of sunburn risk. It is normally 10-18°C lower than fruit surface temperature. In grapes, hot and dry summers typically result in early harvests of mature fruits associate with lower-yield harvests.

Sunburn protectants should be considered when forecast air temperatures are 30°C or above particularly where relative humidity is likely to be low and trees are water stressed. Increased irrigation, of course, mitigates growing stress due to excessive heat. Metabolism is quick under optimal environmental conditions that include warm dry days and cool nights. Humidity acts to slow down metabolic processes in grapes at least (without considering direct biochemical influences) by disproportionately increasing nighttime temperature when daytime temperature is already at detrimentally high levels. Under such extreme conditions, new-world wine growers and makers, preferring to enhance the grapes' maturity and the character of the vintage apparently prefer to harvest later even at the risk of losing a significant portion of the crop to heat stress.

2. Frost

In clear sky and minimum wind, radiational cooling forces temperature inversions to occur where air temperatures 10-20 m above the field are at least 2-3°C warmer than at 1 m above ground. For formation of inversions, wind speeds must be less than 7 km/h; for inversions to break up, wind speeds must be greater than 7 km/h. This is because the wind mixes the air above the field with the air at ground level. Frost occurs when water vapour in the air freezes upon contact with an object that has a surface temperature below 0°C. It is commonly seen as ice crystals on tree branches, grass stems, and car windscreens. Radiation frost occurs at night in cold temperatures when heat radiates into the atmosphere and leaves the ground and ambient air cooler. Drought years are conducive to radiant frost due to the lack of cloud cover and the increased difference between day and night time temperatures whereas advective frost occurs when a mass of very cold air moves over an area, replacing the warmer air in that area. It is not affected by cloud cover

When plant tissues freeze, ice crystals form and rupture cell walls. The resulting dead tissue leads to the familiar water-soaked and scorched appearance of frosted plants. Late, heavy frosts during spring can kill fruitlets at any stage of development. The severity of the damage varies with the phenological stage and type of fruit tree affected. The duration of exposure to frost and the rate of tissue defrosting also influence the amount of damage. Developing fruit can withstand relatively low temperatures early in the season, but become more susceptible to frost damage. Fruitlets which are killed shed early in the season. Fruit can also suffer sub-lethal damage which distorts growth during maturation and makes it unmarketable. This damage is often expressed as banded areas of russetted, corky growth or constricted bands leading to misshapen fruit. Severe, late frosts can also kill the buds responsible for the following year's crop and have an on-going effect on orchard productivity.

Frost prevention and protection

Protection has three basic objectives i.e. to reduce heat loss from the surface, stirring the air and break up the temperature inversion layer and adding heat to maintain the temperature above the danger level.

Site selection

Where advection frosts occur, frost-sensitive crops need to be planted on slopes so that cold air can flow downhill and escape the orchard. Belts of trees, weedy fence lines and buildings can trap cold air. Low branches on wind breaks at the bottom of slopes need to be pruned to allow cold air to drain. Windbreaks uphill of an orchard can prevent cold airflows reaching the crop. Orchards located next to dams, lakes or rivers are usually less prone to frost damage. Air masses over water cool less rapidly at night than over land. Where radiant frost occurs, orchards need to be well irrigated. Moist, compact soil absorbs heat during the day and radiates this heat during the night. This heat may be enough to prevent damage from light frost. Dry, loose soils do not warm during the day and offer no protection during the night. Mulches need to be applied after flowering and kept moist. Inter-rows need to be slashed as low as possible to trap radiant heat during the day for release in the early mornings to reduce frost damage.

Netting

Netting provides some protection from frost damage in light to moderate frost events. Nets hold heat radiating from the ground and raise the temperature in the orchard. Nets also block much of the cold air flow associated with advection frosts. Frost protection is really only effective for radiant frosts when winds are light or calm, and in low-lying, frost-prone areas. Protection efforts need to focus on the most frost-prone blocks and in response to forecasts of the critical temperatures needed for substantial crop damage.

Wind machines

Wind machines can raise air temperatures around plants by about a third to half the temperature inversion difference. One protection method is to use wind machines in orchards of high value crops. Wind machines are tall, fixed-in-place, engine-driven fans that pull warm air down from at least 15 m above ground during strong temperature inversions, blowing it down and out, pushing away and replacing cold air near target crops. This raises air temperatures around cold-sensitive perennial crops such as grapes. Wind machines break up micro-scale air boundary layers over plant surfaces, improving sensible heat transfer from the air to the plants. Wind machines transfer heat by forced convection. The area protected covers 3-5 ha, depending on topography, field layout, strength of temperature inversion, time of year and drift due to slight winds.

Overhead irrigation and use of chemicals

This method protects plant tissue. It does not raise the air temperature or warm plant parts. There is likely to be a build-up of ice on branches and the resulting weight can cause breakages. Sprinklers need to be turned on with the onset of freezing conditions and maintained at a rate of 1.5 to 4 mm per hour until temperatures have risen above freezing point. If irrigation is turned off prematurely, heat will transfer from the plant tissues to melt the ice, so freeze damage is likely. Abscisic acid has been demonstrated to reduce chilling-induced injury in some crops (Wang 1993).

3. Hailstorm

Hailstorms are the result of four atmospheric factors which are characterized as strong convective instability creating strong updrafts, abundant moisture at low levels feeding into the updrafts, strong wind shear aloft, usually veering with height, enhancing updrafts and some dynamical mechanisms that can assist the release of instability such as air flow over mountain ridges. The development of hailstones typically occurs 5 to 7 km/hr above the earth's surface. Its size and shape depend on how fast the storm is moving and how strong the updrafts are inside the storm. A typical hail-streak is about 1.5 km wide and 8 km in length. However, these may vary from a few acres to large belts, about 16 km wide and 160 km long. The volume of hail reaching the ground falls at a speed of about 40 m/sec, and is usually less than 10 per cent of the volume of rain produced by a thunder storm (Bal *et al.,* 2014).

Hail control mechanisms

Artificial hail control is an important measure in disaster prevention and mitigation. Cloud seeding for hail suppression is based on the cloud microphysical concept in which seeding is postulated to reduce hail severity. The natural and artificial ice crystals compete for the available super-cooled liquid cloud water within the storm. Hence, the hailstones that are formed within the seeded cloud volumes will be smaller and produce less damage if they should survive the fall to the surface. If sufficient nuclei are introduced into the new growth region of the storm, then the hailstones will be small enough to melt completely before reaching the ground. Another concept is to create shock waves which can prevent the formation and growth of hail by melting altogether. Shockwaves are produced using hail guns/cannons. The super-cooled water situated on the external layer of hailstone is transformed from liquid state to solid state. Therefore the hail nuclei are not able to melt anymore and remain at a small size which thus minimizes the damage when they hit the ground. It is be reiterated that hail control mechanisms cannot eliminate hail completely but the cloud seeding can be beneficial (Bal *et al.,* 2014).

Preventive measures

Protective screens termed as anti-hail nets above the crop can be appropriately utilised especially for high value crops. The hail climatology, microclimatic effects of the cover, its durability properties and installation cost are useful in evaluation of the merit of such screens. These anti-hail nets are not effective against strong hail storms. Tree shelterbelts can markedly reduce hail damage in their immediate vicinity since hails are usually associated with strong winds. Some hail is intercepted directly by the trees protecting crops immediately downwind. The trees also create a change in the air flow so that the area in the lee of them is particularly sheltered with hail deflected laterally. Wind speeds will also be less in the lee of the shelter so the total hail kinetic energy, which results both from the vertical fall speed of the hail and the wind speed will be less (Bal *et al.,* 2014).

Post-hail management options

For orchard crops, it is advised to remove all the broken branches and twigs. Fruit thinning of hail damaged fruits will improve yield and quality of remaining fruits. Contingency crops like green gram and black gram may be taken to utilize soil moisture and to recover a part of losses. Pruning may be necessary to retrain young trees and optimize new growth. Bud breaking chemicals and growth regulators may be applied to induce the vegetative growth along with fertilizers. In fruit crops like pomegranate, grapes, guava, etc. heavy flower and fruit drop has occurred due to hail damage. The damaged flowers and fruits started rotting due to cold shocks and infection by bacterial and fungal pathogens like grey sooty mould. Hence, such fruits, flowers should promptly be collected and destroyed to avoid further spread of infection. This will help in preventing spread of insect pests like fruit borers that are hiding inside damaged fruits. Since the severely damaged barks on branches and stems become very susceptible to attack by insect pests like stem borers, bark borers, and infection by fungal diseases like blight, rot, wilts, etc, the damaged/ broken branches of plants should be pruned immediately. The spraying of 1 % Bordeaux mixture was beneficial in healing of wounds. Alternatively, spraying of copper oxychloride at 2-2.5 g/ litre can also be done. Due to cloudy weather coupled with increased atmospheric humidity and wet foliage conditions, the plants suffered from hail damage were likely to be affected severely with sucking pests like thrips, jassids, mites, and diseases like powdery mildew, leaf blights, fruit and flower rot, etc. For the control of sucking pest attacking new foliage/ growth, spraying of systemic insecticides/ acaricides like Spinosad (at 0.3 ml/litre), fipronil (at 1.5 ml/ litre) should be done. For controlling powdery mildew, spraying of Hexaconazol (at 1 ml/litre), propiconaole (at 1 g/litre) was useful. Spraying of Mancozeb (at 2.5 g/ litre) or Carbendazim (Bavistin) (at 1 g/ litre) helped to prevent infection of blights and rot diseases (Bal *et al.,* 2014).

4. Severe storms

It is difficult to make projections about thunderstorms because of its localized nature. As per the IPCC report, the warming climate is likely to increase the length and frequency of severe thunderstorm environments especially during summer storm season. Storm damage in orchards is determined by storm severity and crop maturity. A severe storm close to harvest is more serious than a less severe storm earlier in the season. Fruit damage depends on fruit maturity, size and texture of hailstones, and the speed and force of hail as it hits fruit. A high velocity impact early in the season can cause deep depressions in fruit which persist to harvest. Later season damage can vary from bruised fruit to broken skin and holes in the fruit flesh. Fruit can also be stripped from the trees by the force of wind. Damage in mature fruit quickly become focal points for brown rot (stone fruit, cherries), blue rot (apples) and grey rot (all fruit). While the disease may initially be limited to damaged tissue, it can quickly spread to intact fruit in warm, humid conditions. Hailstones and wind can damage tree trunks and limbs and create entry points for pests and diseases.

Prevention

Storm preparation is critical because little can be done to reduce losses after a storm passes through. Netting is expensive, but the reduced risk of crop loss from hail, and from bird, bat and sunburn damage justifies the expenditure. Structured netting (with posts and tensioning wires) is the most expensive option and requires seasonal checking and adjustment. Throw-over netting can provide similar levels of protection but makes orchard management difficult and may need to be removed in summer for pruning and harvest. Both systems have advantages and orchardists need to investigate their options.

Post storm management

Removing damaged material allows trees to divert resources to potentially good fruit and minimizes the risk of disease. Severely damaged limbs need to be summer pruned and removed from the orchard. Large wounds caused by pruning and hail damage need to be sealed as soon as possible with water-based paint or a commercial wound dressing to prevent bacterial and fungal diseases and insect pests. If not done immediately after the storm there is risk of sealing in disease organisms. Hail-damaged and split fruit needs to be removed and mulched. Damaged fruit can become a reservoir of disease and infect intact fruit resulting in increased loss close to harvest. Diseases and pests are also likely to be a more substantial issue in subsequent seasons if damaged fruit is left to rot in-situ. Use of fertilizers and irrigation also allows trees to recover from storm stress. Applications of calcium increase fruit firmness, but the timing is critical and effectiveness is influenced by irrigation.

5. Meteorological drought

Drought results in wilting, reduction in photosynthesis, disturbances in physiological processes, cessation of growth, or even death of the plant or plant parts which ultimately reduce the potential yield. The occurrence of drought conditions during production of fruit and vegetable crops is becoming more frequent with climate change patterns (Whitmore, 2000). Water stress during the production phase of crops may affect their physiology and morphology in such a manner as to influence susceptibility to weight loss in storage (Kumar *et al.*, 2013). Another negative effect associated with water deficit is that, pre-harvest water stress (watering to 25-75% of soil water field capacity) can weaken the cells, resulting in higher membrane leakage (i.e. cell damage) and consequently greater weight loss in storage. Therefore it is critical to avoid water stress until the fruit has reached maximum size in order to minimize incidence of chilling-induced injury in storage. Any trait can thus have different (positive, negative, or neutral) impacts on yield, depending on the drought scenario. Water conservation traits have beneficial effects in the most severe drought conditions or in terminal drought scenarios, in which it is useful to save water for the end of the crop cycle and to decrease the numbers of sinks for assimilates in order to obtain larger seeds. Traits that serve to conserve water (conservative traits) include low stomatal conductance, low leaf growth rate, high WUE, or deep but sparse root systems. Conversely, growth 'maintenance' traits have beneficial effects under milder drought conditions, in which the soil profile is periodically re-watered.

Selection of suitable crops and varieties

Varieties having good root system and capacity to recoup after the alleviation of stress need to be selected and depending upon situation, it is recommended to use short duration varieties. The fruit crop selected for arid region must be such that its maximum growth period synchronizes with the period of maximum water availability and low vapour pressure deficit in the atmosphere. Fruits like ber, custard apple, phalsa, *Cordiamyxa* are suitable for such condition. Otherwise crop selected should be such that their reproductive cycle can be monitored to synchronize with maximum moisture availability periods e.g. Guava, pomegranate and acid lime which bear fruits in distinct *bahars* and the *bahar* which coincides with rainy season can be encouraged. Since water is a limiting factor in arids, the crop selected should have drought tolerance mechanism like deep root system to draw water from deeper soil (ber, date palm), leaf shedding (ber, gonda), water binding mechanism (fig), presence of thorns (caronda), stomata at lower side (anona), wax coating (ber), leaf orientation (*aonla*), hairyness, sunken and covered stomata (fig, phalsa, ber and gonda).

In sugarcane, drought stress occurring when the leaf canopy is already well established has more serious impacts on final yield in terms of total biomass, stalk biomass, and stalk sucrose than when drought occurs earlier during the season (Robertson *et al.*, 1999). Periods of soil water deficits can occur at any time during the growing season, and not all stages of plant development are equally sensitive in terms of impact on ultimate crop yield. For example, crop species such as maize are particularly sensitive to drought at the heading stage, which is just before tassel flowering.

For more severe water deficit scenarios, breeding strategies tend to focus on reducing the risk of total yield loss. This can be achieved by decreasing cumulative transpiration, which often incurs a penalty in terms of yield potential and performance under mild or moderate water deficit conditions. Such strategies therefore often increase yield stability at the expense of yield potential. These strategies include

(a) shortening the duration of the crop cycle (i.e. phenological adjustment) to escape from the drought period. This strategy has proved to be most successful for C_3 cereals such as wheat or barley in environments, which experience drought at the end of the growing season, here referred to as 'terminal drought' (Araus *et al.,* 2002)

(b) reducing leaf area or stomatal conductance, traits which frequently increase the WUE (i.e. the amount of harvested biomass per unit transpired water).

Biomass production is tightly linked to transpiration, WUE, and nitrogen accumulation. Blum (2011) has argued that breeding for high WUE under drought conditions will ultimately result in low-yielding genotypes with reduced drought tolerance because WUE is increased by reduced transpiration and water use. Therefore, biomass production under most drought conditions (i.e. except for marginal drought-prone areas) can only be enhanced by maximizing soil moisture capture for transpiration, which also involves minimized water loss by soil evaporation. Collectively, this property has been called effective use of water (Blum, 2011).

The spatial distribution of the roots in the soil, and not total root mass *per se*, defines the ability of a root system to take up water. An 'ideal' root system, with an even distribution of <-1 cm root m^{-3} of soil (root length density or RLD) is considered to be most suitable under field conditions (Tardieu *et al.,* 1992). Higher RLDs are adequate when evaporative demand increases and soil water reserve decreases. However high RLD values might ultimately prove to be wasteful of photo assimilate and other resources without giving appreciable increase in water uptake.

Improved method of seedling production: Improved method of seedling production such as Portray grown seedling using coco peat, nylon net protection and bio-fertilizers/bio-pesticide inoculation at nursery stage has good potential for obtaining sturdy, uniform and healthy seedlings. These seedlings when transplanted in the main-field will establish better with less root damage and fare better in overcoming biotic and abiotic stresses particularly during water stress conditions.

Adoption of soil and moisture conservation techniques: Contour cultivation, contour trip cropping, mixed Cropping, tillage, mulching, zero tillage, are some of the agronomical measures for the *in-situ* soil moisture conservation.

Mechanical measures like contour bunding, graded bunding, bench terracing, vertical mulching etc. also need to be followed for effective soil and moisture conservation in dry lands. Another technology for efficient utilization of runoff is water harvesting recycling. Rainwater harvesting includes collecting runoff water into dug out ponds or tanks in small depressions, gullies and into storage dams of earth or masonry structures. Rain water harvesting is possible in areas having rainfall as little as 500 to 800 mm. Depending on the rainfall and soil characteristics, 10-50 % of the runoff can be collected in farm pond. Surface run off thus collected in a farm pond can be used to provide protective irrigation in the period of prolonged dry spell. Trench planting (0.5-1.0 m) is recommended for ber, aonla and custard apple to conserve moisture. The trenches collect rain water along with silt and organic matter and thus promote tree growth. Planting in trenches is common for pineapple under dry conditions. Contouring, contour bunds, graded bunding, terraces, broad bed furrow system, curved land ploughing, loose boulders are some of the other agronomical measures used to conserve moisture.

Enhancing soil organic matter content: Incorporation of crop residues and farm yard manure to soil improves the organic matter status, improves soil structure and soil moisture storage capacity. Organic matter content of the soil can also be improved by fallowing alley cropping, green manuring, crop rotation and agro forestry. Vegetable being short duration crop and having faster growth phases, the available organic matter needs to be properly composted. Vermi composting can be followed for quicker usage of available organic matter in the soil and improving the soil moisture holding capacity.

Water saving irrigation method

In the midst of increasing urban and environmental demands on water, agriculture must improve water use efficiency generally. Adding climate change to this mix only intensifies the demands on water use in agriculture. With hotter temperatures and changing precipitation patterns, controlling water supplies and improving irrigation access and efficiency will become increasingly important. Climate changes will burden currently irrigated areas and may even outstrip current irrigation capacity due to general water shortages, but farmers with no access to irrigation are clearly most vulnerable to precipitation volatility.

Across the Middle East and North Africa, Central Asia and Southern Africa, water availability is projected to decline dramatically with climate change and population growth in the next several decades. It is no exaggeration that the future of agriculture in these regions hinges primarily on improving the efficiency of existing irrigation systems and, where profitable, extending irrigation infrastructure. In places with limited access to irrigation, well timed 'deficit irrigation' can make a substantial difference in productivity. With dwindling water supplies, such deficit irrigation techniques will become increasingly important. In non-irrigated areas, water conservation and water harvesting techniques may be farmers' only alternative to abandoning cultivation agriculture all together. Drip irrigation has proved its superiority over other conventional method of irrigation, in horticulture due to precise and direct application of water in root zone.A considerable saving in water, increased growth, development and yields of fruits and vegetables and control of weeds, saving in labour under drip irrigation are the added advantages. Drip irrigation can be adopted in fruit crops and also to all vegetable crops including closed spaced crops like onions and beans. The saving in water is to the tune of 30-50 % depending on the crop and season.

The following methods may be adopted under limited water conditions to save water:

(a) *Water saving irrigation method*: Under limited water situations, water-saving irrigation methods like alternate furrow irrigation or widely spaced furrow irrigation and drip irrigation systems can be adopted. Studies conducted on methods of irrigation in capsicum, tomato, okra and cauliflower indicated that adopting alternate-furrow irrigation and widely-spaced furrow irrigation saved 35 to 40 per cent of irrigation water without adversely affecting yield.

(b) *Mulching Practices* - The technique of covering the soil with natural crop residues or plastic films for soil and water conservation is called mulching (Suresh Kumar *et al.,* 2012). Mulching can be practiced in fruits and vegetable crops using crop residues and other organic material available in the farm. Recently plastic mulches have come into use due to the inherent advantages of efficient moisture conservation, weed suppression and maintenance of soil structure. Wide variety of vegetables can be successfully grown using mulches. In addition to soil and water conservation , improved yield and quality, suppression of weed growth , mulches can improve the use efficiency of applied fertilizer nutrients and also use of reflective mulches are likely to minimize the incidence of virus diseases. For vegetable production generally polyethylene mulch film of 30 micron thick and 1 to 1.2 m width is used. Generally raised bed with drip irrigation system is followed while laying the mulch film. Depending upon situation and availability of water, this technology can be used for fruits and vegetable crops. The cost of initial establishment is lower compared to drip system. Further in summer the sprinkling of water helps in reducing the microclimate temperature and increasing the humidity, thereby improving the growth and yield of the crop. The water saved is to the tune of 20 to 30 per cent.

Drip irrigation has proved its superiority over other conventional method of irrigation, in horticulture due to precise and direct application of water in root zone. A considerable saving in water, increased growth, development and yields of fruits and vegetables and control of weeds, saving in labour under drip irrigation are the added advantages. Drip irrigation can be adopted in fruit crops and also to all vegetable crops including closed spaced crops like onions and beans. The saving in water is to the tune of 30 to 50 per cent depending on the crop and season. The cost of initial establishment sprinkler is lower compared to drip system. Further in summer the sprinkling of water helps in reducing the microclimate temperature and increasing the humidity, thereby improving the growth and yield of the crop. The water saved is to the tune of 20 to 30 per cent. Water-saving irrigation methods like alternate furrow irrigation or widely spaced furrow irrigation can be adopted. Studies conducted on methods of irrigation in capsicum, tomato, okra and cauliflower indicated that adopting alternate-furrow irrigation and widely-spaced furrow irrigation saved 35 to 40 per cent of irrigation water without adversely affecting yield.

Protected cultivation: Protective structure is a facility to protect crops from biotic and abiotic constraints. Structures for protected cultivation include green houses, plastic/net houses and tunnels. Commonly used protected structures are poly-houses and net or shade houses. Rain-shelter is a simple structure covered with polyethylene sheet which helps in producing the crops which are affected by excessive rainfall. The productivity of tomato, onions and melons are adversely affected in the event of high rainfall due to difficulty in managing the foliar diseases, lack of proper soil aeration and drainage and also depending on the nature of crop physical damage of the foliage and flower drop. Net house cultivation and shade net cultivation provide better microclimate especially during summer in minimizing the high temperature effect and improving the relative humidity.

Greenhouse Gas

One of the pre-eminent and incontrovertible manifestations of the global climate change is the increase of greenhouse gases (GHGs) in the earth's atmosphere. Although the majority of emissions are produced from energy supply, industry, transport, and buildings, agriculture and forestry (including land use change) together add up to over 30% of current emissions. According to MoEF, Govt. of India estimates, agriculture sector in India contributes about 28% of total GHGs emissions and this is likely to increase in the future due to increasing population and therefore need for more production to meet the food demand. Emissions from soils account for about 12% while manure management and crop residues burning account for 6% of the emissions. Over 90% of N_2O emissions constituting 4% of India's GHGs emissions come from agriculture sector, largely due to fertilizer use. Recently the carbon dioxide (CO_2) and methane (CH_4) concentrations have reached from 280 ppm and 0.7 ppm in the pre-industrial period to 395.4 ppm and 1.76 ppm at present, respectively. Nitrous oxide (N_2O) concentration is increasing at the rate of 0.22% every year to attain a value of 310 ppb (IPCC, 2001) and attained 324 ppb in 2014 (Blasing, 2014). The predicted consequence of rise in CO_2 and other GHGs is elevation of temperature of the earth's surface. Though on global level, emphasis is being given to restrict the CO_2 emission, the beneficial effect of its increase in concentration can be captured.

Capturing the beneficial effects of CO_2 fertilization effect

Higher concentrations of atmospheric CO_2 due to increased use of fossil fuels, deforestation and biomass burning, can have a positive influence on photosynthesis under optimal growing conditions of light, temperature, nutrient and moisture supply, biomass production can increase, especially of plants with C_3 photo-synthetic metabolism above and even more below ground.

C_3 species respond more to increased CO_2; C_4 species respond better than C_3 plants to higher temperature and their water-use efficiency increases more than for C_3 plants. There are some indications that enhancements can decline over time ('down-regulation'). A total of 10 to 20% of the approximate doubling of crop productivity over the past 100 years could be due to this effect (Tans *et al.*, 1990; Ainsworth and Long, 2005) and forest growth or regrowth may have been stimulated as well. Further productivity increases may occur in the coming century, in the order of 30% or more where plant nutrients and moisture are adequate. Higher CO_2 values would also mitigate the plant growth damage caused by pollutants such as NOx and SO_2 because of smaller stomatal openings. Higher percentages of starch in grasses improves their feeding quality, implying less need for feed mixes when silaging.

CO_2 fertilization effect on quality of produce

For rice, the amylose content of the grain, a major determinant of cooking quality is increased under elevated CO_2 (Conroy *et al.*, 2001). Cooked rice grain from plants grown in high-CO_2 environments would be firmer than that from today's plants. However, concentrations of iron and zinc, which are important for human nutrition, would be lower. Moreover, the protein content of the grain decreases under combined increases of temperature and CO_2. Studies using FACE have shown that increases in CO_2 lead to decreased concentrations of micronutrients in crop plants. This may have knock-on effects on other parts of ecosystems as herbivores will need to eat more food to gain the same amount of protein. Studies have shown that higher CO_2 levels lead to reduced plant uptake of nitrogen (and a smaller number showing the same for trace elements such as zinc) resulting in crops with lower nutritional value. This would primarily impact on populations in poorer countries less able to compensate by eating more food, more varied diets, or possibly taking supplements. Reduced nitrogen content in grazing plants has also been shown to reduce animal productivity in sheep, which depend on microbes in their gut to digest plants, which in turn depend on nitrogen intake.

Earlier flowering in response to increased CO_2 was reported in most crop species, but delayed and unchanging flowering time at elevated CO_2 has also been reported (Springer & Ward, 2007). Elevated CO_2 (+300 ppm over ambient concentration) may even increase sterility in rice and thus depress crop yield at higher temperature, possibly by further increasing temperature within the plant canopy. The negative effects of elevated CO_2 on forage N and crude protein concentrations are usually greater than for fiber (e.g., celluloses, lignin) fractions, which can result in lower digestibility under high CO_2 (Idso & Idso, 2001; Milchunas *et al.*, 2005)

Ways to Tackle Atmosphere Related Stresses in Future Climae

Substituting cultivars

In contrast to developed counties, which have seen dramatic yield gains in the past century through investments in agricultural innovation and operate close to the technological frontier, much of developing country agriculture is far from this frontier. Profitable adaptation and farmer adoption of suitable varieties and crops could spark substantial yield gains. These productivity gains could confer a substantial mitigation benefit in the form of foregone land conversion or even reversion of some sensitive lands to grass or forests. Since land use changes, including deforestation and conversion to agricultural production account for 17% of global CO_2 emissions (World Bank, 2009), productivity gains represent a significant mitigation mechanism in agriculture. New cultivars and traits can also lead to less intensive use of other inputs such as fertilizers and pesticides and the associated equipment.

In addition to increasing productivity generally, several new cultivars and traits offer farmers greater flexibility in adapting to climate change, including traits that confer tolerance to drought and heat, tolerance to salinity (e.g., due to rising sea levels in coastal areas), and early maturation in order to shorten the growing season and reduce farmers' exposure to risk of extreme weather events. These promising new traits and cultivars, which are mostly still in development, can emerge from traditional breeding techniques that leverage existing ones that are well suited to vagaries of the local production environment as well as from more advanced biotechnology techniques such as marker assisted selection and genetic modification.

In many places, new traits for the crops farmers have traditionally cultivated will confer sufficient scope for adaptation. In other places, shifting to a totally different mix of crops will be required to cope with dramatic changes in rainfall or temperature, and cropping systems will fundamentally change as a result. Climate change will also lead to new pest and disease pressures. The nuances of temperature changes – e.g., higher low temperatures and fewer freezes – could shorten dormant periods, speed pest and disease growth and change the dynamics of these populations and their resistance. Crops, cultivars and traits that are resistant to pests and diseases will improve producers' ability to adapt to climate change. To the extent that these varieties reduce the need for pesticides, they also reduce carbon emissions by decreasing pesticide demand as well as the number of in-field applications. Breakthroughs in nitrogen use efficiency could substantially mitigate emissions in agriculture. The mitigation potential of new crops and varieties extends to direct carbon sequestration and perhaps to second generation biofuel crops. There are several second generation biofuel crops (i.e., beyond sugar cane and maize) that appear promising as fuel sources (e.g., *Miscanthus*). The damage to fruit trees was relatively more in low-lying areas where cold air settled and remained for a longer time on ground. But temperate fruits such as apple, peach, plum and cherry gave higher yield due to extended chilling. One of the best ways to reduce the impact of the cold wave is to go in for proper selection of fruit species and varieties as per the site conditions. Apart from that, wind breaks or shelter belts should be provided. There should be frequent irrigation, smoking, covering young fruit plants with thatches or plastic sheets.

For the development of traits and cultivars that help to mitigate and adapt to climate change, agricultural biotechnology stands out as an especially promising set of tools. While it remains controversial in some policy arenas and public fora, agricultural biotechnology has produced dramatic improvements in yield and reductions in production costs and input use intensity. Many of the promising traits and varieties discussed above owe their existence to biotechnology, including genetically modified crops with pest resistance (Bt) and herbicide tolerance (Roundup Ready) and conventionally bred varieties that benefit from breeding tools such as marker selection and tissue culture.

Manipulating Through Agronomic Practices

Adoption of Conservation agriculture

The basic principles of conservation agriculture (CA) are soil cover, particularly through retention of crop residues on the soil surface as mulch or cover crops; a minimum level of soil disturbance, e.g., reduced or zero tillage practices; and sensible, profitable crop rotations. Soil moisture can be conserved through mulching either with black polythene or locally available mulches, growing cover crops or inter- culturing in the orchards to check soil erosion and runoff rain water. If area around tree basin remains covered with mulches, like grass mulch (10-12 kg/ basin) and black polythene mulch (100 micron) throughout the growth period, it helps in conserving the soil moisture, saving water for more critical stages during summer and reduced the weed population by 60 % with grass mulch and 100% with black polythene mulch. The polythene mulch maintained 29% more soil moisture compared to un-mulched trees on soil available water content basis. Surface-applied mulches provide several benefits to crop production by controlling evaporation from the soil surface, heat energy and nutrient status in soil, buffering drastic changes in soil temperature. Soil temperature, especially at the surface layer is reported to be important for the translocation of photo- assimilates. Straw mulch ameliorates environmental stresses and improves the product quality and safety.

CA is one way of improving soil moisture management. Considerable saving in water (up to 20-30%) and nutrients are achieved with zero-till planting and particularly in laser levelled and bed planted crop. In maize-wheat cropping system economic yield of maize was increased by 19 and 29% with furrow-irrigated raised-bed (FIRB) planting (5.66 t ha^{-1}) compared to flat no-till and conventional-till planting systems, respectively. The productivity of wheat was higher by 7 and 9%, respectively, under flat no-till (5.56 t ha^{-1}) compared to FIRB and conventional till flat planting. The water productivity (kg grain m^{-3} water) of maize and wheat was higher in FIRB planting (2.79 and 1.98) followed by flat no-till (1.74 and 1.89) and the lowest (1.36 and 1.38) in conventional till system. (Gangwar, 2012).

Permanent soil cover, either plant residues or growing crops, protects the soil surface from the negative effect of raindrop impacts. It reduces crust formation and susceptibility to erosion, and enhances porosity on the soil surface. It also reduces direct water loss through evaporation from the upper layers of the soil and establishes better conditions. Reducing or stopping tillage means that the soil is not disturbed and that the moisture loss and soil compaction that follows tillage is avoided. This increases the infiltration and percolation of water through the soil, leading to better root development and crop growth. The overall impact of CA is a productive soil system, able to carry crops through dry conditions because of the enhanced soil water store, the deep rooting of the crops, the biological activity and the high content of organic matter.

Adoption of agroforestry system

CA system (includes planting on raised beds, cropping sequences/rotations and agroforestry systems) when adopted in appropriate spatial and temporal patterns can further enhance natural ecological processes which contribute to system resilience and reduced vulnerability to yield. RCTs could very well be supplemented by rainwater conservation, increased bund height, irrigation scheduling, land levelling etc. for in-situ moisture/water conservation and conjunctive use of rain, surface and groundwater to enhance water productivity. Moreover, agroforestry systems can be useful in maintaining production during drier years, a common phenomenon in arid regions. In drought situations, deep root systems of trees are able to explore a large volume of water and nutrients, which help to maintain depleting soil moisture conditions to some extent which act as an insurance against drought (Roy *et al.,* 2011).

Shifting in sowing /planting dates

Jalota *et al.,* 2012 simulated the impact of climate change scenarios on crop yields in rice-wheat cropping system in central India Punjab followed by assessment of shifting transplanting/sowing dates as an adaption strategy to mitigate that impact. The simulation analysis using CROPSYST model showed decline in crop yields depending upon changed levels of temperature and CO_2 in different climate scenarios. Under the projected scenario of 2080 (CCCMA model) shifting trans-/planting date from 171st to 178th for rice and from 309th to 324th Julian day for wheat resulted in least reduction in crop yields. The simulated results of dynamic crop growth model WTGROWS (Kalra *et al.,* 2008) also clearly indicates that optimal sowing date of wheat can be delayed by 3-8 days per degree rise in temperature to sustain yield under elevated temperature conditions.

Plant nutrition management

The trends of nutrient absorption, translocation and assimilation are likely to worsen under changing climatic condition. An urgent effort is required to improve crop nutrition and soil fertility management in poor nations, integrating agroecological and socio-cultural aspects of the problem, to avert worsening of a situation that is already desperate. To address this challenge it is urgent that the fertility and productivity of agro-ecosystems in developing regions be maintained and even improved to keep pace with population growth. The conceptual framework for this effort should be integrated soil fertility management (ISFM), consisting of three primary components: 1) judicious use of fertilizers and soil amendments, 2) soil conservation to reduce erosion, maintain soil organic matter, and enhance water and nutrient bioavailability, and 3) cultivation of crop species, genotypes and cropping systems that make optimal use of soil resources for food production while conserving soil fertility.

Application of foliar nutrition and plant growth regulators

The foliar application of nutrients during water stress conditions helps in the better growth by quick absorption of nutrients. The spraying of K and Ca induces drought tolerance in arable (cereals, pulses and vegetables) crops. Spraying of micronutrients and secondary nutrients improves crop yields and quality. Anti-transpiration coatings have been shown to be effective for maintaining quality through control of water loss. Several chemicals have been successfully used like Acropyl in grapes, polycot in banana and kaolinite (3-8%) in different fruit plants. The energy input can be reduced by increasing plant reflectivity by using effective chemicals like zinc oxide, kaolinite, chalk etc alone or in combination with other anti-transpirants. Film forming compounds like wiltpruf, mobileaf, clear spray, vapour guard and folicoat can be used to reduce transpiration and water loss.

Foliar sprays of 50 mMIAA, KNO_3, GA_3, or benzylaminopurine (BAP) partially counteracted the effect of water deficit on photosynthesis and transpiration. Growth regulator applications can also potentially enhance stress resistance in crops which are prone to show accelerated ripening or senescence in response to drought stress. As consequence anti-ethylene products such as aminovinylglycine (AVG) and 1-methylcyclopropene (1-MCP) could be used to mitigate drought stress. While foliar cytokinin (CK) application can prevent ABA-induced photosynthetic limitation, the effects can be transient and of little consequence in the long term. Paclobutrazole (10 mg/ lit) is used to avoid moisture stress in mango. Accelerated ripening or senescence is most often mediated by ethylene production in response to the stress. As consequence anti-ethylene products such as aminovinylglycine (AVG) and 1-methylcyclopropene (1-MCP) could be used to mitigate drought stress.

Proper water management

Drip irrigation has proved its superiority over other conventional methods of irrigation in wider spaced crops due to precise and direct application of water in root zone, while PRD emerged as a potential technology for increasing water productivity in horticultural crops by creating partial moisture stress to produce hormones for minimizing stress impact. A considerable saving in water, increased growth, development and yields of cereals, fruits and vegetables and control of weeds, saving in labour under drip irrigation are the added advantages. The saving in water is to the tune of 30 to 50 per cent depending on the crop and season. The cost of initial establishment sprinkler is lower compared to drip system. Further in summer, the sprinkling of water helps in reducing the microclimate temperature and increase humidity, thereby improving the growth and yield of the crop. The water saved to the tune of 20 to 30 per cent. Water-saving irrigation methods like alternate furrow irrigation or widely spaced furrow irrigation can be adopted. Studies conducted on methods of irrigation in capsicum, tomato, okra and cauliflower indicated that adopting alternate-furrow irrigation and widely-spaced furrow irrigation saved 35 to 40 per cent of irrigation water without adversely affecting yield.

Beneficial microbes

Microbial activity in soils is stimulated by the release of carbon-rich material in the form of root border cells, and/or the selective exudation of specific sugars, carboxylic acids, or amino acids that encourage the development of cultivar-specific, plant-beneficial, microbial communities. Some of these microorganisms benefiting from the availability of fresh energy are also capable of enhancing plant growth and development, which are generally called as Plant growth promoting (PGP) rhizosphere microorganisms. Plant growth-promoting rhizobacteria (PGPR) and fungi (mycorrhizae) can facilitate plant growth directly by facilitating the uptake of nutrients from the environment, by influencing phytohormone production (e.g. auxin, cytokinin, or giberallin), and/or by enzymatic lowering of plant ethylene levels. In addition to facilitating the growth of plant, these microorganisms can protect plants from the deleterious effects of flooding and drought. The beneficial plant-microbe interactions in the rhizosphere are the primary determinants of plant health and soil fertility. Rhizobacteria include mycorrhization helper bacteria (MHB) and plant growth promoting rhizobacteria (PGPR), which assist AMF to colonize the plant roots. The most common bacteria in the mycorrhizosphere are *Pseudomonas*. Some extremely mycorrhizal dependent plants, including grapes, citrus, melons, oaks and pines may quite literally starve to death in soils that lack this helpful fungi. These benefits of mycorrhizal symbioses in vegetables, fruits and tree species, both agronomically by increased growth and yield as well as ecologically by improved fitness, indicate that mycorrhizal plants are often more competitive and better able to tolerate environmental stresses.

Wind break and shelterbelts

To overcome the adverse effect of high temperature and dry winds, tall growing trees need to be planted all along the boundary of the farm. Windbreaks help to deflect and to filter through the wind current thereby reducing the velocity of wind resulting lower displacement of wind around tree and cause reduction in transpiration and evaporation. The orientation of windbreak should more or less be at right angles to the prevailing winds. It is believed that the windbreak is effective for a distance equivalent to 3 to 4 times of tree height. *Acacia tortalis, Cassia siamea* and *Prosopis julifera* are some of the useful tree species suitable as wind break under less water conditions. Inter cropping of vegetable crops of the area can be practiced in orchards during summer months. Maize/ Sorghum can be grown all along the border of the plot to mitigate the effect of desiccating winds. Crop rotation, cover crops, strip cropping, mulching (Suresh Kumar *et al.*, 2011), contour hedge row intercropping system are used to conserve moisture.

Conclusion

Agriculture is our primary source of food and has significant potential for mitigation of global GHG emissions, and is particularly sensitive to climate change. Almost certainly, climate change will be severe in most developing countries and will directly – and, in some cases, dramatically – hurt agricultural production in these countries. Yet, development cannot be taken for granted and the dual burden of climate change adaptation and mitigation may make economic transformation more difficult. As climate change will amplify the atmospheric stressors, especially temperature and water availability in many places, input use efficiency and appropriate adaptive measures must be taken to meet the future productivity demands. It is a fool's errand to attempt to fully catalogue in any comprehensive way agricultural technologies with potential for climate change mitigation and adaptation over the next seven decades. If history is any guide, such technologies will be developed with the passage of time and with experiences. Innovations in agriculture have always been an important and even more vital in the context of climate change.

References

Ainsworth E.A. and Long S.P. (2005) What have we learned from 15 years of free air CO_2 enrichment (FACE)? A meta-analytic review of the responses of photosynthesis, canopy properties and plant production to rising CO_2. New Phytologist, 165: 351–372.

Bal S.K., Saha S., Fand B.B., Singh N.P., Rane J. and Minhas P.S. (2014) Hailstorms: Causes, Damage and Post-hail Management in Agriculture. Technical Bulletin No. 5, ICAR-National Institute of Abiotic Stress Management, Malegaon, Baramati, Pune, Maharashtra. p.44.

Beck E.H, Fetitig S. and Knake C. (2007) Specific and unspecific responses of plants to cold and drought stress. Journal of Biological Sciences, 32(3): 501-510.

Blasing T.J. (2014) Recent Greenhouse Gas Concentrations. DOI: 10.3334/CDIAC/atg.032, http://cdiac.ornl.gov/pns/current_ghg.html.

Blum A.J. (2006) Rising carbon dioxide concentrations and the future of crop production. Journal of the Science of Food and Agriculture, 86: 1289–1291.

Craufurd P.Q. and Wheeler T.R. (2009) Climate change and the flowering time of annual crops. Journal of Experimental Botany, 60: 2529–2539.

Gangwar B. (2012) An over view of PDFSR and Winter School. In: Singh, K.K., Singh, V.K. and Mishra, R.P. (Eds.), System Based Conservation Agriculture for Sustained Productivity and Soil Health, Project Directorate for Farming Systems Research, Modipuram, Meerut, Uttar Pradesh, p. 408.

Idso S.B. and Idso K.E. (2001) Effects of atmospheric CO2 enrichment on plant constituents related to animal and human health. Environmental and Experimental Botany, 45: 179–199.

IPCC (2001) Climate Change 2001: The Scientific Basis - Contribution of Working Group I to the Third Assessment Report of the Intergovernmental Panel on Climate Change (IPCC), Cambridge University Press, Cambridge.

IPCC (2007) Climate Change 2007: Synthesis Report. Contributions of Working Groups I, II, and III to the Fourth Assessment Report of the Intergovernmental Panel on Climate Change. Geneva: IPCC.

Jalota S.K., Kaur H., Ray S.S., Tripathi R., Vashisht B.B. and Bal S.K. (2012) Mitigating Future Climate Change Effects by Shifting Planting Dates of Crops in Rice-Wheat Cropping System. Regional Environmental Change, 12(4): 913-922.

Jalota S.K., Kaur H., Ray S.S., Tripathi R., Vashisht B.B. and Bal S.K. (2013) Past and General Circulation Model-driven future trends of climate change in Central Indian Punjab: ensuing yield of rice-wheat cropping system. Current Science, 104(1): 105-110.

Kalra N., Aggarwal P.K., Chander S., Pathak H., Choudhary R., Choudhary A., Mukesh S., Rai H.K., Soni U.A., Anil S., Jolly M., Singh U.K., Owrs A., Hussain M.Z. (2003) Impacts of climate change on agriculture. In: Shukla P.R., Sharma S.K., Ravindernath N.H., Garg A., Bhattacharya S. (eds) Climate change and India: vulnerability assessment and adoption. Orient Longman Private Ltd., Hyderabad, pp 191-226.

Kumar P.S., Minhas P.S., Govindasamy, V. and Choudhary R.L. (2014) Influence of Moisture Stress on Growth, Development, Physiological Process and Quality of Fruits and Vegetables and Its Management Strategies. *In*. (Eds. R.K. Gaur and Pradeep Sharma) Approaches to Plant Stress and their Management, Springer publications., New Delhi, p.125-148

Lehmann J., Gaunt J. and Rondon M. (2006) Bio-Char Sequestration in Terrestrial Ecosystems – a review. Mitigation and Adaptation Strategies for Global Change, 11(2): 395-419.

Milchunas D.G., Mosier A.R., Morgan J.A., LeCain D.R., King J.Y. and Nelson J.A. (2005) Elevated CO2 and defoliation effects on a short grass steppe: Forage quality versus quantity for ruminants. Agriculture, Ecosystems and Environment, 111:166–184.

Porter J.R. and Semenov M.A. (2005) Crop responses to climatic variation. Philosophical Transactions of the Royal Society B: Biological Sciences, 360: 2021–2035.

Roy M.M., Tiwari J.C. and Ram M. (2011) Agroforestry for climate change adaptations and livelihood improvement in Indian hot arid regions. International Journal of Agriculture and Crop Sciences, 3-2: 43-54.

Suresh Kumar P., Choudhary V.K. and Bhagawati R. (2012) Influence of mulching and irrigation level on water-use efficiency, plant growth and quality of strawberry (*Fragaria xananassa*). Indian Journal of Agricultural Sciences, 82(2):127-133.

Vara Prasad, P.V., Craufurd, P. Q., Summerfield, R. J. and Wheeler, T. R. (2000) Effects of short-episodes of heat stress on flower production and fruit-set of groundnut (*Arachis hypogaea* L.) Journal of Experimental Botany, 51: 777–784.

Waggoner P.E. (1983) Agriculture and a climate changed by more carbon dioxide. Changing climate. National Academy Press, Washington, DC, pp 383–418.

Wang C.Y. (1993) Approaches to Reducing Chilling Injury of Fruits and Vegetables. Horticultural Review, 15: 63-95.

CLIMATE CHANGE MANAGEMENT STRATEGIES IN AGRICULTURE

5

Water Management in Relation to Climate Change

Anil Kumar Singh

India has shown to the world that with 2.4 % of the land and 4 % of the water resources available at the global level, it can provide the food security to its millions. A record production of almost 265 million tonnes in 2013-14 is a testimony to the strides taken in the field of agriculture. However, the agriculture sector faces many new challenges in the future and climate change (global warming) is perhaps the most formidable one.

The new emerging demands of the relatively more-affluent Indian population, particularly its middle class, coupled with a net cultivated area unlikely to exceed 143 Mha in 2050 as well as an estimated rainfed agriculture to cover around 45 % of the net sown area, are further compounded with the harsh reality that highly productive agricultural land is being continuously lost out to the industry and urban sectors. How will the country meet the target of 400 Mt of food grain production is the mute question, in a situation where natural resource base is continuously degrading and climate change with its attendant impact is adversely affecting the agricultural production system (Singh, 2014).

Undoubtedly, one of the pillars of the Green Revolution has been the Government of India's focus on development of water resources in the country. Today at 65.26 Mha (2011-12), India has the second largest net irrigated area in the world but if one examines the productivity of irrigated areas at national level, it is only around 3 t ha^{-1}. The efficiency of surface irrigation systems is less than 40% implying that at least 60% of the water being supplied is being lost at various stages in the system.

Status of Water Resources (Singh, 2013; 2014)

Total utilizable water resource in the country has been estimated to be about 1123 billion cubic metres (BCM) (690 BCM from surface and 433 BCM from ground water), which is just 28% of the water derived from precipitation

(Table 1). About 80% of the water (688 BCM) is being diverted for irrigation, which may increase to 1072 BCM by 2050. On the basis of the available water resources, the total irrigation potential from surface and ground water resources is estimated to be 139.9 Mha. The major source for irrigation is groundwater. Annual groundwater recharge is about 433 BCM of which 212.5 BCM is used for irrigation and 18.1 BCM for domestic and industrial use (CGWB 2011). By 2025, demand for domestic and industrial water usage may increase to 29.2 BCM. The sector-wise estimated demand is given in Table 2.

Water is a finite resource the availability of which is declining with each passing day. If the international yardstick of 1700 m³ per capita availability of water is taken as a criterion, then the whole country became "Water stressed" in the year 2007 with per capita water availability declining to 1656 m³ from 5200 m³ in 1951 (Figure 1). The reported per capita availability of water was 1545 m³ in 2011. It is estimated that the per capita availability of water is likely to reach a level of around 1100 m³ in 2050 but in reality the threshold may be reached much earlier considering the wasteful ways of water usage. It may be noted that per capita availability of <1000 m³ indicates "Water scarcity". However, if the per capita availability is scrutinized basin-wise than a significant number of them in the country were either in the "water scarcity zone" or worse in 2010 (Figure 2).

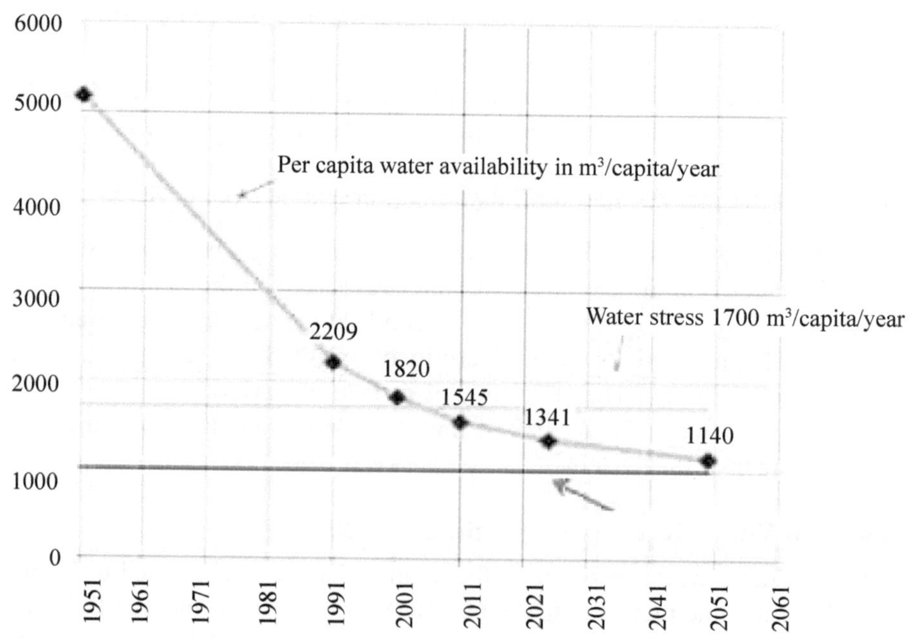

Fig. 1: Per capita availability of water in India

The statistics related to the per capita storage or the days of average flows are equally disturbing. The data available indicate that the per capita storage in India at 220 m^3 is one fifth that of China (1110 m^3). The scenario in terms of "days of average flow" at the national level is around 20 days compared to Colorado in USA which is 850-900 days.

Table 1: Water resources of India.

Estimated Annual Precipitation (including snow fall)	4000 BCM
Estimated Utilisable Water	
Surface	690 BCM*
Ground	433 BCM
Total	1123 BCM
Irrigation Potential	
***Major and Medium (Surface Water)**	
Ultimate	58.5 Mha
Created (upto 2009-10)	44.4 Mha
Utilized (upto 2009-10)	34.4 Mha
***Minor Irrigation**	
Ultimate	81.4 Mha
Created (upto 2009-10)	62.8 Mha
Utilized (upto 2009-10)	52.5 Mha
Total (Major and Medium+Minor)	
Ultimate	139.9 Mha
Created (upto 2009-10)	107.2 Mha
Utilized (upto 2009-10)	96.9 Mha
Area Irrigated (2009-10)	
Gross	85.08 Mha
Net	61.94 Mha

Source: CGWB (2011)

Table 2: Projected water demand in India by different sectors.

S.No.	Sector	Water Demand in km^3 (or BCM*)									
		Standing Sub-Committee of MOWR			NCIWRD						
		2010	2025	20250	2010		2025		2050		
					Low	High	Low	High	Low	High
1.	Irrigation	688	910	1072	543	557	561	611	628	807
2.	Drinking water	56	73	102	42	43	55	62	90	111
3.	Industry	12	23	63	37	37	67	67	81	81
4.	Energy	5	15	130	18	19	31	33	63	70
5.	Other	52	72	80	54	54	70	70	111	111
	Total	813	1093	1447	694	710	784	843	973	1180

*BCM: Billion cubic metres

Source: CGWB (2011)

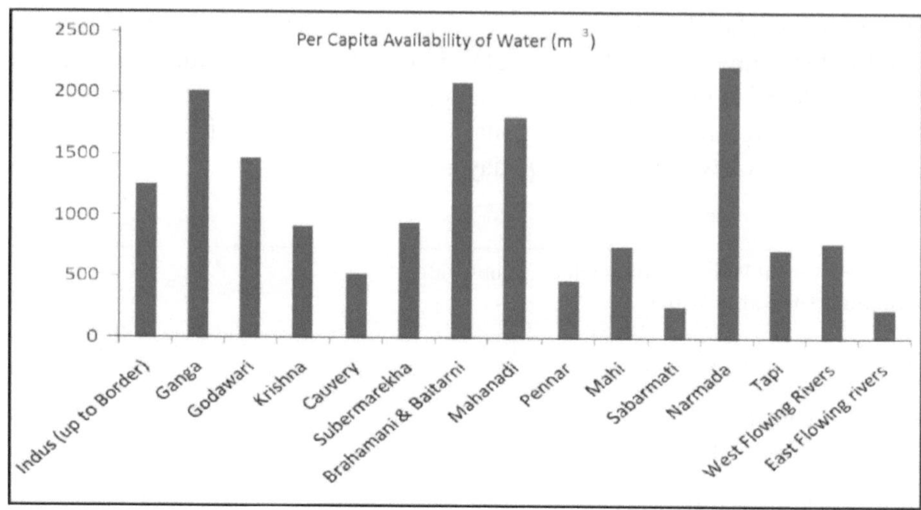

Fig. 2: Per capita water availability of water in the river basin of India in 2010

A study completed in late 2011 revealed that the gap between the irrigation potential created and utilized in these projects was substantial and is growing. It is not only the demand–supply mismatch in the major and medium irrigation projects but also the fact that the gap between the irrigation potential created and utilized has been widening continuously and today it stands at more than 25 Mha (Figure 3). Top most priority should be given to bridge this gap considering the fact that the Government of India currently spends more than Rs. 1.25 lakhs to develop one hectare of irrigated land in the major and medium projects. This is a huge amount of investment. Major reasons are low water discharge, insufficient water distribution mechanism, unequal water distribution across farmers located at different points, loss of water during distribution, incorrect recording of irrigated area and diversion of cultivable land to other purposes within the command area. The conclusion of a study carried out on 28 major/ medium irrigation projects in the country have indicated that the overall water use efficiency was 38% with the conveyance efficiency pegged at 69% and on-farm application efficiency at 52%.

The most important initiative for bridging the gap between irrigation potential created (IPC) and irrigation potential utilized (IPU) is the Command Area Development Programme (CADP) that has been running since 1974-75. The statistics indicate that up to 2012, CADP could only manage 20.02 Mha of field channels and construct only 2.25 Mha of field drains, again leaving a lot to be desired in terms of execution.

Ground Water Development

Development of ground water resources which was a low key affair in the earlier stages, is now contributing to more than 60% of the irrigation demand and playing a critical role in ensuring food security to the millions in the country. However, it's over exploitation is now a very serious cause of concern and the potential of judicious exploitation of ground water resources is now confined only to some states in the eastern part of the country. Out of the 5842 assessed blocks, 4277 are safe while the rest are in the semi-critical (523), critical (169) and a large number of over-exploited (802) blocks. India has a dubious distinction of having the highest ground water abstraction (Table 3) considering the availability of the ground water resources, Government of India (GoI) has rightly decided to focus on Eastern India and has implemented a programme aptly titled "Bringing Green Revolution to Eastern India (BGREI)". In view of the seriousness of the problem, the Government is, therefore, putting a lot of emphasis on artificial recharge of ground water. A project for setting up 225 thousand recharge structures at an estimated cost of approximately Rs. 20,000 crores has been launched with this objective and a clear defined goal that it has to be implemented in a participatory mode with involvement of local communities, locally elected bodies like Panchayats, and Non Government Organizations (NGOs).

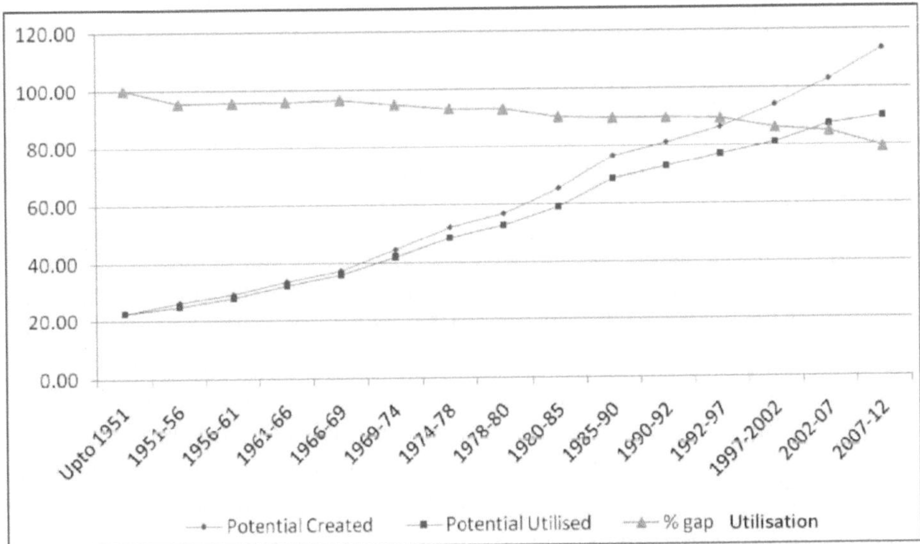

Fig. 3: Plan-wise irrigation potential created and utilized over the years.

Table 3: Groundwater-Abstracting Countries as of 2010

Country	Abstraction (km^2/year)
India	251
China	112
United States of America	112
Pakistan	64
Iran	60
Bangladesh	35
Mexico	29
Saudi Arabia	23
Indonesia	14
Italy	14

Rainfed Agriculture

Although India has the largest irrigated area in the world, it has a very sizeable area under rainfed agriculture (nearly 80 Mha) which caters largely to pulses, oilseeds, minor millets, cotton and even rice (~45%) but the productivity of which is very low (<1 t ha^{-1}). The average annual rainfall in India is 117 cm which is higher than the global average of 110 cm but there is a wide spatial and temporal variation in its distribution and intensity. The ratio of the minimum, average and maximum is of the order of 1:10:100 with the maximum being around 11000 mm in Cherrapunji (Meghalaya) to 100 mm in the Thar desert. The number of rainy days can vary from 4 days to over 300 days a year and the intensity can vary anywhere from <1 cm hr-1 up to 15 cm hr-1. It has also been estimated that almost 80% of the rain occurs in only 100 hr in a year. Soil and nutrients loss due to run off is of the order of 5.34 billion tonnes and 6 million tonnes, respectively, resulting in production losses estimated to be in excess of Rs. 115 billion.

A comprehensive assessment of district level water harvesting potential by Sharma *et al.* (2010) revealed that the potential to realize the rainfed agriculture lies in the harvest of small part of available surplus runoff and reutilize it for supplemental irrigation at critical crop growth stages. The study identified about 28.5 Mha of potential rainfed area covering a large number of districts in central and eastern India that can generate sufficient runoff (114 BCM) for harvesting and reutilization. It is possible to raise the rainfed crop production by a total of 28-36 Mt from an area of 20-25 Mha during normal monsoon years which could result in about 12% increase over the present production. With adoption of improved technologies (much likely possible once 'critical water requirements' are assured), the benefits could be still higher as shown (Figure 4). Extensive area coverage rather than intensive irrigation needs to be followed in regions with higher than 750 mm yr^{-1} rainfall, since there is larger possibility of alleviating the inseason drought spells and ensuring a second crop with limited water application. This may be made an integral component of the on-going and new development schemes in the identified rural districts.

Wastewater Management

With the projected scenarios of water availability compounded by global warming, increasing urbanization (circa 55% of India's estimated 1.6 billion would be urban by 2050) and industrialization, huge volumes of waste waters will be generated. These waste waters would be of varying qualities depending upon the source. In developing countries like India, the problems associated with wastewater reuse arise from its lack of treatment. On the whole only 60% of industrial water and 26% of domestic water is treated in India. Considering the size, its vast population and various sectoral demands, India is a growing market for water and wastewater treatment. It includes both sewage treatment and effluent treatment. Industrial and municipal sectors account for almost 90% of the estimated US$ 2000-2100 million water treatment market in India. Currently, 75% of the rural population and 85% of the urban population have access to the public water supply. However, municipal agencies in many Indian towns and cities are unable to increase their water supply capacities to match population growth, especially in the urban areas. The challenge, thus, is to find lowcost, low-tech and user friendly methods, which on one hand avoid threatening our substantial wastewater -dependent livelihoods and on the other hand protect degradation of our valuable natural resources.

As stated earlier, more than half of the country would be living in cities and towns by 2050, and generally, sizeable amount of domestic as well as industrial waste waters. Since these wastewaters also contain a significant amount of nutrients and their disposal is an environmental issue, they should be considered as resource for irrigation and systematic efforts should be made for their primary treatment as it is cost effective. Peri-urban agriculture needs to be given high priority to meet the specific demands of the urban population particularly vegetables and flowers.

Therefore, it is time to declare these wastewaters as a "resource" so that one can start to scientificantly develop cost-effective and eco-friendly solutions to meet the domestic, industrial, agricultural and environmental demands of the country which would be the most populous nation of the world in a couple of decades.

For agricultural purposes the use of constructed wetlands is now being recognized as an efficient technology for wastewater treatment (Kaur *et al.*, 2011). Compared to the conventional treatment systems, constructed wetlands need lesser material and energy, are easily operated, have no sludge disposal problems and can be maintained by untrained personnel. Further these systems have lower construction, maintenance and operation costs as these are driven by natural energies of sun, wind, soil, microorganisms, plants and animals. Attempts sould be made to upscale this technology with suitable fine tuning for specific locators.

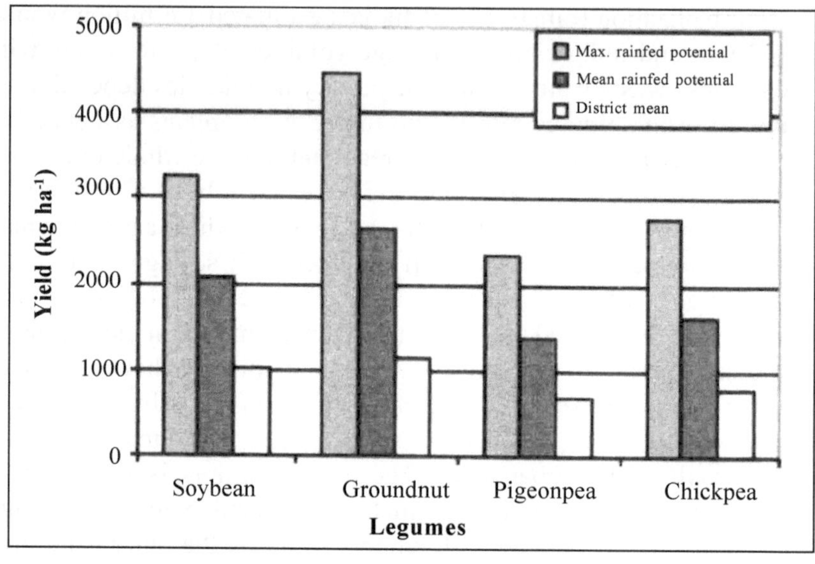

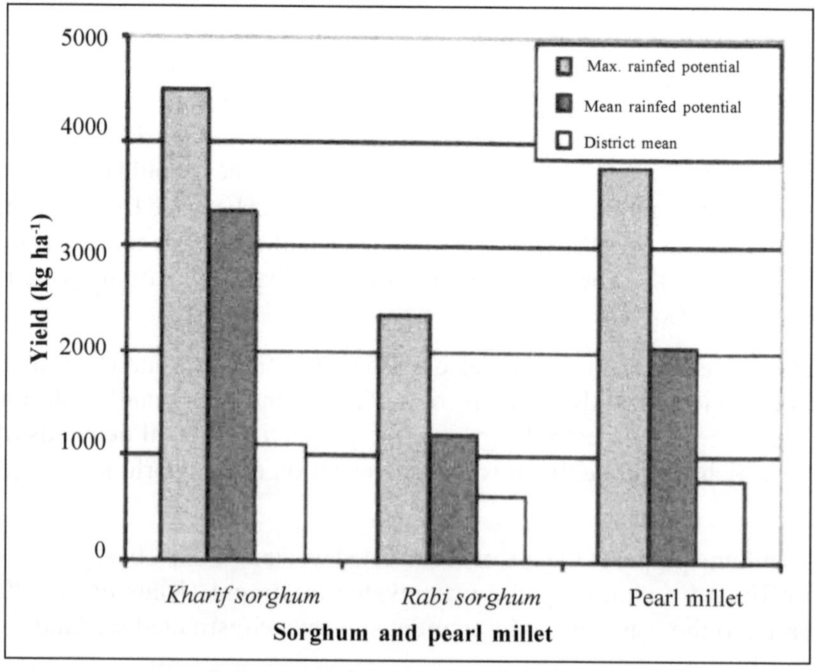

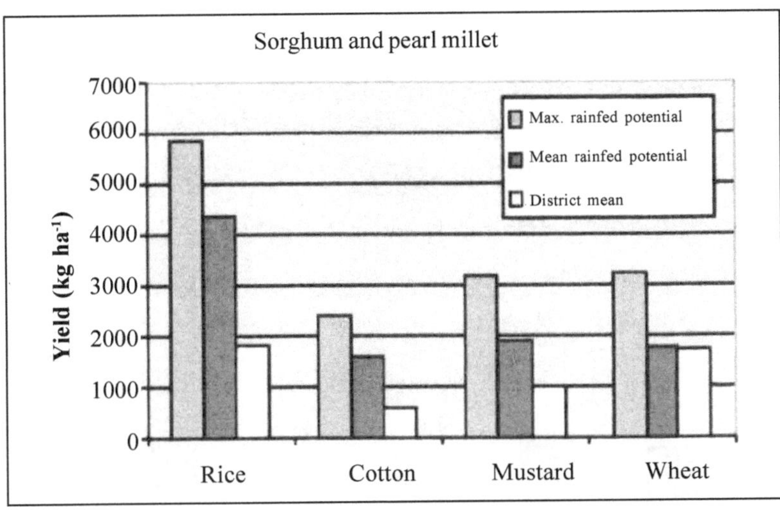

Fig . 4: Rainfed potential yield and yield gap of crops in India (van-Rooijen *et al.* 2009)

Hence, for planned, strategic, safe and sustainable use of wastewaters there is a need for policy decisions and coherent programmes encompassing low-cost decentralized waste water treatment technologies, nano/bio-filters, efficient microbial strains, and organic/ inorganic amendments, appropriate crops/ cropping systems, cultivation of remunerative non-edible crops and modern sewage water application methods.

Multiple uses of water

Since the scarcity of water is going to worsen in the future, the emphasis has to shift from conventional land productivity approach to enhancing water productivity on a systems basis. Multiple use of water provides opportunities for increasing water productivity, enabling diversification to high value products, reducing risk, optimizing resource use and a regular flow of income as well as judicious utilization of water of poor/marginal quality. Integrated farming system (IFS)-based agriculture can enhance the water productivity, reduce the amount of external inputs to be purchased by the farmers and increase his income (Figure 5).

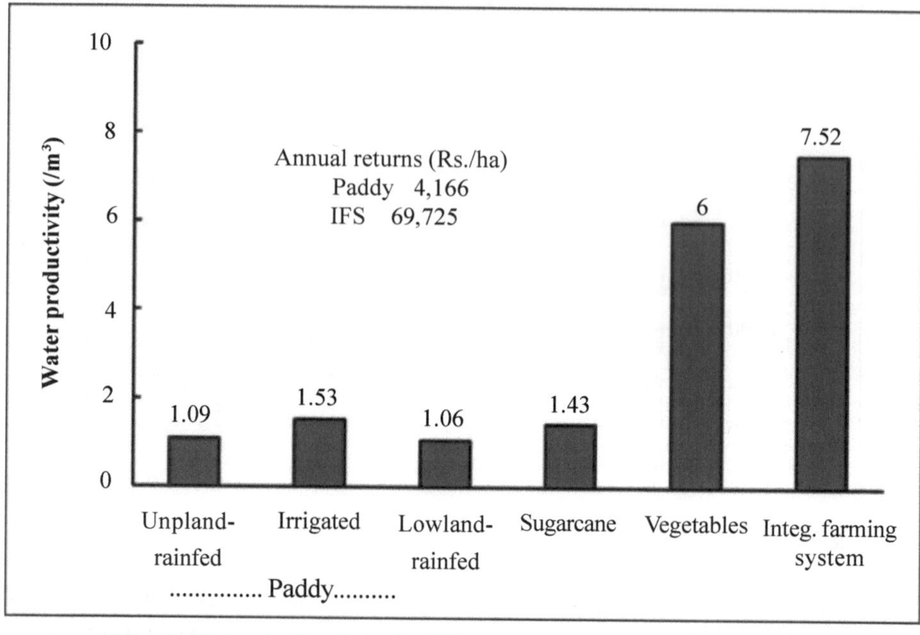

Fig. 5: Water productivity in different cropping / farming systems

Use of Modern Tools in Water Management

The stupendous progress in information technology coupled with the rapid advancements made in Geographical Information Systems (GIS), simulation tools, sensors, precession farming and remote sensing have opened up new vistas for water resources development and management. These tools should be an integral system of scientific management of irrigation networks, water distribution, crop planning and related operational activities as they will enable the system managers to take correct and timely decisions. These tools can be equally, if not more effectively utilized in assessment and monitoring of watershed related development studies. Decision Support Systems (DSS) for real time monitoring and decision-making with inputs from remote sensing and ground based inputs can contribute significantly in improving water use efficiencies in large irrigation projects and integrated watershed management.

Developments in biotechnology have reached a level where gene flow has no boundary. Marker Assisted Selection (MAS), gene pyramiding etc. can enable identification as well as introduction of genes (single/multiple) that can enhance water use efficiency as well as increase tolerance to waterlogging, soil salinity or heavy metal toxicity. The potential of this technology needs to be utilized fully.

Impact of Climate Change (Gautam and Singh 2015)

Greenhouse gases

The increasing levels of green house gases (GHGs) in the atmosphere have been attributed as one of the major driving force behind the rapid climate change phenomenon. Though the increase in the level of CO_2 is expected to produce some beneficial effects on crop dry maater production, it may soon be nullified by associated water and thermal stresses leading to overall deterioration of agro-climatic conditions for food production systems. The CO_2 concentration has reached 400 ppm and it has been rising continuously since the last several decades (Fig. 6). This indicates that the impact of measures taken by the various countries has not made any impact. At the global scale, the historical temperature-yield relationships indicate that global warming is very likely to offset some of the yield gains from technological advance, rising CO_2 and other non-climatic factors (Lobell and Field, 2007).

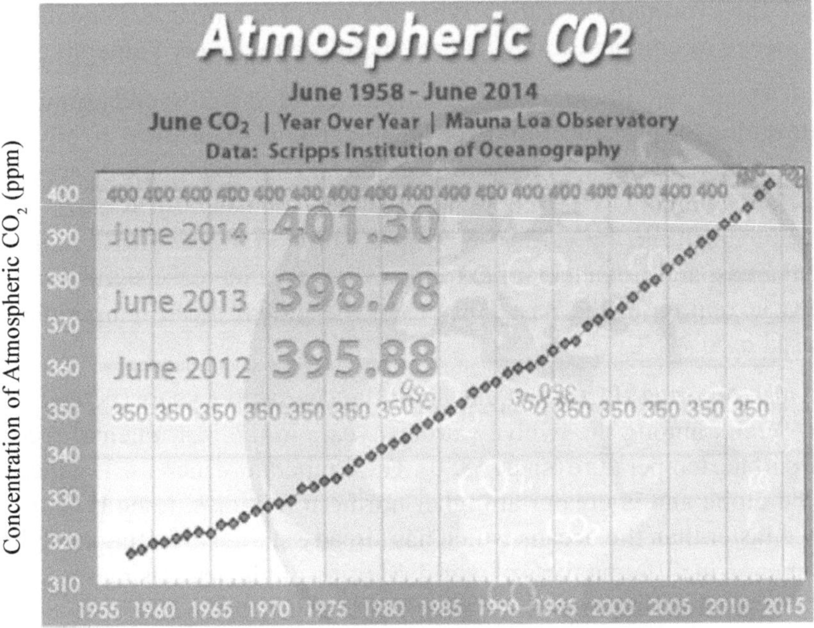

Fig. 6: Current global CO_2 concentration levels, *Source*: NOAA-ESRL

Role of carbondioxide

Research suggests that atmospheric carbon dioxide levels may affect water availability through its influence on vegetation. Controlled experiments indicate that elevated CO_2 concentrations would increase the resistance of plant "pores," that is stomata, to water vapor transport. Experiments suggest that a doubling of CO_2 would increase stomatal resistance and reduce the rate of transpiration vis-à-vis the passage of water vapor from plants by about 50 percent on average. The resulting decrease in transpiration would tend to increase runoff. On the other hand, CO_2 also has been demonstrated to increase plant growth, leading to a larger area of transpiring tissue and a corresponding increase in transpiration. Other factors that might offset increases in plant water-use efficiency associated with a CO_2 enriched atmosphere are a potential increase in leaf temperatures caused by reduced transpiration rates and species changes in vegetation communities. The net effect of opposing influences on water supplies would depend on the type of vegetation and other interacting factors, such as soil type and climate.

The management and labour costs of farm production will rise to a great level owing to increased incidences of pests and diseases as well as weeds. Additionally, extreme events like hails and frosts will also negatively impact crop production. As of now, most of the developing countries, including India, are not fully prepared to deal with the adverse impacts expected as a consequence of climate change and are therefore relatively vulnerable.

The 4th Assessment Report (IPCC, 2007 has reported almost change on earth's climate during the past 400,000 years, but this change accelerated greatly during the last century, due to rapid industrialization and indiscriminate destruction of natural environment. The report has projected that by 2100 earth's mean temperature will rise by 1.4 to 5.8 °C, precipitation will decrease in the sub-tropical areas, and frequency of extreme events will increase significantly. As of now, in reality, in the past 100 years, the global mean temperature has increased by 0.74 °C.

It was also reported IPCC, that eleven of the last twelve years (between 1995 to 2009) rank among the twelve warmest years in the instrumental record of global surface temperature (since 1850). The temperature increase is widespread over the globe and is greater at higher northern latitudes. Land regions have warmed faster than the oceans which has already started affecting the climatic phenomenon in different parts of world. Melting of glaciers, rising sea level are some of the most important manifestations of it. IPCC (2007) have further reported change in their frequency and /or intensity of extreme events over the last 50 years. The report has made the following significant observations:

However, climate change impacts are already being felt, as the last 60 years were the warmest in the last 1000 years. Since the beginning of 21st Century, India has experienced droughts in quick succession, of which the 2009 one was the most recent one that significantly affected *kharif* crop production. It was the 2nd largest all India monsoon rainfall deficit since 1972 (23% below normal).

Incidentally, 2009 also achieved the distinction of being the warmest year in past several centuries across the world. However, after 2009, 2010 proved to the warmest year on record since 1850. 2011 is now the 11th warmest year on record since 1850. Apart from that, 1998 was one of the warmest years, 2003 experienced unprecedented heat and cold waves across the globe, occurrence of high temperature in March 2004 adversely affected crops like wheat, apple, potato etc. across northern India; 2005 witnessed destructive hurricanes/cyclones across the globe. Again, 2007 was as warm as 1998 in the entire northern hemisphere and unusual summer rains and floods were experienced in many parts of India. Besides that, the amount and distribution of rainfall is becoming more and more erratic which is causing greater incidences of droughts and floods globally. The increase in frequency of heavy rainfall events in last 50 years over Central India (Fig. 7) points towards a significant change in climate pattern in India (Goswami, 2006).

In 2005, the conference (Defra, 2005) "Avoiding Dangerous Climate Change" raised concerns that climate change was occurring more quickly than previously anticipated. Large-scale, irreversible system disruption and the destabilization of the Antarctic ice sheets are serious risks: changes to polar ice, glaciers and rainfall regimes have already occurred.

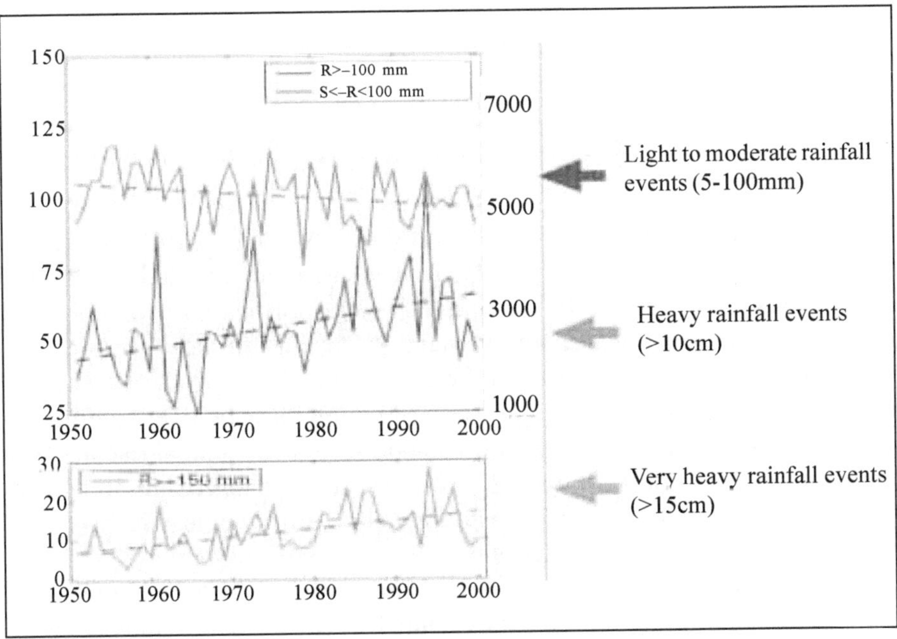

Fig. 7: Change in intensity of rainfall over Central India in the last 50 years
Source: (Goswami, 2006)

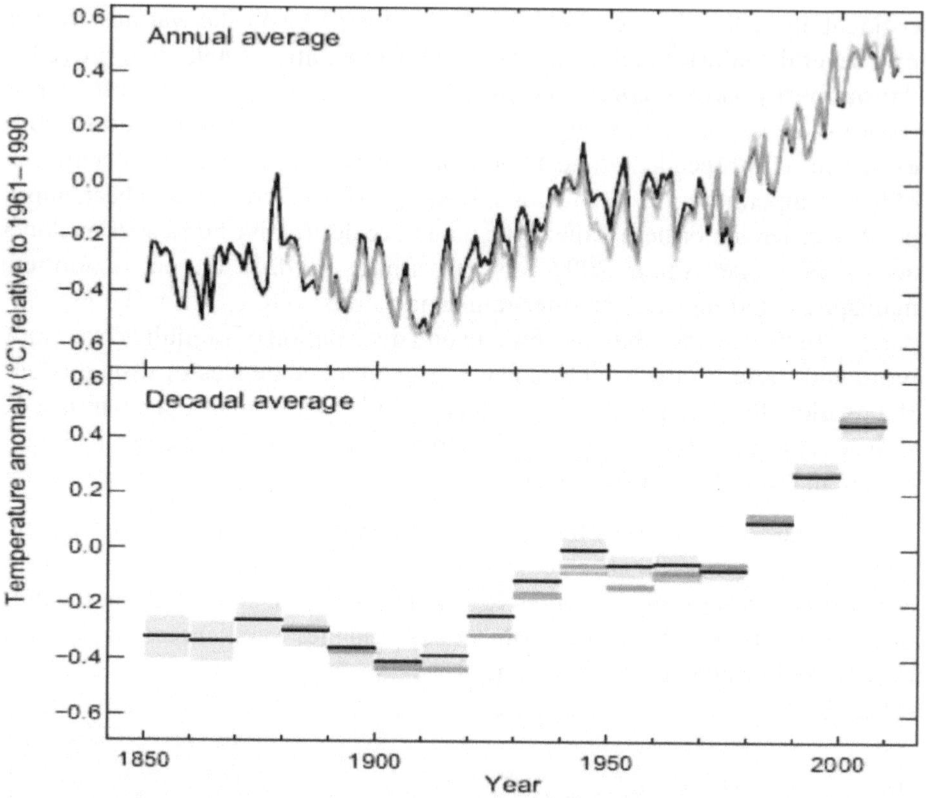

Fig. 8: Observed globally averaged combined land and ocean surface temperature anomaly 1850–2012. The year 2013 ties with 2003 as the fourth warmest year globally since records began in 1880

The rainfall analysis of India (Rao *et al.*, 2009) showed that significant negative trends of rainfall were observed in the eastern parts of Madhya Pradesh, Chhattisgarh and parts of Jharkhand, U.P., and northeast India. Lal (2001) reported that annual mean area-averaged surface warming over the Indian sub-continent is to likely to range between 3.5 and 5.5 °C by 2080s (Table 4). These projections showed more warming in winter season over summer. The spatial distribution of surface warming suggests a mean annual rise in surface temperatures in North India by 3 °C or more by 2050. The study also indicated that during winter, the surface mean air temperature could rise by 3 °C in Northern and Central parts, while it would rise by 2 °C in Southern parts by the same period. In case of rainfall, a marginal increase of 7-10% in annual rainfall is projected over the sub-continent by 2080. Nevertheless, the study suggests a fall in rainfall by 5-25% in winter, while it would increase by 10-15% in summer, Marked variability is seen even in the onset and withdrawal of monsoon over the period. The glaciers and the snowfields in the Himalayas are on the decline as a result of climate variability. The rate of retreat of the snow of Gangotri glacier demonstrated a sharp rise in the first half of the 20th century. Glacial melt would lead to increased summer river flow and floods over the next few decades, followed by a serious reduction in flows thereafter.

Table 4: Projected mean temperature and rainfall changes over the Indian sub continent (Lal, 2001)

Year	Season	Temperature change (°C)		Rainfall change (%)	
		Lowest	Highest	Lowest	Highest
2020s	Annual	1.00	1.41	2.16	5.97
	Rabi	1.08	1.54	-1.95	4.36
	Kharif	0.87	1.17	1.81	5.10
2050s	Annual	2.23	2.87	5.36	9.34
	Rabi	2.54	3.18	-9.22	3.82
	Kharif	1.81	2.37	7.18	10.52
2080s	Annual	3.53	5.55	7.48	9.90
	Rabi	4.14	6.31	-24.83	-4.50
	Kharif	2.91	4.62	10.10	15.18

Impact on water resources

A warmer climate will modify the hydrologic cycle, altering rainfall, magnitude and timing of run-off. Warm air holds more moisture and increases evaporation of surface moisture. With more moisture in the atmosphere, rainfall and snowfall events tend to be more intense, increasing the potential for floods. However, if there is little or no moisture in the soil to evaporate, the incident solar radiation will lead to increase in the temperature, contributing to longer and more severe droughts (Trenberth, 1999). Therefore, change in climate will affect the soil moisture, groundwater recharge and frequency of flood or drought episodes and finally groundwater level in different areas.

At present, available statistics on water demand show that the agriculture sector is the largest consumer of water in India using more than 80% of the available water. The quantity of water used for agriculture has increased progressively through the years as more and more areas were brought under irrigation. Since 1950-51, the net irrigated area in India rose from 21 mha to 65.26 million ha up to 2011-12. Contribution of surface water and groundwater resources for irrigation has played a significant role in India attaining self-sufficiency in food production during the past three decades, but water is likely to become more scarce and critical input in future. By judicious utilization, the demand for water from farm sector can be pegged at 68% by the year 2050, but agriculture will still remain the largest consumer. In order to meet this demand, augmentation of the existing water resources by development of additional sources of water or conservation of the existing resources through impounding more water in the existing water bodies and their conjunctive use will be needed (Mall *et al.*, 2006). The enhanced surface warming over the Indian subcontinent by the end of the next century would result in an increase in pre-monsoonal and monsoonal rainfall and no substantial change in winter rainfall over the central plains. This would result in an increase in the monsoonal and annual run-off in the central plains, with no substantial change in winter run-off and increase in evaporation and soil wetness during the monsoon on an annual basis.

In another study made by IITM, Pune in collaboration with DEFRA, UK reported that there will be an increase in annual rainfall and annual flow in Godavari and Ganga basins by 2071-2100 period.

In the recent report submitted by IPCC (2013), it has been three decades was warmer than the previous one but the rate of warming per decade during the period 1998-2012 was less than the period 1951-2012 (Fig. 8). The report has also stated that the contributions made by anthropogenic sources are responsible for global warming.

Goswami and Rao (2003) used a combination of a hydrological model (SWAT: Soil and Water Assessment Tool) and weather data generated by the Hadley Centre for Climate Prediction (HadRM2) at a resolution of 0.440 X 0.440 on a GIS framework for quantifying the impact of climate change on the water balance components in 12 river basins of India. The results of their study have been presented in Fig. 9. The projected impact is likely to vary from basin to basin. There would be basin which will have an increase in rainfall, runoff, and evapotranspiration (ET), e.g., Mahanadi, while in some basins there would be a decrease in all the three components, e.g., Sabarmati.

Chaturvedi *et al.* (2012) used the representative concentration pathways (RCPs) along with the Coupled Model Intercomparison Project 5 (CMIP5) described by IPCC (2013) to generate scenarios of temperature and precipitation for India. They concluded that the ensemble model's output was closer to the observed climate than any individual model. They reported that under the "business" as usual scenario" the mean warming is likely to between 1.7-2 °C by 2030s and 3.3-4.8 °C by 2080s compared to preindustrial era. The precipitation is likely to increase by 4-5 % by 2030s and 6-14 % by 2080s. Temperature projections are more robust than precipitation projections. Another important project impact is the likely increase in extreme rainfall events of >40 mm/day for 2060s and beyond.

The climate is expected not only to affect input (recharge) and output (discharge), but also to inûuence the quantity and quality of the groundwater. For example, drying up of millions of wells/bore wells, poor quality of groundwater, salinity ingress, drying up of rivers like Swarnamukhi, Impact of waste effluents into river (river-aquifer interaction), Over exploitation of freshwater from coastal aquifer, seawater Intrusion, sea level rise. Groundwater quality water recharged during an arid period may have a higher concentration of salts and hence higher TDS, while during a wet period the converse may occur (Sukhija *et al.* 1998). However, to appreciate such changes long-term monitoring of rainfall and groundwater quality is required. It is also possible to link the occurrence of certain ions in groundwater to particular water–rock processes that occurred during specific past climatic periods.

Crop Water Requirements

In a study carried out by Rao *et al.* (2011), crop water requirements of four important crops being cultivated in the various agro-climatic zones of the country, CROPWAT 4 Windows 4.3 (Smith, 1992; 1993) for the years 2020 and 2050 using the weather data generated by HadCM3 Global Circulation Model. On the basis of their results summarized in Table 6, it can be concluded that the lowest crop water requirement (260 – 300 mm) was observed for wheat at northwest Indian stations like Sirsa and Ganganagar. In case of perecentage increase in crop water requirement and reference crop evapotranspiration, the highest increase was noticed at Solapur for sorghum, Sirsa and Ganganagar for wheat, Karimnagar and Begusarai for maize in 2020 and 2050 respectively, and at Ongole and Gulburga for pearl millet. It is important to note that though the actual values of crop water requirement and reference evaportranspiration were low at Ganganagar and Sirsa in 1990, the percentage increase is more in 2020 and 2050 at these places compared to other wheat growing stations. These stations come under water scarcity zone and adequate measures to augment irrigation water availability are needed in order to sustain wheat production.

In another study, Rao *et al.* (2012) used FAO-CROPWAT for estimating crop water requirement and irrigation demand for four major crops *viz.* groundnut, soybean, wheat and *rabi* sorghum for ten selected locations in the country. NATCOM projected climate change scenarios were used for estimating water requirements for 2021-2050 and 2071-2098 (NATCOM, 2009). They assessed the water requirements both for sustaining current productivity levels and future requirements i.e. 10 and 20 per cent increase in yield. Their results indicated that soybean did not require any additional irrigation but *rabi* sorghum and wheat would require additional irrigation requirement of different magnitudes to sustain the current production levels. Groundnut irrigation requirements differed according to location with no additional requirement for middle Gujarat region for current production levels. However, net irrigation requirement for enhanced production in future climate scenarios were higher in wheat and *rabi* sorghum crops compared to groundnut and soybean.

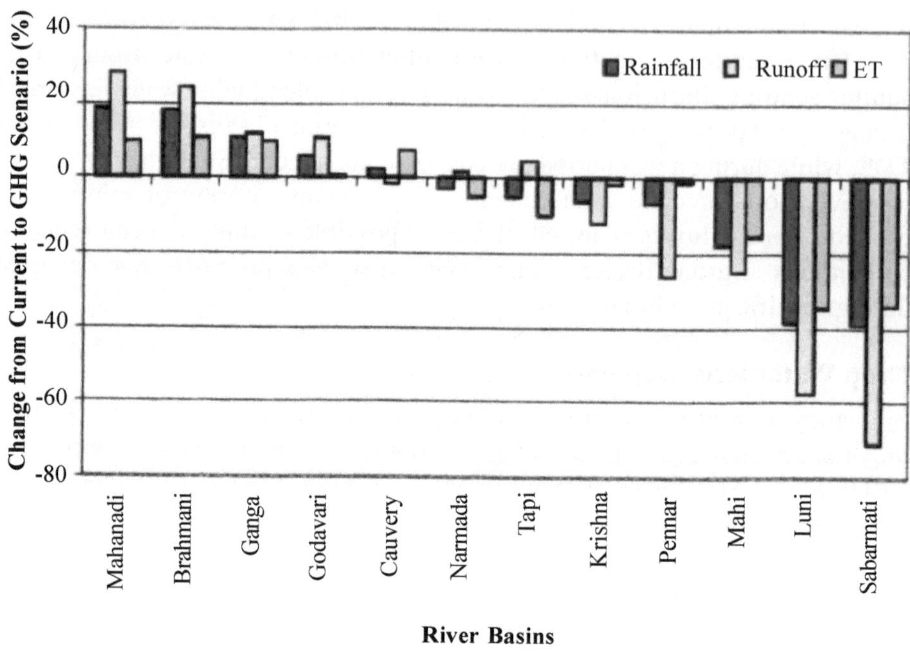

Fig. 9: Percent change in mean annual water balance for control and CHG climate scenarios

Conclusion

It is suggested there is an urgent need to improve understanding and modeling of climate changes related to the hydrological cycle at scales relevant to decision making. Information about the water-related impacts of climate change is inadequate, especially with respect to water quality, aquatic ecosystems and groundwater - including their socio implies economic dimensions. Agrarian water crisis can be solved through identification of input use efficient crops and cropping systems, integrated use of rain, surface and ground waters, judicious use of polluted and poor quality waters, awareness among farmers about the value and scarcity of water and negative fallouts of improper use, participatory water management, recharging the ground water aquifers, multiple uses of water, enhancing income per unit quantity of water consumed, speedy development and transfer of cost - effective and eco-friendly technologies.

References

Ageta, Y. and Kadota, T. (1992). Predictions of changes of glacier mass balance in the Nepal Himalya and Tibetea Plateau: a case study of air temperature increase for three glaciers. Ann. Glaciol.16: 89–94.

Bahadur, J., Dash, S. K. and Bahadur, J. (1999). The Himalayan Environment, New Age International (P) Ltd, New Delhi, pp. 258–267.

Central Ground Water Board (2006). Dynamic Ground Water Resources of India. New Delhi.

CGWB (2011) Ground Water Year Book – India 2010-11. Central Ground Water Board, Ministry Water Resources, Giovernment of India. http://www.cgwb.gov.in/documents/ground%20 Water%20Year%20Book-2010-11.pdf.

Chaturvedi RK, Joshi J, Jayaraman M, Bala G, Ravindranath NH, (2012). Multi – model climate change projections for India under representative concentration pathways, Current Science 103:791-802.

Fushimi, H., In (eds Domoto, A. et al., 2000). Tsukiji-Shokan Pub Co Ltd., Japan and The World Conservation Union, Gland, Switzerland, pp. 42–45.

Gautam, M. and Singh, A.K. (2015). Impact of climate change on water resources. In Climate Change Modelling, Planning and Policy for Agriculture, Eds. Singh, A.K. et al. Springer India, pp. 219-231.

Goswami, B. N. (2006). Increasing trend of extreme rain events and possibility of extremes of seasonal mean Indian monsoon in a warming world (http://saarc-sdmc.nic.in/pdf/workshops/ kathmandu/pres16.pdf)

IPCC (2007). Inter-governmental Panel on Climate Change, 4th Assessment Report-Synthesis Report.

IPCC (2013). Summary for policy makers. In: Stocker TF, Qing D, Platter GK, Tignor M, Allen SK, Boschung J, Nauels A, Xia Y, Bex V, Midgley PM (eds) climate change 2013: pshycal science basic contribution of Working Group I too the fifth assessment report of the Intergovernmental Panel on Climate Changes, Cambridge University Press, Cambridge/New York.

Ives, J. D. and Messerli, B (1989). The Himalayan Dilemma: Reconciling Development and Conservation, Routledge, London.

Lal, M. (2001). Future climate change: Implications for Indian summer monsoon and its variability. Current Science, 81 (9):. 1205.

Lobell, D. B. and Field, C. B. (2007). Global scale climate-crop yield relationships and the impacts of recent past. Environmental Research Letters, 2, p. 7.

Mall, R. K., Akhilesh Gupta, Ranjeet Singh, R. S. Singh and L. S. Rathore (2006). Water resources and climate change: An Indian perspective. Current Science, 90 (12): 1610-1626.

NATCOM.(2009). India's National Communication to UNFCCC. Data extraction tool for Regional Climate Scenarios (PRECIS) for India. Ministry of Environment and Forests, Government of India, New Delhi.

Rao., GGSN., G.G.S.N. Rao, A.V.M.S. Rao and Rao, V.U.M. (2009) Trends in Rainfall and Temperature in Rainfed India in Previous Century In Global Climate Change and Indian Agriculture-case studies from ICAR Network Project (PK Aggarwal ed.), ICAR, New Delhi Pub., pp.71-73.

Rao, V.U.M., Rao, A.V.M.S., Rao, G.G.S.N., Satyanarayan, T, Manikandan, N. and Venkateshwarlu, A.(2011). Impact of Climate Change on Crop Water Requirements and Adaptation Strategies. In "Challenges and Opportunities in Agrometeorology". Eds. Attri, S.D. et al. Springer-Verlag, Berlin Heidelberg: 311-319 p.

Rao, B.B., Sandeep, V.M., Rao, V.U.M. and Rao, A.V.M.S. (2012). Climate Change and Crop Water Requirements: An assessment for future climates. Journal of Agrometeorology, 14: 125-129

Romani, Saleem (2006), Groundwater Management – Emerging challenges. Groundwater Governance – Ownership of Groundwater and its Pricing. Proceedings of the 12th National Symposium on Hydrology. November 14-15, 2006. New Delhi.

Sharma, B.R., Rao, K.V., Vittal, K.P.R., Ramakrihana, Y.S. and Amarasinghe, U. (2010) Estimating the potential of rainfed agriculture in India: Prospects for water productivity improvents. Agricultural Water Management 97: 23-30.

Singh, A.K. (2013).Water Management: Issues and Strategies for Higher Productivity. In "Fertiliser Sector at the Crossroads", FAI Annual Seminar 2013, New Delhi, p. SII-3/1-15.

Singh, A.K. (2014). Sustainable Management of Water resources: Issues and Strategies. Bulletin of the Indian Society of Soil Science, 29: 1-26.

Smith M (1992) CROPWAT. A computer program for irrigation planning and management. Irrig. & Drain. Paper 46 FAO, Rome.

Smith M (1993) CLIMWAT for CROPWAT. A climate database for irrigation planning and management. Irrig. & Drain, paper, 49, Food and Agricultural Organisation, Rome.

Sukhija BS, Reddy DV, Nagabhusanan P, Chand R (1998) Validity of the environment chloride method for recharge evaluation of coastal aquifers, India. J Hydrol 99:349-366.

Trenberth, K. E. (1999). Conceptual framework for changes of extremes of the hydrological cycle with climate change. Climatic Change, 42: 327–339.

Upadhyay, D. S. (1995). Cold Climate Hydrometeorology, New Age International (P) Ltd, New Delhi, p. 345.

6

Climate Change Adaptation and Mitigation Strategies in Rainfed Agriculture

Ch. Srinivasa Rao, K.A. Gopinath, G. Venkatesh and M.P. Jain

Introduction

Rainfed agro-ecosystems occupy a considerable place in Indian agriculture, covering 80 M ha in arid, semi-arid and sub-humid climatic zones; constituting nearly 57% of the net cultivated area. Rainfed areas contribute almost 100% of forest products, 84-87% of coarse-grain cereals and pulses, 80% of horticulture, 77% of oilseeds, 60% of cotton and 50% of fine cereals including rice, wheat, maize, sorghum etc (Srinivasarao *et al.*, 2010; 2011). Rainfed regions support 60% of livestock, 40% of human population and contribute 40% of food grains and several special-attribute commodities such as seed spices, dyes, herbs, gums, etc. (Srinivasarao *et al.*, 2011). It is estimated that even after achieving the full irrigation potential, barring for successful completion of river linking project, nearly 45 to 50% of the total cultivated area will remain dependent on rain. Undoubtedly, the rainfed agriculture would continue to occupy a prominent place in Indian agriculture for a long time to come (Venkateswarlu *et al.*, 2011). Despite the progress made so far, rainfed agriculture in India encounters multiple risks and constraints relating to biophysical and socio-economic issues. Agricultural productivity in rainfed areas continues to remain low and unstable due to weather aberrations, nutrient disorder and poor socio-economic status of farmers.

Features of rainfed ecosystems

***Frequent droughts*:** Droughts and famines are the general features of rainfed agriculture in India. The risk involved in successful cultivation of crops depends on the nature of drought (chronic and contingent); it's probable duration, and periodicity of occurrence within the season. In the arid region where the mean annual rainfall is less than 500 mm, drought is almost an inevitable phenomenon in most of the years (Ramakrishna, 1997). In semi-arid regions (mean annual rainfall 500-750 mm), droughts occur in 40 to 60% of the years due to deficit seasonal rainfall or inadequate soil moisture availability between two successive rainfall events. Even in dry sub-humid regions (annual rainfall 750-1200 mm), contingent drought situations occur due to break in monsoon conditions. Very high incidence of drought (>20%) is observed in a few districts in Rajasthan and Gujarat. The incidence is relatively low in the Western ghats, Eastern and North-Eastern India (Fig. 1; Rama Rao *et al*., 2013).

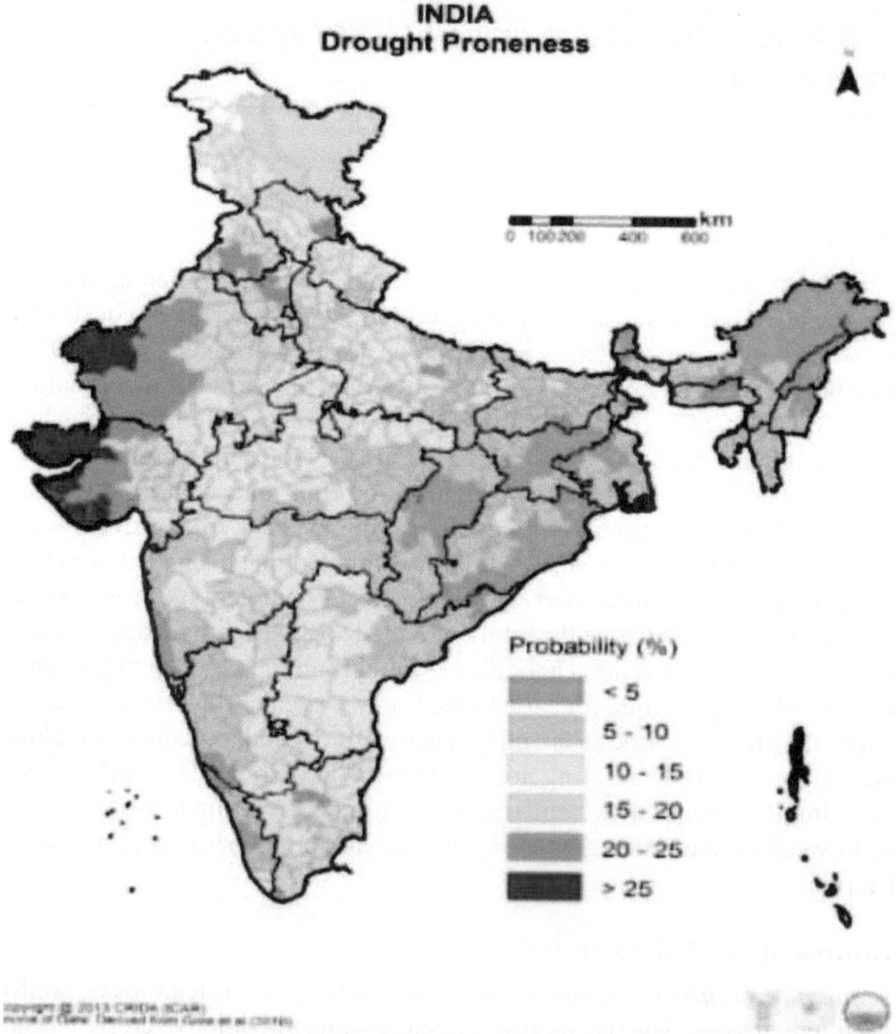

Fig. 1: Probability of drought occurrence in different parts of India (*Source*: ICAR CRIDA)

Degraded soils: There have been several estimates of area of degraded/waste lands by different organizations. These estimates have been harmonized by adopting spatial data integration with geographical information system for different environment (Maji, 2007). The total degraded area is estimated at 120.7 Mha, of which 104.2 Mha (86.3%) is arable land, and 16.5 Mha (13.7%) is open forest land. Of the total degraded land area, 73.3 Mha (60.7%) is caused by water erosion, 12.4 Mha (10.3%) by wind erosion, 5.4 Mha (4.5%) by salinization, and 5.1 Mha (4.2%) by soil acidification. Some areas are affected by more than one degradation progress (Maji, 2007). In many states in India, anywhere between 40 to 80% of the land area is classified as degraded in some form or the other. Soils in rainfed areas are subject to a prolonged double exclusion, being unable to gain from chemical fertilizers and receiving no support for locally validated fertility enhancing practices like crop residue incorporation, composting, farm yard manure application, etc.

Low soil organic carbon content: Soils of rainfed areas are highly degraded besides having low soil organic carbon. Most of the rainfed soils are low in organic carbon due to rapid oxidation process in dry regions of the country. Besides, low biomass generation and erosion of top organic carbon rich soil under intensive rainfall are other important factors leading to low soil organic carbon status in tropical regions (Srinivasarao, 2011). Severe soil organic carbon (SOC) depletion is a major constraint in rainfed agro ecosystems because it directly influences soil quality, crop productivity and sustainability. Vertisols, Inceptisols and Alfisols comprise a major share of soil organic carbon (SOC) stocks in the top 30 cm depth. Organic carbon stocks in the soil profiles across the country showed wide variations and followed the order Vertisols> Inceptisols> Alfisols>Aridisols (Srinivasarao *et al*., 2009, 2011).

Multi-nutrient deficiencies: The increased use of fertilizers alone, often in an unbalanced manner has resulted in deteriorating soil health and multiple nutrient deficiencies in various areas of the high potential rainfed regions. Indian soils not only show deficiency of NPK but also of secondary nutrients (S, Ca and Mg) and micro nutrients (B, Zn, Cu, Fe, Mn etc.) in most parts of the country. Besides the three primary nutrients (NPK), deficiency of S and micro nutrients like Zn and B in many of states, and of Fe, Mn and Mo in some states, has become a limiting factor in increasing food productivity. In rainfed India, results of analyses of soil samples from farmers' fields in several districts of Andhra Pradesh, Karnataka, Tamil Nadu, Rajasthan, and Madhya Pradesh and Junagarh district, Gujarat showed that almost all farmers' fields had low organic carbon (C), low to moderate available phosphorus (P), and generally adequate extractable potassium (K). However, the widespread deficiencies of sulfur (S), boron (B), and zinc (Zn) were most revealing; their deficiencies varied with nutrient, district, and state (Sahrawat *et al.* 2010).

***Low input use*:** Use of production inputs like fertilizers, supplemental irrigation, good quality seeds, pesticides, and herbicides are lower in rainfed than in irrigated crops. Thus, productivity of rainfed crops has remained low. Although it has been amply demonstrated that soils in rainfed regions are multi-nutrient deficient, balanced use of these inputs in rainfed crops is rarely achieved. There are wide disparities of inputs between irrigated and rainfed regions due to numerous uncertainties in the crop production. The vast majority of rainfed farmers in remote areas still practice low external input or no external input farming which is well integrated with livestock, particularly small ruminants. Based on several surveys and reports, it is estimated that up to 30% of the rainfed farmers in many remote areas of the country do not use chemical fertilizers and pesticides (Venkateswarlu, 2008).

***Low investment capacity*:** Indian agriculture is the home of small and marginal farmers. The share of marginal and small farmers accounted for around 81% of operational holdings in 2002-03 as compared to about 62% in 1960-61. Similarly, the area operated by small and marginal farmers has increased from about 19% to 44% during the same period. According to National Sample Survey (NSS) 2003 data, the monthly consumption of marginal farmers was Rs. 2482 and monthly income was Rs. 1659. It shows that they have dis-savings of Rs. 823. The dis-savings for small farmers were Rs. 655. Hence, small holdings need credit for both consumption and investment purposes.7 The indebtedness for the small & marginal farmers from formal institutional sources is lower than large farmers and the reverse is true in the case of informal sources. The dependence on money lenders is the highest for sub-marginal and marginal farmers. Dependence of small and marginal farmers on informal sources is high even in states like Andhra Pradesh, Punjab and Tamil Nadu. They need a level playing field with large farms in terms of accessing land, water, inputs, credit, technology and markets.

***Poor market linkages*:** Most of the rural areas are characterized by subsistence economy. The surplus farm produce is offered for sale after meeting the family requirements. Further, individual production units (families) mostly operate separately making it difficult to pool the produce for efficient marketing. The present marketing system in most of the villages is fraught with many constraints viz. unorganized, unregulated and unprofitable systems, middlemen-dominated, and unreliable marketing channels (Dixit *et al.*, 2013). Facilitating market linkage involves understanding of the demand and supply situation, transient storage opportunities, transport infrastructure and easy access to markets. The idea is to intervene at any of these steps in the value chain to enhance farmers' share in the retail market price.

Climate change and rainfed agriculture

While climate change impacts agriculture sector in general, rainfed agriculture is likely to be more vulnerable in view of its high dependency on monsoon and the likelihood of increased extreme weather events due to aberrant behavior of south-west (SW) monsoon (Venkateswarlu, 2010). About 74% of annual rainfall occurs during SW monsoon (June-September). This rainfall exhibits high coefficient of variation particularly in arid and dry semi-arid regions. Skewed distribution has now become more common with reduction on numbers of rainy, days. Aberrations in SW monsoon which include delay in onset, long dry spells and early withdrawal, all of which affect the crops, strongly influence the productivity levels (Lal, 2001). These aberrations are likely to further increase in future. A long term analysis of rainfall trends in India (1901 to 2004) using Mann Kendall test of significance by CRIDA indicates significant increase in rainfall trends in West Bengal, central India, coastal regions, south western Andhra Pradesh and central Tamil Nadu (Venkateswarlu, 2010). Rainfall is likely to decline by 5 to 10% over southern parts of India whereas 10 to 20% increase is likely over other regions. There is a probable decrease in the number of rainy days over major part of the country the country pointing at likely increase of extreme events. The recent ensemble models projected that the frequency of extreme precipitation days (e.g. 40mm/day) may likely rise. Changes in annual average temperature are expected to increase by 2 to 2.5°C. Warming is likely to be more over northern parts of India. A rise in night temperatures is also likely over India except some small pockets in the peninsular region (CRIDA, 2013). This is a cause of concern for agriculture as increased night temperatures accelerate respiration, hasten crop maturity and reduce yields. Besides hastening crop maturity and reducing crop yields, increased temperatures will also increase the crop water requirement. The risk of crop failure and poor yields always influence farmers' decision on investing on new technologies and level of input use (Pandey *et al*., 2000). Numerous technological (e.g. cropping patterns, crop diversification, soil health management, rainwater harvesting, shifts to drought/ salt tolerant varieties etc) and socio-economic (e.g. ownership of assets, access to services, infrastructural support etc) factors will come into play in enhancing or constraining the current capacity of rainfed farmers to cope with climate change.

Adaptation and mitigation strategies

The purpose of mitigation and adaptation measures is to attempt a gradual reversal of the effects caused by climate change and sustain development under the inescapable effects of climate change. Here is it important to note and understand the subtle difference between mitigation and adaptation. Mitigation and adaptation are related to the temporal and spatial scales on which they are effective. The benefits of mitigation activities carried out today will be evidenced in several decades because of the long residence time of greenhouse gases in the atmosphere, whereas the effects of adaptation measures should be apparent immediately or in the near future (Kumar and Parikh, 2001). Besides, mitigation has global in addition to local benefits, whereas adaptation typically takes place on a local or regional scale.

Successful adaptation to climate change requires long-term investments in strategic research and new policy initiatives that mainstream climate change adaptation into development planning. As a first step, we need to document all the indigenous practices farmers have been following over time for coping with climate change. Secondly, we need to quantify the adaptation and mitigation potential of the existing best bet practices for different crop and livestock production systems in different agro-ecological regions of the country. Thirdly, a long-term strategic research planning is required to evolve new tools and techniques including crop varieties and management practices that help in adaptation (Venkateswarlu, 2010). Technical adaptation measures range from change in production systems like adjusting planting or fishing dates, rotations, multiple cropping/species diversification, crop-livestock pisciculture systems, agroforestry, soil, water and biodiversity conservation and development by building soil biomass, restoring degraded lands, rehabilitating rangelands, harvesting and recycling water, planting trees, developing adapted cultivars and breeds, protecting aquatic ecosystems to maintain long-term productivity (Venkateswarlu and Shanker, 2009). Adaptation measures also take account of establishing disaster risk management plans and risk transfer mechanisms, such as crop insurance and diversified livelihood systems. Mitigation options include carbon sequestration in agriculture and forestry. Mitigation of climate change is a global responsibility. Agriculture, forestry, fisheries/aquaculture provide in principle, a significant potential for greenghouse gases mitigation.

Strategies based on natural resource conservation and management

There are a number of options in soil, water and nutrient management technologies which contribute to both adaptation and mitigation. Much of the research done in rainfed agriculture in India relates to conservation of soil and rainwater and drought proofing which is an ideal strategy for adaptation to climate change (Venkateswarlu and Shanker, 2009). The key technologies are *in situ* moisture conservation, rainwater harvesting and recycling, efficient use of irrigation water, conservation agriculture, energy efficiency in crop production and irrigation and use of poor quality water. The suggested strategies are: characterization of bio-physical and socio-economic resources utilizing GIS and remote sensing; integrated watershed development; developing strategies for improving rainwater use efficiency through rainwater harvesting, storage, and reuse; contingency crop planning to minimize loss of production during drought/flood years (Kapoor, 2006).

Rainwater conservation and harvesting

The strategy of rainwater management in arid and semi-arid regions consists of selection of short duration and low water requiring crops and conserving as much rainwater as possible so that crops can escape moisture stress during the growing period. In addition to *in-situ* conservation, efforts need to be made to divert the surplus water into storage structures which can be used either as standalone resource or in conjunction with groundwater for meeting the critical irrigation requirements. In relatively high rainfall regions, the strategy is to conserve as much rainwater as possible and to harvest the surplus water for life saving irrigation and also for enhancing the cropping intensity and to maximize returns from the harvested water. Apart from enhancing the availability of water by various methods, increasing the water use efficiency should be the focus by arresting various kinds of losses associated with utilization of water.

Rainwater conservation: In-situ soil and water conservation (SWC) practices improve soil structure and soil porosity, increase infiltration and hydraulic conductivity, and consequently increase soil water storage and promote crop growth and build adequate soil cover. Several studies have indicated that *in-situ* SWC practices directly or indirectly influence soil quality as well as soil fertility. These measures are more feasible and practical proposition under most situations and can be adopted by an individual farmer with less draft and amenable even for a small holder. The suitability of a practice depends on the topography of the field, temporal and spatial distribution of rainfall, type of soil, crop, etc. Based on extensive research conducted at various locations in the country, several soil and water conservation measures have been recommended depending on the rainfall regimes (Table 1) and some of these practices are being adopted in a significant area in the country.

Rainwater harvesting: The importance of rainwater harvesting has increased in recent years due to the increased rainfall variability, depletion of groundwater levels and raising temperatures. Besides, due to substantial soil loss, sedimentation of reservoirs is happening thus resulting in loss of storage capacity at a faster rate. Rainwater harvesting not only reduces runoff and soil loss but also facilitates groundwater recharge and prevents early sedimentation of the reservoirs. The rainwater harvesting and groundwater recharge enables the farmers to provide supplementary irrigation to crops during monsoon season and may go for the second crop depending on the availability of water. Therefore, it is desirable to harvest rainwater which has multiple uses (Mishra, 2011). It has been estimated

Table 1: Recommended soil and water conservation measures for various rainfall zones in India

Seasonal rainfall (mm)

< 500	500-700	750-1000	> 1000
Contour cultivation with conservation furrows	Contour cultivation with conservation furrows	BBF (Vertisols) conservation furrows	BBF (Vertisols)
Ridging	Ridging	Sowing across slopes	Field bunds
Sowing across slopes	Sowing across slopes	Tillage	Vegetative bunds
Mulching	Scoops	Lock and spill drains	Graded bunds
Scoops	Tied ridges	Small basins	Level terrace
Tied ridges	Mulching	Field bunds	
Off-season tillage	Zing terrace	Vegetative bunds	
Inter row water harvesting systems	Off-season tillage	Graded bunds	
Small basins	BBF	*Nadi*	
Contour bunds	Inter-row water harvesting system	Zing terrace	
Field bunds	Small basins		
Khadin	Modified contour bunds		
	Field bunds		

Source: Vittal *et al.* (2003)

that about 27.5 M ha of rainfed area of the country, excluding the very arid and wet areas, can contribute an amount of 114 billion M^3 of water for water harvesting which is adequate to provide one supplementary irrigation of 100 mm depth to 20 M ha during drought years and 25 M ha during normal years which can increase yields by 12% over the present production from these areas (Table 2). With the adoption of improved technologies, production can be further increased. The benefits could be still higher if improved irrigation techniques like drip and sprinklers are used for the use of harvested water. In order to realize the above projected benefits, an amount of Rs 63.0 billion per annum for 20 years (a total of Rs 1260.0 billion) is needed to develop 50 million ponds in rainfed areas spread across the country (Sharma *et al.*, 2010).

Table 2: Potential of rainwater harvesting in different rainfed crops

Crop	Rainfed area (M ha)	Surplus runoff (BM^3)	Irrigable area (M ha)	
			During normal monsoon years	During drought years
Rice	6.32	41.21	6.32	6.21
Finger millet	0.30	1.53	0.26	0.22
Maize	2.44	7.71	2.25	1.68
Pearl millet	1.81	3.59	1.37	0.83
Sorghum	2.93	7.71	2.62	1.85
Total coarse cereals	7.50	20.57	6.51	4.60
Cotton	3.17	7.57	2.65	1.72
Castor	0.02	0.14	0.02	0.02
Groundnut	1.66	3.42	1.09	0.71
Sesame	1.05	4.16	0.91	0.74
Soybeans	2.84	13.29	2.84	2.66
Sunflower	0.09	0.11	0.05	0.03
Total oilseeds	6.27	24.21	4.94	4.17
Chickpea	3.00	13.04	2.92	2.56
Pigeonpea	1.83	6.59	1.71	1.37
Total pulses	5.28	20.44	4.63	3.93
Total (all rainfed crops)	28.56	114.02	25.07	20.64

Source: Sharma *et al.* (2010)

Watershed management: Watershed development programs are silently revolutionalizing the rainfed areas and can become growth engine for inclusive and sustainable development in vast tracts of rainfed areas in India. Meta analysis of 311 watershed case studies from different agro-eco regions in India revealed that watershed programs benefitted farmers through enhanced irrigation areas by 33.5%, increasing cropping intensity by 63%, reducing soil loss to 0.8 t/ha and runoff to 13% and improved groundwater availability. Economically the watershed programs were beneficial and viable with a benefit-cost ratio of 1:2.14 and the internal rate of return of 22% (Joshi *et al.*, 2005).

Soil management

Soils hold the key to productivity and resilience to climate vagaries in rainfed agriculture in India. Loss of fertile soil as erosion, depletion of soil organic matter, emerging secondary and micronutrient deficiencies, soil compaction, surface crusting, loss of soil biodiversity have become potential limiting factors for productivity enhancement in these regions.

Site specific nutrient management: Integrated nutrient management and site specific nutrient management (SSNM) has the potential to mitigate effects of climate change. Demonstrated benefits of these technologies are: increased rice yields and thereby increased CO_2 net assimilation and 30-40% increase in nitrogen use efficiency. This offers important prospect for decreasing GHG emissions linked with N fertilizer use in rice systems. Judicious fertilizer application, a principal component of SSNM approach, thus has twofold benefits, i.e. reducing GHG emissions and at the same time improving crop yields. In Andhra Pradesh, under the NAIP program, SSNM was demonstrated in different districts to address nutrient deficiencies which exist within farmer's fields. The net return from cotton due to balanced nutrition in Warangal, Adilabad and Khammam districts showed an increase of the order of 20, 60 and 30%, respectively over farmers' practice. For groundnut, the net return due to balanced nutrition in Kadapa and Anantapur showed an increase of the order of 40 and 25%, respectively over farmers' practice (Table 3).

Conservation agriculture: Conservation agriculture (CA) is gaining importance as an alternative strategy to sustain agricultural production due to the growing resource degradation problems, particularly under rainfed conditions. Sufficient information is available that indicates that CA practices contribute towards saving in time, reduce cost of production and contribute substantially towards profitability and sustainability of carbon in soils. Results from 4-year old CA experiment with maize-horsegram cropping system on Alfisols at Gunegal Research Farm of CRIDA, Hyderabad showed that there were no significant differences in CA and conventional tillage treatments on maize seed and stover yield. However, pooled data from 2010 and 2011 showed that seed mean yield (3.0 t/ha) and stover mean yield (5.5 t/ha) in CA were on par with conventional system. It was also observed that nutrient uptake was more in CA compared to conventional system. There was an increase in use efficiency of N, P, K, S and Zn through adoption of CA practices compared to conventional system. However, less biomass productivity and competing uses of crop residues constraints the scope of using crop residues for conservation agriculture in dryland ecosystems. Central Research Institute for Dryland Agriculture (CRIDA), Hyderabad has shown that in dryland ecosystems, where only a single crop (maize) is grown in a year, it is possible to raise a second crop (horsegram) with residual soil moisture by covering soil with crop residues.

Table 3: Economics of balanced nutrition in rainfed cotton and groundnut cultivation in farmers fields in Andhra Pradesh during 2008-2010.

Village/ District	Crop	Mean gross return (Rs/ha)		Mean net return (Rs/ha)		Mean B:C ratio	
		SSNM	FP	SSNM	FP	SSNM	FP
Jaffergudem, Warangal (10)*	Cotton	31543(536)	27048(460)	15091(256)	12569(214)	1.95	1.84
Seethagondi, Adilabad (10)*	Cotton	54655(929)	38631(656)	37907(644)	23724(403)	3.26	2.72
T. Cheruvu, Khammam (10)*	Cotton	55273(939	43973(748)	38557(655)	29741(505)	3.35	3.05
B. Yerragudi, Kadapa (14)*	Groundnut	9647(163)	7290(124)	3977(67)	2885(49)	1.78	1.65
Pampanur, Anantapur (12)*	Groundnut	9668(164)	8157(139)	4668(79)	3723(63)	1.82	1.66

*Number of farmer's fields; *Source*: Srinivasarao *et al.* (2011)

Biomass recycling: Indian agriculture produces about 500-550 million tonnes (Mt) of crop residues annually. These crop residues are used as animal feed, soil mulch, manure, thatching for rural homes and fuel for domestic and industrial purposes and thus are of tremendous value to farmers. However, a large portion of these crop residues, about 90-140 Mt annually, is burnt on-farm primarily to clear the fields to facilitate planting of succeeding crops (NAAS, 2012). Burning of crop residues is environmentally unsafe as it leads to release of soot particles and smoke causing human health problems and emission of greenhouse gases adding to global warming and climate change. It also results in loss of plant nutrients such as N, P, K, S and carbon, which are so important for soil health. A large amount of energy is used in cultivation and processing of crops like sugarcane, food grains, vegetables and fruits, which can be recovered by utilizing residues for energy production. This can be a major strategy of climate change mitigation by avoiding burning of fossil fuels and recycling crop residues (Venkateswarlu, 2010). Lack of availability of proper chipping and soil incorporation equipment is one of the major reasons for the colossal wastage of agricultural biomass in India. Increased cost of labour and transport is another reason for lack of interest in utilizing the biomass. Many technologies like briquetting, anaerobic digestion, vermicomposting and biochar etc exist, but they have not been commercially exploited.

There is renewed interest in the use of anaerobic digestion processes for efficient management and conversion of cattle dung and other agro-industrial wastes into clean renewable energy and organic fertilizer source. Similarly, few Maize-horsegram cropping sequence on Alfisols. Rainy season maize (left) and post-rainy season horsegram on stored moisture (right) organizations in India have initiated research on efficient use of bioresidues for biochar production and its use in agriculture. At CRIDA, biochar from castor, cotton and pigeonpea stalks was produced by using a portable kiln (Venkatesh *et al.*, 2013ab) and initial

Maize-horsegram cropping sequence on Alfisols. Rainy season maize (left) and post-rainy season horsegram on stored moisture (right)

results from field experimentation suggest that biochar application to soil helps in improving soil properties and crop yields (CRIDA, 2014).

Tank silt application: Community tanks in peninsular India serve the purpose of collecting and storing not only the rain water, but also the nutrient-rich top soil eroded from their catchment areas. Analysis of tank silt collected from several tanks of 100 districts showed the potential of tank silt in supplying organic carbon and several nutrients. Impact of tank silt application in degraded Alfisols of Andhra Pradesh indicate strong response of maize and castor yields besides improving soil physical, chemical and biological properties

Mulch cum manuring: Shortage of rainfall coupled with its erratic distribution during rainy season causes severe water deficit conditions inducing various intensities of moisture stress. Hence, improving soil surface conditions to increase infiltration and water holding capacity, and to reduce evaporation losses are basic requirements in rainfed areas for realizing higher crop yields. Mulching has been shown to decrease the evaporation of soil moisture and thus helps moisture retention in soil. However, making available enough material to mulch the soils under field conditions still remains a challenge. Crop residue recycling, green manure crops, green leaf manuring, brown manuring etc are some of the important options for adopting mulch cum manuring.

Carbon sequestration: Maintaining or improving SOC concentration in rainfed dryland agroecosystems is a major agronomic challenge. Yet, the data from long-term experiments show that increasing SOC concentration by C sequestration and stabilization positively affects yields of several crops. Agronomic efficiency (AE) of added nutrients and partial factor productivity (PFP) of crops are maintained or enhanced with INM practices including application of organics in conjunction with fertilizers, but decline with application of only chemical fertilizers because of declining SOC concentration and soil quality with continuous cropping (Srinivasarao *et al.*, 2013). The SOC stocks in the soil profiles across the country show wide variations and follow the order Vertisols>Inceptisols>Alfisols>Aridisols. The rate of depletion of SOC can range from 0.15 Mg C/ha/y in rice based system to 0.92 Mg C/ha/y in groundnut-finger millet system. To arrest this depletion C input of 1.10-3.47 Mg C/ha/y is required as a maintenance dose. Further, potential of tropical soils to sequester more C in soil can be harnessed by identifying appropriate production systems and management practices for sustainable development and improved livelihoods in the tropics.

Crop based strategies

Crop based approaches include growing crops and varieties that fit into changed rainfall and seasons, development of short duration varieties, varieties for heat stress, drought and submergence tolerance; evolving varieties which respond positively in terms of growth and yield under high CO_2. In addition, adoption of intercropping systems, crop diversification, improved planting methods, agro-forestry and integrated farming systems help to cope with climate variability and climate change.

Efficient crops and varieties

Selecting suitable crops and varieties will not only help increase production of a single crop but also will help increase cropping intensity. Many criteria have been set out for selecting a crop variety to meet the requirement of rainfed areas; however, the capacity to produce a fairly good yield under limited soil moisture conditions is the most desirable one. Generally, the crop for rainfed areas should be of short duration, drought tolerant and high yielding. Based on the analysis of long-term climatic data in terms of probability of the onset of monsoon, withdrawal of monsoon, and occurrence of dry spells, effective cropping seasons have been worked out for different regions of the country. These will serve as a good guide in selection of efficient crops and their sowing time.

In arid regions of the country, where rainfall is above 300 mm, the length of humid period is about 1-4 weeks. Under these conditions, short duration drought resistant pulses such as mungbean, mothbean, cowpea, and cereals of 10-12 weeks duration such as pearl millet and minor millets are suitable. In semi-arid regions, where the length of humid period is around 6 weeks, rainy season crops are grown in soils that have a capacity to hold less than 150 mm of water. Additional post-rainy season crops can be grown on conserved soil moisture, in soils that can hold more than 200 mm. in soils with 150-200 mm capacity, intercropping is possible. In sub-humid areas, where humid period is more than 12 weeks and the rainfall is twice that of PET, paddy based cropping system is suitable as other crops cannot tolerate water stagnation.

Similarly, Short-duration varieties/hybrids with higher harvest indices that mitigate the effects of drought periods (often occurring at the beginning and the end of the growing season) have been evolved for different agro-climatic conditions (Table 4). Such cultivars are considered a key component of management strategies particularly in the drought-prone areas. Furthermore, choice of suitable crop varieties assumes importance when consideration is given to the normal and late sowing times in a cropping season. The experience has shown that the varieties that are suitable for normal sowing are not necessarily suitable or

Table 4: Improved varieties/hybrids released for rainfed cultivation

State/Region	Crop	Variety	Yield advantage (%) over check	Special features
Maharashtra	Pigeonpea	Vipula	22	Resistant to sterility mosaic and tolerant to wilt, suitable for intercropping with sunflower/ pearl millet
		BSMR-853	31	Suitable for sole cropping in deep soils
	Pearl millet	Shradha	28	Resistant to downy mildew
	Sunflower	Bhanu	34	Tolerant to bud necrosis
	Chickpea	Vijay	20	Drought tolerant and wilt resistant
		Digvijay	25	Drought tolerant and wilt resistant
	Safflower	Bhima	30	Drought tolerant
		Phule Kusuma	30	Highly responsive to protective irrigation in deep black soils
	Sorghum	Phule Vasudha	20	Responsive to protective irrigation
Jharkhand	Rice	Birsa Vikas Dhan-108 & Birsa Vikas Dhan-109	45	Both varieties are tolerant to drought, insect-pests and diseases
	Black gram	Birsa Urid-1	65	Tolerant to drought
	Pigeonpea	Birsa Arhar-1	65	Tolerant to wilt
	Mustard	Shiwani	60	Tolerant to insect-pests
Uttar Pradesh	Pigeonpea	Narendra Arhar-1	45	Wilt and drought tolerant, survives during high rainfall, and suitable for late planting conditions
	Lentil	Malaviya Vishwanath	85	Resistant to rust and moderately resistant to wilt
Kandi region of Punjab	Maize	PMH-2 & JH-3459	85	Both hybrids are tolerant to lodging, mid and late season drought
	Chickpea	PBG-1	70	Resistant to blight and tolerant to drought
South western dry zone of Haryana	Pearl millet	HHB-67	20	Early maturing, resistant to downy mildew, and tolerant to drought

Contd.

State/Region	Crop	Variety	Yield advantage (%) over check	Special features
Odisha	Rice	ZHU-11-26	90	Short duration, resistant to lodging, and tolerant to moisture stress and blast
	Cowpea	SGL-1	75	Non-trailing, short duration variety
	Green gram	PDM-54	100	Early maturing, resistant to yellow mosaic virus, and moderately resistant to powdery mildew
	Turmeric	Sudarshan, Suguna & Subarna	37-38	All varieties have high curcumin content (7-8%) and high yielding
	Yam bean	Rajendra Mishri Kanda-1	45	High yielding

Source: Compiled by the authors

efficient when grown late in the season (Singh, 1995). This is particularly true in case of cereal crops. In case of sequential cropping systems, it is advisable to choose a short duration variety for rainy season cropping, followed by a longer duration variety for post-rainy season. This will facilitate good establishment of post-rainy season crop during the period when soil moisture conditions are favourable. In case of intercropping systems, the choice of variety of one component crop changes with the change in the variety of other crops.

Intercropping

Mixed cropping involving growing more than one crop of different durations on the same field is a widespread traditional practice in rainfed agriculture. This helps to distribute the risk due to adverse climate. But the system is hardly productive. The important contribution of rainfed farming research has been to scientifically modify the system to the level of intercropping where in two crops of different durations are planted in definite row ratios as a system (Umrani, 1999). Research on intercropping systems has focused on identification of potential crop combinations for increased productivity, profitability and risk minimization. Further, production technologies involving defining of base and companion crop combinations, plant population, nutrient management, etc. have been standardized for intercropping systems suitable for different agro-climatic conditions.

Amongst the several combinations tested, pearl millet and pigeonpea proved to be the best combination across several regions (Umrani and Subba Reddy, 1999). The combination has done exceedingly well in regions where late rains are likely to occur in September/October (Bijapur and Solapur) or where deep and moisture retentive soils exist (Akola). Cotton and mungbean/black gram/ soybean is yet another combination in which the relative advantage of intercropping is very good. Groundnut + pigeonpea intercropping system in 7:1 proportion is one of the most successful intercropping system adopted by farmers in scarce rainfall zone of Andhra Pradesh. About 70% of the total groundnut area of 0.85 million ha in this zone is under groundnut + pigeonpea intercropping system. The system is also adopted on a large scale by farmers in southern Karnataka. Similarly, sunflower + pigeonpea system is practiced in more than 6000 ha covering 30% of cropped area during rainy season in Solapur district and pearl millet + pigeonpea is practiced in 80000 ha covering 35% of cropped area during post-rainy season in Ahmednagar, Pune and Dhule districts of Maharashtra (Venkateswarlu *et al.*, 2009).

Contingency crop planning

Contingency crop planning refers to implementing a plan for making alternate crop or cultivar choices in tune with the actual rainfall situation and soils in a given location. In rainfed areas, as a general rule early sowing of crops with the onset of monsoon is the best-bet practice that gives higher realizable yield. Major crops affected due to monsoon delays are those crops that have a narrow sowing window and therefore cannot be taken up if the delay is beyond this cut-off date. Crops with wider sowing windows can still be taken up till the cut-off date without major yield loss and only the change warranted could be the choice of short duration cultivars. Beyond the sowing window, choice of alternate crops or cultivars depends on the farming situation, soil, rainfall and cropping pattern in the location and extent of delay in the onset of monsoon. Breaks in monsoon cause prolonged dry spells and are responsible for early, mid and terminal droughts. These aberrant situations often lead to poor crop performance and or total crop failures. While early season droughts have to be combated with operations like gap filling and re-sowing, mid and late season droughts have to be managed with appropriate contingency measures related to crop, soil nutrient management and moisture conservation measures. Drought also causes loss in livestock productivity due to shortage in fodder production. Appropriate location-specific fodder production strategies are essential for reducing the adverse impact on livestock which is the major source of livelihood in dryland areas (Srinivasarao *et al.*, 2013).

Any contingency measure, either technology related (land, soil, water, crop) or institutional and policy based, which is implemented based on real time weather pattern (including extreme events) in any crop growing season is considered as Real Time Contingency Planning (RTCP). If done timely and effectively, RTCP contributes to household and village food and fodder security (Srinivasarao *et al.*, 2010).

Agro-forestry

Agro-forestry systems provide multiple products such as fodder, fuel wood, fruit, nutrients, commercial wood, etc. depending on the kind of tree species. Integration of perennials in to the arable systems not only provides additional income to the farmers but also reduces the risk during the years of low rainfall. Trees and perennial shrubs in farm lands or on boundaries help in conserving soil and water and contribute towards improvement of micro climate. Tree/perennial systems improve soil fertility and contribute towards amelioration of saline soils and diversify production for increased social, economic and environmental benefits. Trees in landscape protect and stabilize ecosystems, provide material to meet basic requirements of rural population such as fuel wood, wood for making farm implements, fodder during the periods of dry period and generate additional employment. Hence, integrating trees into landscapes is an approach for sustainable intensification of arable systems and contributes towards enhancing productivity in unit time and area and have multifarious benefits and meets diversified needs of communities in comparison to arable cropping under rainfed situations.

Integrated farming systems

A common characteristic of integrated farming systems (IFS) is that they invariably have a combination of crop and livestock enterprises and in some cases may include combinations of poultry, agro-forestry, horticulture, apiary etc. The farming systems approach is considered as important and relevant especially for the small and marginal farmers as location-specific integrated farming systems will be more resilient and adaptive to climatic variability (Gopinath *et al.*, 2012) and bring in much needed year round employment and income to farmers. Farming systems are highly location specific depending on the natural resources available in a given agro-climatic situation. The success of an IFS and its sustainability depends entirely on strengths and weaknesses of the internal factors like natural resources (soil and water), farm size, availability of recyclable residues, labour and external factors such as market prices, infrastructure and the level of post-harvest processing. On-station and on-farm research in different regions of the country has resulted in identification of a number of sustainable and profitable IFS models for rainfed regions. These models need to be validated and replicated in respective areas.

Way forward

Rainfed agriculture is synonymous with risk-prone agriculture due to erratic monsoon. The low and highly variable productivity levels are an impediment to investment by resource-poor farmers. Climatic risks like droughts and floods and edaphic constraints like poor water and nutrient retention capacity, low soil organic matter (SOM) make rainfed agriculture highly vulnerable, less resilient requiring a different outlook and strategy. The following activities should be undertaken by researchers, extension personnel and policy makers (Venkateswarlu *et al.*, 2012) for enhancing the adaptation and mitigation potential of rainfed agriculture to climate change.

- Community level seed banks with buffer stocks of seed material of diverse crops appropriate for the village/area need to be maintained. Seed banks should be controlled and maintained by organized farmer groups.
- Programmes promoting farming systems should have a built in component of improving soil organic matter. Composting methods with high biomass-

to-dung ratio should be targeted to overcome the limitation of availability of dung. There is scope for integrating this service with Mahatma Gandhi National Rural Employment Guarantee Scheme (MNREGS).

- Rainwater harvesting and its efficient utilization should be promoted on a priority basis. Efforts should be made to provide incentives to farmers owning groundwater sources to share water with other farmers to provide protective irrigation during *Kharif* season. Subsidized energy or renewable energy systems may also be used as an incentive for water sharing and social regulation.
- Isolated patches of forest land under ownership of the forest department within the villages / watersheds needs special consideration. Regenerating these lands with people's participation for providing biomass for livestock and livelihoods should be the core purpose of managing these forest areas.
- A specific and viable system has to be put in place for developing feed and fodder banks at strategic places in the rainfed areas. Biomass intensification specially targeting the small ruminants should receive highest priority; much of the shrub/tree-biomass for goats and sheep can be enhanced easily with little effort/resources.
- Post-harvest processing and establishing market linkages add further value to the farming systems approach. Collective procurement and marketing, if directly handled by the producers, will significantly enhance the profitability. With the growth in Self Help Group (SHG) movement across the country, options for collective marketing and value addition are opening up.
- Establishment of agro-service centers in the villages can save the cost of inputs and can also get timely farm advisory services for higher profitability. An effective rural knowledge society and ICT system involving various stakeholders; farmers, development agents and agencies, knowledge generators and distributors (universities, and public and private institutions) should be established for bridging the information and knowledge gaps on production and marketing of different commodities.
- Formation of commodity groups and self-help groups of farm women can help to promote off-season income generation activities which lead to livelihood improvement. Imparting training in emerging skills and crafts as per diversified demands in the market will go a long way for realizing all inclusiveness of landless, asset-less, small and marginal farmers.
- Greater emphasis has to be laid down by the extension department for upscaling exposure visits, trainings and demonstration of location-specific

farming system models, for creating greater awareness and capacity building of the farming community.

References

CRIDANewsletter. 2013. Changing climate and implications to development planning in agriculture. Central Research Institute for Dryland Agriculture, Hyderabad, India, p 2.

CRIDA. 2014. Annual Report 2013-14, Central Research Institute for Dryland Agriculture Hyderabad, India. pp 37-38.

Dixit, S, Gopinath, K.A, Uday Kiran, L. and Anuradha, B. 2013. Linking markets for better incomes. LEISA (India) 15(2): 9-11.

Gopinath, K.A, Sreenath Dixit, Srinivasarao, Ch, Raju, B.M.K, Ravindra Chary, G, Osman, M Ramana, D.B.V, Nataraja, K.C, Gayatri Devi, K, Venkatesh, G, Grover, M, Maheswari, M. and Venkateswarlu, B. 2012. Improving the existing rainfed farming systems of small and marginal farmers in Anantapur district, Andhra Pradesh. Indian J. Dryland Agric. Res. Dev. 27(2): 43-47.

Joshi, P.K, Jha, A.K, Wani, S.P, Laxmi Joshi and Shiyani, R.L. 2005. Meta analysis to assess impact of watershed program and people's participation. Comprehensive assessment research report 8. Colombo, Sri Lanka: Comprehensive Assessment Secretariat, IWMI. pp. 18.

Kapoor, A. 2006. Mitigating natural disasters through preparedness measures. Proc. International Conference on Adaptation to Climate Variability and Change, 5-7 January 2006, New Delhi, organized by Institute for Social and Environmental Transition and Winrock International India. 184 p.

Kumar, K.S.K. and Parikh, J. 2001. Indian agriculture and climate sensitivity. *Glob. Environ. Chang.* 11(2): 147-154.

Lal, R. 2001. Future climate change: Implications for India summer monsoon and its variability. Curr. Sci. 81: 1205-1207.

Maji, A.K. 2007. Assessment of degraded and wastelands of India. *J. Indian Soc. Soil Sci.* 55: 427-435.

Mishra, A. 2011. Rainwater conservation and management in high rainfall region for increasing land and water productivity. In: 'Techniques of Water Conservation & Rainwater Harvesting for Drought Management' (Eds. P.K. Mishra, M. Osman, Satendra and B. Venkateswarlu). 714 p. (SAARC Training Program, 18-29 July, 2011, Central Research Institute for Dryland Agriculture, Hyderabad, India).

NAAS, 2012. Management of crop residues in the context of conservation agriculture. Policy Paper No. 58, National Academy of Agricultural Sciences, New Delhi. pp.12.

Pandey, S, Behura, D.D, Villano, R. and Naik, D. 2000. Economic costs of drought and farmers coping mechanisms: A study of rainfed rice systems in eastern India. Discussion Paper Series 39, IRRI, Phillipines. 35 p.

Rama Rao, C.A, Raju, B.M.K, Subba Rao, A.V.M, Rao, K.V, Rao, V.U.M, Kausalya, R Venkateswarlu, B. and Sikka, A.K. 2013. Atlas on vulnerability of Indian agriculture to climate change. Central Research Institute for Dryland Agriculture, Hyderabad, Andhra Pradesh, India. pp. 29.

Ramakrishna, Y.S. 1997. Climate features of the Indian arid zone. In: 'Desertification control in the arid eco-system of India for sustainable development' (Eds. S. Singh and A. Kar). (Agro Botanical Publishers, Bikaner, India) pp. 27-35.

Sahrawat, K.L, Wani, S.P, Pardhasaradhi, G. and Murthy, K.V.S. 2010. Diagnosis of secondary and micronutrient deficiencies and their management in rainfed agroecosystems: Case study from Indian semi arid tropics. Comm. Soil Sci. Plant Anal. 41(3): 346-360.

Sharma, B.R, Rao, K.V, Vittal, K.P.R, Ramakrishna, Y.S. and Amarasinghe, U. 2010. Estimating the potential of rainfed agriculture in India: Prospects for water productivity improvements. Agric. Water Manage. 97(1): 23-30.

Singh, R.P. 1995. Efficient crops and cropping systems for dryland areas. In: 'Sustainable development of dryland agriculture in India' (Ed. R.P. Singh). pp. 21-283. (Scientific Publishers, Jodhpur, India).

Srinivasarao, Ch. 2011. Nutrient management strategies in rainfed agriculture: constraints and opportunities. Indian J. Fert. 7(4): 12-25.

Srinivasarao, Ch, Ravindra Chary, G, Mishra, P.K, Nagarjuna Kumar, R, Maruthi Sankar, G.R Venkateswarlu, B. and Sikka, A.K. 2013a. Real time contingency planning: Initial experiences from AICRPDA. All India Coordinated Research Project for Dryland Agriculture (AICRPDA),Central Research Institute for Dryland Agrciulture (CRIDA), ICAR Hyderabad - 500 059, India. pp. 63.

Srinivasarao, Ch, Venkateswarlu, B, Dixit, S, Kundu, S, and Gayatri Devi, K. 2011b. Livelihood impacts of soil health improvement in backward and tribal districts of Andhra Pradesh. Central Research Institute for Dryland Agriculture, Hyderabad, Andhra Pradesh, India. pp.119.

Srinivasarao, Ch, Venkateswarlu, B, Dixit, S, Wani, S.P, Sahrawat, K.L, Kundu, S, Gayatri Devi K, Rajesh, C. and Pardasaradhi, G. 2010. Productivity enhancement and improved livelihoods through participatory soil fertility management in tribal districts of Andhra Pradesh. Indian J. Dryland Agric. Res. Dev. 25(2): 23-32.

Srinivasarao, Ch, Venkateswarlu, B, Rattan Lal, Singh, A.K. and Kundu, S. 2013b. Sustainable management of soils of dryland ecosystems of India for enhancing agronomic productivity and sequestering carbon. In: 'Advances in Agronomy' (Ed. Donald L. Sparks). Vol. 121 pp. 253-329. (Academic Press, Burlington).

Srinivasarao, Ch, Venkateswarlu, B, Sreenath Dixit, Veeraiah, R, Rammohan, S, Sanjeev Reddy, B Kundu, S. and Gayatri Devi, K. 2010. Implementation of contingency crop planning for drought in a tribal village in Andhra Pradesh: Impact on food and fodder security and livelihoods. Indian J. Dryland Agric. Res. Dev. 25(1): 23-30.

Srinivasarao, Ch, Vittal, K.P.R, Venkateswarlu, B, Wani, S.P, Sahrawat, K.L, Marimuthu, S, Kundu, S. 2009. Carbon stocks in different soils under diverse rainfed production systems in tropical India. Comm. Soil Sci. Plant Anal. 40(15): 2338-2356.

Umrani, N.K. 1999. Development of dryland technologies across five foundation locations: Historical perspective. In: 'Fifty years of dryland agricultural research in India' (Eds. H.P. Singh, Y.S. Ramakrishna, K.L. Sharma, and B. Venkateswarlu). pp. 1-9. (Central Research Institute for Dryland Agriculture, Hyderabad, India).

Umrani, N.K. and Subba Reddy, G. 1999. Cropping systems research in drylands – a review. In: 'Fifty years of dryland agricultural research in India' (Eds. H.P. Singh, Y.S. Ramakrishna, K.L. Sharma, and B. Venkateswarlu). pp. 105-114. (Central Research Institute for Dryland Agriculture, Hyderabad, India).

Venkatesh, G, Srinivasarao, Ch, Venkateswarlu, B, Gopinath, K.A, Prasad, J.V.N.S, Sanjeeva Reddy, B, Sasikala, Ch, Rajeshwara Rao, G. and Ramesh Babu, P.V. 2013a. Operational process for biochar preparation from castor bean stalk and its characterization for soil application. Indian J. Dryland Agric. Res. Dev. 28(2): 21-26.

Venkatesh, G, Venkateswarlu, B, Gopinath, K.A, Srinivasa Rao, Ch, Korwar, G.R, Sanjeeva Reddy, B, Prasad, J.V.N.S, Grover, M, Raju, B.M.K, Sasikala, Ch. and Venkanna, K. 2013b. Biochar production technology for conversion of cotton stalk bioresidue into biochar and its characterization for soil amendment qualities. Indian J. Dryland Agric. Res. Dev. 28(1): 48-57.

Venkateswarlu, B, Singh, A.K, Prasad, Y.G, Ravindra Chary, G, Srinivasa Rao, Ch, Rao, K.V, Ramana, D.B.V. and Rao, V.U.M. 2011. District level contingency plans for weather aberrations in India. Central Research Institute for Dryland Agriculture, Natural Management Division, Indian Council of Agricultural Research, Hyderabad – 500 059 India. pp. 136.

Venkateswarlu, B. 2008. Organic farming in rainfed agriculture: Prospects and limitations. In: 'Organic farming in rainfed agriculture: Opportunities and constraints' (Eds. B. Ventateswarlu, S.S. Balloli and Y.S. Ramakrishna). pp. 7-11. (Central Research Institute for Dryland Agriculture, Hyderabad, India).

Venkateswarlu, B. 2010. Climate change: Adaptation and mitigation strategies in rainfed agriculture. The 21st Dr. S.P. Raychaudhuri Memorial Lecture, delivered on 25th September, 2010 at BAU, Ranchi.

Venkateswarlu, B. and Arun K. Shanker. 2009. Climate change and agriculture: Adaptation and mitigation strategies. Indian J. Agron. 54(2): 226-230.

Venkateswarlu, B, Mishra, P.K, Ravindra Chary, G, Maruthi Sankar, G.R. and Subba Reddy, G. 2009. Rainfed farming - A compendium of improved technologies. All India Co-ordinated Research Project for Dryland Agriculture, Indian Council of Agricultural Research, Hyderabad, India. pp 132.

Venkateswarlu, B, Singh, A.K, Srinivasa Rao, Ch, Gauranga Kar, Ashwani Kumar and Virmani, S.M. 2012. Natural resource management for accelerating agricultural productivity. Stadium Press (India) Pvt. Ltd., New Delhi. pp 234.

Vittal, K.P.R, Singh, H.P, Rao, K.V, Sharma, K.L, Victor, U.S, Chary, G.R, Sankar, G. R.M, Samra J.S. and Singh, G. 2003. Guidelines on drought coping plans for rainfed production systems. All India Coordinated Research Project for Dryland Agriculture, Central Research Institute for Dryland Agriculture, Indian Council of Agricultural Research, Hyderabad. pp.39.

7

Weed Management in Conservation Agricultural Systems

A.R. Sharma and K.K. Barman

Introduction

Adoption of green revolution technologies during 1960s led to increased productivity and elimination of acute foodgrain shortages in India. These technologies primarily involved growing of high-yielding dwarf varieties of rice and wheat, increased use of chemical fertilizers and other agrochemicals, and spread of irrigation facilities. This was also accompanied by other modern methods of cultivation, which included maximum tilling of land, virtually clean cultivation with complete removal of crop residues and other biomass from the field, fixed crop rotations mostly involving cereals, and elimination of fertility-restoring pulses and oilseed crops in the highly productive north-western plain zone of the country.

The transformation from 'traditional animal-based subsistence farming' to 'intensive chemical- and tractor-based conventional agriculture' have led to multiplicity of issues associated with sustainability of these production practices. Conventional crop production technologies are characterized by: (i) intensive tillage to prepare fine seed- and root-bed for sowing to ensure proper germination and initial vigour, faster absorption of moisture, control of weeds and other pests, mixing of fertilizers and organic manures; (ii) monocropping systems; (iii) clean cultivation involving removal or burning of all residues after harvesting leading to continuous mining of nutrients and moisture from the soil profile, and bare soil with no soil cover; (iv) indiscriminate use of insecticides and pesticides, and excessive and imbalanced use of chemical fertilizers leading to decline in input- use efficiency and factor productivity, and increase in pollution of environment, ground water, streams, rivers and oceans; and (v) energy-intensive farming systems.

Over the last 4 decades of adoption of these technologies, there have been emerging concerns about the natural resource degradation. It is realized that soils are getting impoverished due to imbalanced use of fertilizers, discontinuation of traditional practices like mulching, intercropping and inclusion of legumes in cropping systems. Further, the use of organic manures, compost and growing of green manure crops has also decreased considerably due to various reasons. Similarly, water resources are under great stress due to their indiscriminate exploitation and also getting polluted due to various human interferences. Burning of fossil fuels, crop residues, excessive tillage including puddling for rice cultivation are leading to emission of greenhouse gases, which are responsible for climate change and global warming. Further, there is now a growing realization that the productivity levels are stagnating and the incomes of the farmers are reducing due to the rising cost of the inputs and farm operations. It is feared that conventional cultivation practices are not sustainable in the long-run, and there is a need to change the way we do crop production in arable lands.

Conservation agriculture - a new paradigm

Widespread resource degradation problems under conventional agricultural production systems, and the need of reducing production costs, increasing profitability and making agriculture more competitive, made the conservation issues more imperative. Globally innovations of conservation agriculture-based crop management technologies are said to be more efficient, use less inputs, improve production and income, and address the emerging problems (Gupta and Seth 2007). Additionally, the secondary drivers as (i) availability of new farm machinery, (ii) availability of new biocide molecules for efficient weed, insect-pest and disease control, (iii) ever-decreasing labour force and ever-increasing labour cost, (iv) increasing production costs, energy shortages, erosion losses, pollution hazards and escalating fuel cost, and (v) residue burning, have accelerated a change in thinking of researchers, policy makers and farmers to adopt modified methods for cultivation of crops aimed at improving productivity and resource-use efficiency.

Adequate food production for ever-increasing global population can only be achieved through the implementation of sustainable growing practices that minimize environmental degradation and preserve resources while maintaining high-yielding, profitable systems. Conservation agriculture (CA) practices are designed to achieve agricultural sustainability by implementation of sustainable management practices that minimize environmental degradation and conserve resources while maintaining high-yielding, profitable systems, and also improve the biological functions of the agro-ecosystem with limited mechanical practices and judicious use of external inputs. It is characterized by three linked principles, viz. (i) continuous minimum mechanical soil disturbance, (ii) permanent organic soil cover, and (iii) diversification of crop species grown in sequence and/or associations. A host of benefits can be achieved through employing components of conservation agriculture, including reduced soil erosion and water runoff, increased productivity through improved soil quality, increased water availability, increased biotic diversity, and reduced labour demands.

Conservation agriculture systems require a total paradigm shift from conventional agriculture with regard to management of crops, soil, water, nutrients, weeds and farm machinery (Table 1).

Historical perspective

Jethro Tull (1671-1741), who is regarded as 'Father of Tillage' carried out numerous experiments dealing with cultural practices. He believed that soil should be finely pulvarised to provide proper pebulum for the growing plant. He propounded a theory that 'Soil particulars are ingested through openings in plant roots due to the processes caused by the swelling of growing roots'. Dust Bowl era between 1931 and 1939 exposed the vulnerability of plough-based agriculture, as wind blew away precious top soil from the drought-ravaged southern plains of US, leaving behind failed crops and farms. It was realized that tillage is the root cause of agricultural land degradation – one of the most serious environmental problems worldwide – which poses a threat to crop production and rural livelihoods, particularly in poor and densely populated areas of developing countries.

In the early 1940s, Edward H. Faulkner (1886-1964) wrote a book 'Plowman's folly', which the *Nature* magazine termed "an agricultural bombshell" when he blamed the-then universally used mouldboard plough for disastrous tillage of the soil. He questioned the use of plough for cultivation of crops, and showed that all standard wisdom used as a rationale for ploughing and working the soil was invalid. His ideas were considered 'mad' and without merit, until after his death when soil experts and other scientists began to admit 'We didn't pay attention to what he said when he was alive'.

Masanobu Fukuoka (1913-2008) worked for more than 65 years at his farm in Japan and developed a system of natural farming. He did not plough his fields, used no agricultural chemicals or prepared fertilizers, did not flood his rice fields as farmers have done in Asia for centuries, and yet his yields equaled or surpassed the most productive farms in Japan. His book 'The One Straw Revolution' contains the spiritual memoir of a man whose innovative system of cultivating the earth reflects a deep faith in the wholeness and balance of the natural world. In the recent times (2007), David R. Montgomery wrote the

Table 1: Some distinguishing features of conventional and conservation agriculture systems

Conventional agriculture	Conservation agriculture
• Cultivating land, using science and technology to dominate nature	• Least interference with natural processes
• Excessive mechanical tillage and soil erosion	• No-till or drastically reduced tillage (biological tillage)
• High wind and soil erosion	• Low wind and soil erosion
• Residue burning or removal (bare soil surface)	• Surface retention of residues (permanently covered)
• Water infiltration is low	• Infiltration rate of water is high
• Use of *ex-situ* FYM/composts	• Use of *in-situ* organics/composts
• Green manuring (incorporated)	• Brown manuring/cover crops (surface retention)
• Kills established weeds but also stimulates more weed seeds to germinate	• Weeds are a problem in the early stages of adoption but decrease with time
• Free-wheeling of farm machinery, increased soil compaction	• Controlled traffic, compaction in tramline, no compaction in crop area
• Mono cropping/culture, less efficient rotations	• Diversified and more efficient rotations
• Heavy reliance on manual labor, uncertainty of operations	• Mechanized operations, ensure timeliness of operations
• Poor adaptation to stresses, yield losses are more under stress conditions	• More resilience to stresses, yield losses are less under stress conditions
• Productivity gains in long-run are in declining order	• Productivity gains in long-run are in incremental order

award winning book 'Dirt – The Erosion of Civilizations', which shows that with any form of tillage including the non-inversion tillage, the rate of soil degradation and soil erosion is greater than the rate of soil formation. According to his research, tillage has caused the destruction of agricultural base and of its productive capacity nearly everywhere, and continues to do so.

Prospects of conservation agriculture

The concept of conservation agriculture has been developed to reverse the process of land degradation and ensure sustainable crop production. This requires a systems approach, i.e. efficient seeding machinery, nutrient, water, weed and pest management. This technology has been adopted globally on more than 150 M ha in about 50 countries, largely in rainfed areas. The major countries are: USA, Australia, Canada, Argentina, Brazil, Paraguay, Uruguay and New Zealand (Table 2). In India, this realization started in early 1990s when some experiments were initiated on zero-till wheat in north-western India, primarily through the efforts of IRRI, CIMMYT and NATP. There was good success obtained in many states, and the area under zero-till wheat reached up to 3 M ha by the beginning of current century. However, the acreage have stagnated now and some farmers have even switched back to minimum or conventional systems because of some practical constraints and lack of technical know-how.

Table 2: Global adoption of conservation agriculture systems

Country	Area (M ha)	% of global area
USA	35.6	22.7
Brazil	31.8	20.3
Argentina	29.2	18.6
Canada	18.3	11.7
Australia	17.7	11.3
China	6.7	4.3
Russian Federation	4.5	2.9
Paraguay	3.0	1.9
Kazakhstan	2.0	1.3
Others	8.2	5.2
Total	157.0	100.0

Source: www.fao.org/ag/ca/6c.html dt. 02.09.2016

Harvesting of major crops like rice and wheat is now being done with combines and the residue management has become a major issue in many states including central India. Burning of residues *in situ* is rampant despite restrictions imposed and incentives offered by the governments. This is the most unhealthy practice as it leads to loss of precious plant nutrients and environmental problems. In recent years, there have been some major developments, which have led to a change in our approach for promotion of conservation agriculture. New generation farm machinery has become available which can place the seed and fertilizer at an appropriate depth in the desired amounts. Further, these machines can work in standing as well as loose crop residues; thus, providing a very effective mulch cover for moisture and nutrient conservation, temperature moderation and weed control. Availability of new herbicide molecules has further necessitated a change in our thinking about weed management. Further, other triggering factors for shift towards conservation agriculture are labour scarcity, deteriorating soil health, declining factor productivity, rising cost and low income. Thus, conservation agriculture systems help in overcoming the problems being experienced in conventional farming systems.

Weeds in conservation agricultural system

Tillage affects weeds by uprooting, dismembering, and burying them deep enough to prevent emergence. Ploughing also moves weed seeds both vertically and horizontally, and changes the soil environment; thereby promoting or inhibiting weed seed germination and emergence. Reduction in tillage intensity and frequency, as practiced under CA, generally increases weed infestation. Compared to conventional tillage (CT), presence of weed seeds is more in the soil surface under zero tillage (ZT), which favours relatively higher weed germination. Mishra and Singh (2012) observed that over the course of time, a ZT– ZT sequence favoured relatively higher weed growth over a CT–CT sequence in a rice–wheat system.

Infestation of *Phalaris minor* is a serious problem in wheat grown after rice in north-western India. This has increased due to continuous adoption of rice-wheat system and use of same or similar herbicides. New herbicides with a different mode of action are being advocated for effective control of *Phalaris*. Modifying cultivation practices may help in reducing its infestation. Results of on-farm trials at several locations in Haryana revealed that population density of *Phalaris minor* was considerably lower and grain yield of wheat was comparatively higher under ZT than CT (Figure 1).

Changes from conventional to conservation farming practices lead to a weed flora shift in the crop field, which in turn dictate the requirements of weed management technologies involving various approaches, viz. preventive measures, cultural practices (tillage, crop residues as mulches, intercropping, competitive crop cultivars, herbicide tolerant cultivars, planting dates, crop rotations etc.), and herbicides. Weed control in CA depends upon use of herbicides and agronomic practices. However, the recent development of post-emergence broad-spectrum herbicides provides an opportunity to control weeds more

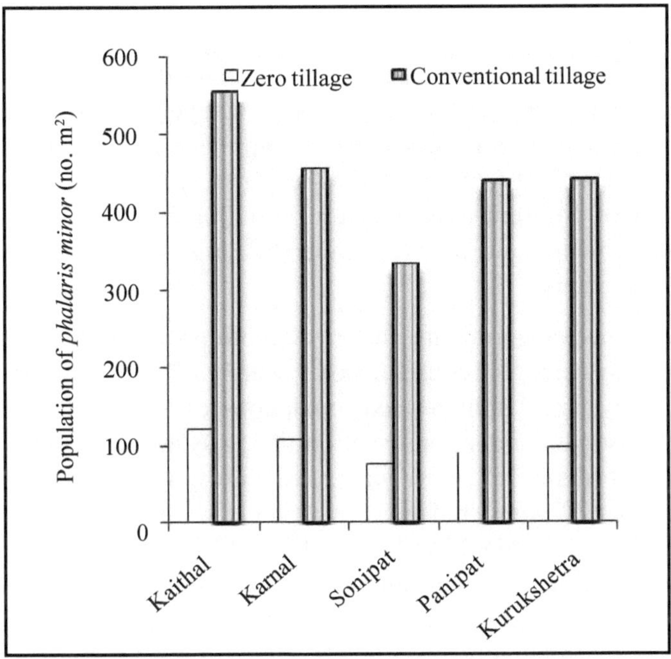

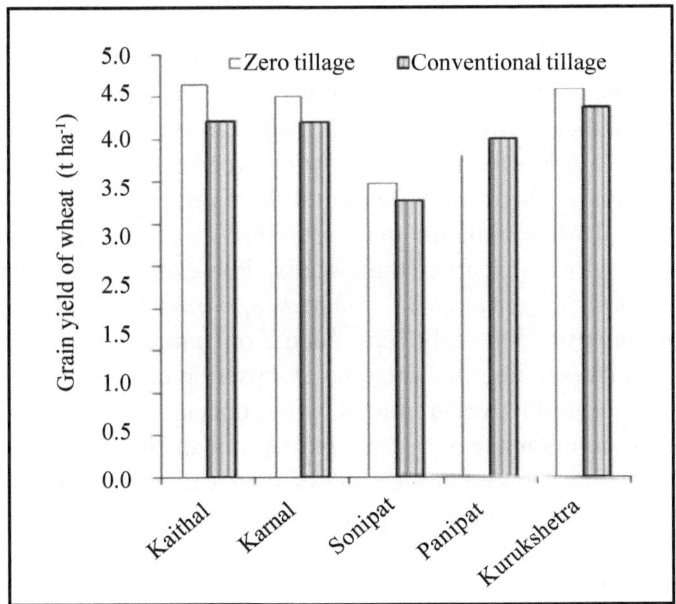

Fig. 1: Effect of tillage on population of *Phalaris minor* and grain yield of wheat at different locations in Haryana (Gupta and Seth, 2007)

effectively in CA, and enabling to have uniform crop stands and yield levels similar to CT systems.

Weed infestations show a drastic change due to continuous adoption of different cultivation practices such as sowing technique, tillage, method of weed control, residue management, cropping system, and application of inputs. Continuous adoption of a particular factor over the years may either increase or decrease the infestation of different weed species. Zero tillage brings about a paradigm shift in weed growth, requiring clear understanding and formulating a strategy for their management.

Certain weed species germinate and grow more profusely than others under continuous ZT system. As a consequence, a weed shift occurs due to the change from CT to ZT system. A shift in weed populations towards small-seeded annuals is generally observed under conservation tillage systems. Contrary to this, in spite of small seed size, *Phalaris minor* has shown a remarkable reduction in population under ZT compared to CT system in the Indo-Gangetic Plains. This may be attributed to: (i) higher soil strength in ZT because of crust development in the absence of tillage, which can mechanically impede seedling emergence (Chhokar *et al.* 2007), (ii) less soil temperature fluctuation under ZT (Gathala *et al.* 2011), or (iii) relatively lower levels of light stimuli, N mineralization and gas exchange under ZT (Franke *et al.* 2007).

Shifts in weed population towards perennials have also been observed in conservation tillage systems. Perennial weeds thrive in reduced or no-tillage systems because the root system is not disturbed and herbicides used to control annual weeds are not effective on perennial weeds. Perennial monocots are considered a greater threat than perennial dicots in the adoption of reduced tillage systems. Unlike annuals, many perennial weeds can reproduce from several structural organs other than seeds. For example, purple nutsedge (*Cyperus rotundus*), tiger grass (*Saccharum spontaneum*) and johnson grass (*Sorghum halepense*) generally reproduce from underground plant storage structures, i.e. tubers or nuts and rhizomes. Conservation tillage may encourage these perennial reproductive structures by not burying them to depths that are unfavourable for emergence or by failing to uproot and kill them. Weed species shifts and losses in crop yield as a result of increased weed density have been cited as major hurdles to the widespread adoption of CA.

Weed management options

Management of weeds is a major issue in agricultural production system, particularly under CA where the infestation is likely to be higher than conventional intensive–tillage system. Understanding ecology, seed bank and dynamics of specific weed flora is essential for developing effective management strategies in divergent situations. Weed control in CA is a greater challenge than in conventional agriculture because there is no weed seed burial by tillage operations. The behaviour of weeds and their interaction with crops under CA is complex and not fully understood. The weed species that germinate in response to light are likely to be more problematic in CA. In addition, perennial weeds become more challenging in this system. In the past, attempts to implement CA have often resulted in a yield penalty because reduced tillage failed to control weed interference. However, the recent development of post-emergence broad-spectrum herbicides provides an opportunity to control weeds in CA. Crop yields can be similar for conventional and conservation tillage systems if the weeds are controlled and crop stands are uniform. Various approaches employed to successfully manage weeds in CA systems includes preventive measures, cultural practices (tillage, crop residue as mulches, intercropping, cover cropping, etc.), use of herbicide-tolerant cultivars, and herbicides.

(a) Preventive measures

Preventive weed control encompasses all measures taken to prevent or arrest the introduction and arrest of weeds. Weed seeds resembling the shape and size of crop seeds are often the major source of contamination in crop seeds. Contamination usually occurs at crop harvesting if the life cycle of crop and weeds is of similar duration. Preventive measures are the first and most important steps to manage weeds, in general and especially under CA, as the presence of even a small quantity of weed seeds may cause a serious infestation in the forthcoming seasons.

(b) Cultural practices

A long-term goal of sustainable and successful weed management is not to merely control weeds in a crop field, but rather to create a system that reduces weed establishment and minimizes weed competition with crops. Further, since environmental protection is a global concern, the age-old weed management practices, viz. tillage, mulching, inter-cultivation, intercropping, cover crops, crop rotation/diversification and other agro-techniques (once labeled as uneconomical or impractical) should be relooked and given due emphasis in managing weeds under CA. One of the pillars of CA is ground cover with dead or live mulch, which leaves less time for weeds to establish during fallow or a turnaround period. Some other common problems under CA include emergence from recently produced weed seeds that remain near the soil surface, lack of disruption of perennial weed roots, interception of herbicides by thick surface residues, and a change in the timing of weed emergence. Shrestha *et al.* (2002) concluded that long-term changes in weed flora are driven by an interaction of several factors, including tillage, environment, crop rotation, crop type, and timing and type of weed management practice.

(i) Tillage

Tillage has long been an essential component of conventional agricultural systems and it is the most important among the traditional means of weed management. The effect of primary tillage on weeds is mainly related to the type of implement used and tillage depth. These factors impact the weed seed and propagule distribution over the soil profile; and therefore directly affect the number of weeds that can emerge in a field. Differential distribution of seeds in the soil profile subsequently leads to changes in weed population dynamics. Weed seeds buried deep germinate but fail to emerge due to the thick soil layer above it, resulting in death of the weed seedling. Tillage stimulates weed germination and emergence of many weed seeds through brief exposure to light. ZT wheat in a rice–wheat system reduces *Phalaris minor* infestation, which is highly competitive and can cause drastic wheat yield reductions under heavy infestation, but it favours the infestation of *Rumex dentatus*, *Malva parviflora* and wild oat (Chhokar *et al.*, 2007).

Furrow Irrigated Raised-Bed System (FIRBS) and ridge tillage systems are the form of reduced and conservation tillage, respectively, that appear to overcome weed control problems associated with conventional and ZT systems. Besides improved weed management, FIRBS has been found to improve input use-efficiency. Chauhan *et al.* (1998) obtained reasonably good control of *Phalaris minor* in wheat on raised beds but broad-leaved weeds in furrows were not controlled. The problem with *Phalaris minor* was less as the weed seeds lying on top of the raised beds failed to germinate because the top of bed dried quickly. This method also facilitated mechanical weeding as the area in the furrows could easily be cultivated and even manual weeding could be done. When crop plants are 40 cm tall, soil is excavated from the furrows and is moved back to the ridge crest, thereby affecting weeds and crop–weed interaction. However, changes in weed communities were influenced more by location and year than by tillage systems. For instance, FIRBS effectively reduced total weed density and biomass at Palampur but was not superior to other tillage practices in Hisar (Table 3).

(ii) Stale seedbed

Seedbed preparation can contribute to weed management by affecting weed seed dynamics and seedling densities at planting. Cultivation for seedbed preparation affects the weeds in two ways: (i) it destroys the emerged vegetation after primary tillage, and (ii) it stimulates weed seed germination and consequent seedling emergence and reallocation of seeds towards the soil surface. This

Table 3: Effect of tillage on total weed density, dry matter of weeds at 60 DAS in different locations

Location	Weed density (no.m^{-2})			Weed dry weight (g m^{-2})			Reference
	CT	ZT	FIRB	CT	ZT	FIRB	
Faizabad	-	-	-	14.40	20.2	-	Yadav *et al.* (2005)
Palampur	270.0	283.3	241.0	131.3	139.4	107.3	Chopra and Angiras (2008a)
Palampur	228.0	245.0	203.0	113.0	126.0	91.0	Chopra and Angiras (2008b)
Karnal	83.2	62.0	-	18.1	20.7	-	Chopra and Chopra (2010)
Delhi	137.9	168.5	-	15.6	19.1	-	Tuti and Das (2011)
Hisar	89.3	87.4	96.1	30.1	26.5	32.4	Jat *et al.*(2013)

phenomenon could be exploited to manage weeds through application of the stale (false) seedbed technique.

Zero-till stale seedbed practice can help to reduce weed pressure in CA systems. In this technique, the field is irrigated 10–15 days prior to actual seeding to favour the germination of weed seeds lying on the soil surface. Emerged weeds are then destroyed by the application of non-selective herbicides like glyphosate, paraquat or glufosinate. It depletes the seed bank in the surface layer of the soil and reduces subsequent weed emergence. Where light rains occur for an extended period before the onset of the monsoon or irrigation is available, it may be possible to kill several flushes of weed growth before planting. To ensure success, cropping should be delayed until the main flush of emergence has passed. However, this practice may not be exploited where the season available for crop growth is short, which may reduce the yield potential of the crop. The main advantage of the stale seedbed practice is that the crop emerges in a weed-free environment, with a competitive advantage over late-emerging weed seedlings. The practice of false seed-bed may decrease weed infestation in crops by 80% or more compared to standard seedbed preparation (Van der Weide *et al.* 2002).

The stale seedbed technique is widely used in many countries to manage weedy rice and awnless barnyard grass in rainfed rice. Stale seedbeds reduce weed populations in direct-seeded rice, and may be especially effective when combined with no-till practices (Chauhan *et al.,* 2006).

(iii) Crop residues

Crop residues present on the soil surface can inûuence weed seed germination and seedling emergence by interfering with sunlight availability and creating physical impedance, as well as improving soil and moisture conservation and soil tilth. Residues on the soil surface can vary greatly in dimension, structure, distribution pattern and spatial heterogeneity. Weed biology, and the quantity, position (vertical or flat, and below- or above ground -weed seeds) and allelopathic potential of the crop residues influence weed germination (Chauhan *et al.* 2006).

Soil cover using crop residues is a useful technique to manage weeds. Weed emergence generally declines with increasing residue amounts on soil surface. However, the emergence of certain weed species is also favoured by some crop residue at low amounts. For example, germination and growth of wild oat and animated oat (*Avena sterilis*) may get stimulated with low levels of wheat residue. High amounts of crop residues have implications for weed management in CA through reduced and delayed weed emergence. The crop gets competitive advantage over weeds due to delayed weed emergence, which results in relatively less impact on crop yield loss. Further, late emerging weed plants produce less number of seeds than the early emerging ones (Chauhan and Johnson 2010). Chauhan and Abugho (2012) reported that 6 t ha^{-1} crop residues reduced the emergence of jungle rice, crowfoot grass and rice flat sedge by 80–95% but reduced the emergence of barnyard grass only by 35% (Figure 2).

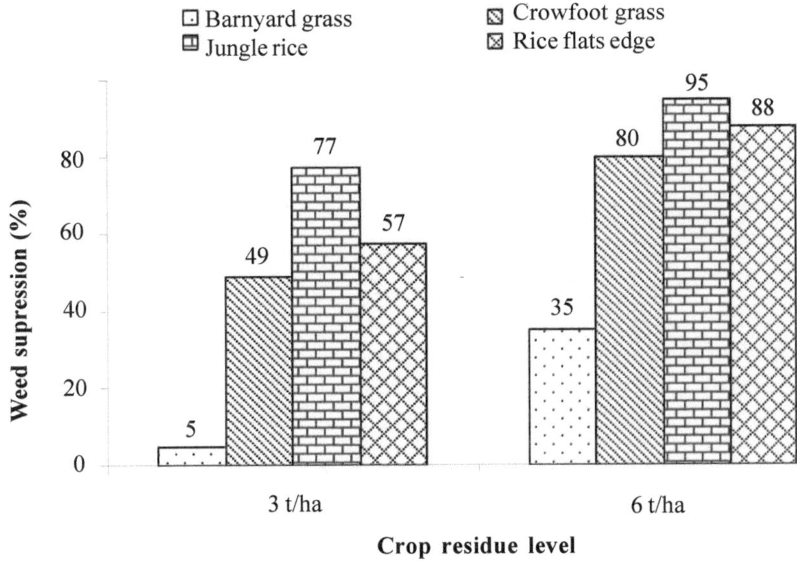

Fig. 2: The effect of rice residues on weed germination (Chauhan and Abugho 2012)

The increased moisture content and decreased temperature of soil due to the presence of crop residue may increase the germination of some weed species. In dryland areas, the amount of available crop residues may be insufficient to substantially suppress weed germination and growth (Chauhan and Johnson 2010). Further, certain crops like oilseeds and pulses produce less biomass than cereals. Therefore, the effects of crop residue on the weed population depend on the region, crop and rainfall. There is a need to integrate herbicide use with residue retention to achieve season-long weed control. In high-residue situations, it is important that residue does not hinder crop emergence.

(iv) Intercropping

Intercropping of short-duration, quick-growing, and early-maturing legume crops with long-duration and wide-spaced crops leads to quick ground cover with higher total weed suppressing ability than sole cropping. This technique enhances weed control by increasing shade and crop competition. Like cover crops, intercrops increase the ecological diversity in a field. In addition, they often compete better with weeds for light, water and nutrients. Success of intercropping relies on the best match between the requirements of the component

Table 4: Effect of intercropping on weed growth and crop productivity (Dubey 2008)

Intercropping system	Weed density (number m^{-2})				Weed dry matter ($g\ m^{-2}$)	Maize equivalent yield($t\ ha^{-1}$)
	Echinochloa colona	*Phylanthus niruri*	*Commelina communis*	Total		
Sole maize	5.3* (27.9)	7.5 (56.1)	3.5 (11.7)	10.4 (107.5)	6.0 (35.4)	2.97
Maize + cowpea (grain)	3.6 (12.9)	6.1 (37.2)	2.9 (8.1)	8.3 (68.4)	4.8 (22.5)	4.07
Maize + cowpea (fodder)	4.1 (16.8)	6.4 (40.6)	3.5 (11.7)	8.6 (73.5)	5.3 (27.8)	3.27
LSD (P=0.05)	0.5	0.6	NS	0.6	0.5	-

*Square-root transformed values, original values are in parentheses

species for light, water and nutrients, which increases resource use. Many short-duration pulses like cowpea, greengram and soybean effectively smother weeds without reducing the yield of the main crop. For instance, total weed growth reduced under intercropping combinations of chickpea + mustard over the sole chickpea crop without losing productivity of the main crop (Rathi *et al.* 2007). Similar observations were also recorded by Dubey (2008) under maize+cowpea intercropping system (Table 4). Compared with the sole crop, increased canopy cover and decreased light availability for weeds in maize–legume intercropping was responsible for the reduction in weed density and dry matter.

One of the principles of CA is to include green manuring, with its bioherbicidal characteristics and weed smothering capabilities, along with an additional benefit of adding biomass to soil. *Sesbania* can be grown with rice as a co-culture to suppress weeds, and in addition to weed control, it can also fix large amounts of N (Ladha *et al.* 2000). *Sesbania* intercropping for 25–30 days in a dry-seeded rice under CA followed by killing of *Sesbania* using 2,4-D or mechanical means was effective in controlling weeds, but the contribution from N_2 fixation was small because of intercropping and short growth duration (Singh *et al.* 2007). This practice was also a highly beneficial resource conservation technology for soil and water conservation, weed control and nutrient supplementation in maize (Sharma *et al.* 2010).

(v) Cover cropping

Ground cover with dead or live mulch, allowing less time for weeds to establish during fallow or turnaround period is an important component of CA technology. Inclusion of cover crops in a rotation between two main crops is a good preventive measure when developing a weed management strategy. Cover crops are fundamental and sustainable tools to manage weeds, optimize the use of natural resources, and reduce water runoff, nutrient leaching and soil erosion. Competition from a strong cover crop can virtually shut down the growth of many annual weeds emerging from seeds. Aggressive cover crops can even substantially reduce growth and reproduction of perennial weeds that emerge or regenerate from roots, rhizomes or tubers, and are more difficult to suppress. Cover crop effects on weeds largely depend upon the species and weed community composition. Weed suppression is exerted partly through resource competition for light, nutrients and water during the cover crop growing cycle, and partly through physical and chemical effects that occur when cover crop residues are left on the soil surface as dead mulch or ploughed down.

Weed pressure in CA can be reduced by including short-duration legume crops e.g. blackgram, cowpea, greengram, and *Sesbania* during the fallow period between harvesting wheat and planting rice. This practice facilitates emergence of weeds during the legume period (stale seedbed effects) and reduces the population during the rice season (Kumar *et al.* 2012). Growing green manure or cover crops in the summer season or as a relay crop to efficiently suppress weed growth is a cost and labour efficient practice. Therefore, green manures are sometimes also called the herbicides of small farmers. Perennial grasses such as *Imperata cylindrica* and *Cynodon dactylon*, and other problem weeds like *Striga* spp. and *Chromolaena odorata* can be suppressed by one or two seasons of cover crops.

(vi) Crop diversification

Crop rotation involves alternating different crops in a systematic sequence on the same land. It limits the build-up of weed populations and prevents weed shifts as the weed species tend to thrive in a crop with similar growth requirements. Different crops require different cultural practices, which help to disrupt the growing cycle of weeds, and prevent any weed species to dominate. For example, johnson grass generally becomes a predominant weed in a continuous maize system but may be controlled by rotating with cotton. In monocropping systems, several weed species persist and expand rapidly. Cropping sequences provide varying patterns of resource competition, allelopathic interference, soil disturbance and mechanical damage, and thus provide an unstable environment that prevents the proliferation and dominance of a particular weed, and discourages growth and reproduction of troublesome weed species. The prolonged cultivation of the rice–wheat system in north-western India has resulted in increased population of sedges and grassy weeds. The diversification of the system even for a short period and intensification by including summer legumes/green manuring decreased the weed menace (Singh *et al.* 2008).

Certain crop-associated weed species e.g. barnyard grass in rice, wild oat and *Phalaris minor* in wheat, and *Cuscuta* in alfalfa may be discouraged by following a rotation of crops with contrasting growth and cultural requirements. Crop rotation is an effective practice for managing *Phalaris minor* because selection pressure is diversiûed by changing patterns of disturbances (Chhokar and Malik, 2002). Changing from rice–wheat to any other sequence not involving rice, reduces the population of *Phalaris minor* in wheat. In case, where sugarcane is taken followed by one ratoon, *Phalaris minor* population goes down considerably. Replacing wheat with other crops like Egyptian clover (*Trifolium alexandrinum*), potato (*Solanum tuberosum*), sunflower (*Helianthus annuus*) and annual rape (*Brassica napus*) for 2–3 years in a rice–wheat cropping system significantly reduces the population of *Phalaris minor*.

A rice–wheat rotation suppressed the establishment and growth of wild oat in wheat, while a maize–wheat rotation resulted in a gradual build-up of wild oat. The inclusion of sesame in several cropping sequences reduced the aerial growth of nutsedge (Varshney 2000). Parasitic weeds can be successfully managed by rotating the host crop with trap crops, as they induce germination of weed seeds but are themselves not parasitized. The added advantage of the crop rotation is that it also allows growers to use new herbicides that may control problematic weeds.

Mechanical measures

Farm mechanization plays a vital role for the success of CA in different agro-ecologies and socio-economic farming groups. It ensures timeliness, precision and quality of field operations; reduces production costs; saves labour; reduces weather risk under the changing climatic scenario; improves productivity, environmental quality and sustainability; and generates rural employment through on-farm and off-farm activities. Reduced labour and machinery costs are economic considerations that are frequently given as additional reasons to use CA practices. Compared to intensive tilled conventional rice–wheat system, ZT systems require much less energy and give higher energy output:input ratio as well as system productivity (Gangwar *et al.* 2006, Kumar *et al.* 2012). Mishra and Singh (2012) reported lower cost of cultivation, higher net returns and wider benefit:cost ratio in a ZT rice–wheat systems. Similarly in a ZT maize–wheat system, minimum energy usage, higher water productivity, higher net returns and enhanced energy input : output ratio were recorded by Ram *et al.* (2010).

Conservation agriculture is essentially machine-driven and suitable farm machinery is required for land leveling, sowing, fertilization, weeding, irrigation, harvesting and other operations. Hence, the availability of suitable farm machineries is of paramount importance for adoption of this technology by farmers. Farooq *et al.* (2007) noticed that access to ZT drills contributed towards the adoption of the ZT wheat technology in Pakistan's Punjab province. Direct drilling with ZT drill is a practice that addresses the issues of labour, energy, water and soil health. However, this machine faces difficulties if crop stubbles are in high quantity, a situation that commonly occurs in CA systems. Harvesting of rice, wheat and many other crops like maize, sorghum, pegionpea, chickpea, greengram and blackgram is being done through combine harvester in many parts of India. In most regions, the crop residue lying on the soil surface is burnt so as to prepare the field for sowing of next crop. This is a very unscientific practice as it leads to environmental problem, loss of C and other essential nutrients. Despite some measures taken by different states and ban imposed by Supreme Court, burning is still the most common means of straw disposal.

'Happy Seeder' technology—an improved version of the no-till seed drill and initially developed for direct drilling of wheat into rice residues (typically 5–9 t ha^{-1} of anchored and loose straw) in north-west India—is a recent novel approach which combines stubble mulching and seed-cum-fertilizer drilling functions. The stubble is cut and picked up in front of the sowing tynes, which engage almost bare soil, and deposited behind the seed drill as surface mulch. In addition to the benefits of direct drilling and retaining organic matter, the mulch also assists in moisture conservation and weed control. Observations from farmers' fields across Indian Punjab showed that the Happy Seeder (zero tillage) and rotavator (reduced tillage) are efficient methods for control of weeds as well as for *in situ* management of paddy straw (Kang 2013). The average reduction in the weed population in the happy seeder-sown wheat crop over the rotavator and farmer's practice was 26.5 and 47.7%, respectively (Singh *et al.,* 2013). Advanced versions of the happy seeder, *viz.* turbo seeder, PCR planter and easy seeder are being developed for more efficient sowing and fertilizer placement.

Laser land leveling, an integral component of CA, provides uniform moisture distribution to the entire field and ensures a proper crop stand and growth with reduced weed infestation. Unleveled fields frequently exhibit patchy crop growth with higher weed infestation. Compared to an unleveled field, weed management in a laser–leveled field is relatively easy and requires less labour for manual weeding operations due to less weed infestation. Lower weed populations in wheat were recorded under precisely leveled fields (200 no. m^{-2}) compared to traditional leveled fields (350 no. m^{-2}) (Jat *et al.* 2003). Precision land leveling may reduce up to 75% of the labour requirement needed for weeding operations (Rickman 2002).

Chemical weed management

The use of herbicides for managing weeds is becoming popular because they are cheaper than traditional weeding methods, require less labour, tackle difficult-to-control weeds, and allow flexibility in weed management. Herbicides are an integral part of weed management in CA. However, to sustain CA systems, herbicide rotation and/or integration of weed management practices is preferred as continuous use of a single herbicide over a long period of time may result in the development of resistant biotypes, shifts in weed ûora and negative effects on the succeeding crop and environment. In CA, the diverse weed flora that emerges in the field after harvesting the preceding crop must be killed using non-selective herbicides like glyphosate, paraquat and glufosinate. Non-selective burn-down herbicides can be applied before or after crop planting but prior to crop emergence in order to minimize further weed emergence.

Unlike in a conventional system, crop residues present at the time of herbicide application in CA systems may decrease the herbicide's effectiveness as the residues intercept herbicide droplets and reduce the amount of herbicide that reaches the soil surface. Proper selection of herbicide formulations for application under CA is necessary to increase their efficacy. For example, pre-emergence herbicides applied as granules may provide better weed control than liquid-forms in zero-till systems. Some herbicides intercepted by crop residues in CA systems are prone to volatilization, photo-degradation and other losses. The extent of loss, however, varies depending upon chemical properties and formulations. Herbicides with high vapour pressure, e.g. dinitroanilines are susceptible to volatilization from the soil surface. Climatic conditions and herbicide application methods significantly affect herbicide persistence under CA systems. Crop residues can intercept 15–80% of the applied herbicides which may result in reduced efficacy of herbicides in CA systems (Chauhan *et al.* 2012). Choosing an appropriate herbicide and timing of its application is critical in CA systems as weed control under zero-till systems varies with weed species and herbicides used.

Pre-emergence herbicides may not be as efficient in controlling weeds in CA systems due to the presence of crop residues which can bind to soil-applied herbicides and favour the weed seedlings to escape the applied herbicides. For example, barnyard grass was fully controlled by pendimethalin and oxadiazon when applied on bare soil (without residue cover); however, some seedlings survived when these herbicides were applied in the presence of residue cover (Chauhan and Abugho 2012).

Several selective post-emergence herbicides, some of which are low-dose and high-potency molecules, are now available to effectively manage weeds in major field crops like rice, wheat and soybean under CA (Table 5). The effectiveness of post-emergence herbicides may be reduced by the presence of crop residues. Wolf *et al.* (2000) observed that the quantity of spray lodged on smooth pigweed (*Amaranthus hybridus*) was reduced by 38–52% by standing wheat stubble depending upon the spray travel speed. Post-emergence herbicides should be applied once the weeds become established, since the timing of weed emergence is less uniform in CA systems than in conventional-tilled systems.

Integrated weed management

Considering the diversity of weed problems in CA systems, no single method of weed control, viz. cultural, mechanical or chemical, provides the desired level of weed control. Therefore, a combination of different weed management strategies should be evaluated to widen the weed control spectrum and efficacy for sustainable crop production. Integrated weed management is not meant to

Table 5: Promising herbicides for weed control in different field crops under conservation agriculture

Herbicide	Dose (g ha^{-1})	Time of application	Remarks
a. Rice			
Pendimethalin	1000–1250	6-7 DAS/DAT	Annual grasses and some broad-leaved weeds. Ensure sufficient moisture at the time of application.
Pyrazosulfuron	25–30	20–25 DAS/DAT	Annual grasses and some broad-leaved weeds
Azimsulfuron	35	20 DAS/DAT	Annual grasses and some broad-leaved weeds.
Bispyribac-sodium	25	15–25 DAS/DAT	Annual grasses and some broad-leaved weeds
Chlorimuron+metsulfuron	4	15–20 DAS/DAT	Annual broad-leaved weeds and sedges
2,4-D	500–750	20–25 DAS/DAT	Annual broad-leaved weeds and sedges
Fenoxaprop-p-ethyl	60–70	30–35 DAS/DAT	Annual grasses especially *Echinochloa* spp.
Fenoxaprop-p-ethyl+2,4-D	60–70 + 500	20–25 DAS/DAT	Annual grasses and broad-leaved weeds
Fenoxaprop-p-ethyl+Almix	60–70 + 20	20–25 DAS/DAT	Annual grasses , broad-leaved weeds and sedges
Bensulfuron+pretilachlor	10000	0–3 DAS/DAT	Annual grasses and broad-leaved weeds
b. Wheat			
Pendimethalin	1000–1250	0–3 DAS	Annual grasses and some broad-leaved weeds. Ensure sufficient moisture at the time of application.
Clodinafop propargyl	60	25–30 DAS	Annual grasses specially wild oat
2,4-D	500–750	20–25 DAS	Annual broad-leaved weeds and sedges
Metribuzin	175–200	30–35 DAS	Annual grasses and broad-leaved weeds
Sufosulfuron	25	25–30 DAS	Annual broad-leaved weeds and grasses
Sufosulfuron +metsulfuron	25 + 2	25–30 DAS	Annual grasses , broad-leaved weeds and sedges
Mesosulfuron+idosulfuron	12 + 24	20–25 DAS	Annual grasses , broad-leaved weeds and sedges
Isoproturon +metsulfuron	1000 + 4	20–25 DAS	Annual grasses and broad-leaved weeds

Contd.

Herbicide	Dose (g ha^{-1})	Time of application	Remarks
c. Maize			
Atrazine	1000-1500	1 - 2 DAS	Broad-spectrum control of weeds
Pendimethalin	1000-1500	1 - 2 DAS	Annual grasses and some broad-leaved weeds
2,4-D	750	30 - 35 DAS	Broad-leaved weeds and few sedges
Topramezone	50-70	25 - 30 DAS	Broad spectrum control of weeds
Tembotrione	100-120	25 - 30 DAS	Broad spectrum control of weeds
d. Soybean			
Metribuzin	35–525	0–3 DAS	Annual grasses and broad-leaved weeds
Chlorimuron ethyl	6–9	15–20 DAS	Annual grasses, broad-leaved weeds and sedges
Fenoxaprop	80–100	20–25 DAS	Annual grasses
Fenoxaprop+chlorimuron	80 + 6	20–25 DAS	Annual grasses and broad-leaved weeds
Imazethapyr	100	20–25 DAS	Annual grasses and broad-leaved weeds

Note: Non-selective herbicides like paraquat and glyphosate should be applied prior to sowing to kill existing weeds

replace selective, safe and efficient herbicides but is a sound strategy to encourage judicious use of herbicides along with other safe, effective, economical and eco-friendly control measures. The use of clean crop seeds and seeders, and weed-free irrigation canals and bunds should be integrated for effective weed management. Weed control efficiency of applied herbicides and crop competitiveness against weeds can be improved by combining good agronomic practices, timeliness of operations, fertilizer and water management, and retaining crop residues on the soil surface. For example, effective ryegrass control (up to 97%) has been observed in a ZT stubble-retained system by using soluble herbicides and minimal disturbance seeders (Crabtree 1999). Similarly, integrating superior genotypes with a high seeding rate and early weed control lead to a 40% yield increase compared with the combination of weaker genotype, low seeding rate and delayed weed control (Harker *et al.* 2003). Approaches such as stale seedbed practice, uniform and dense crop establishment, use of cover crops and crop residues as mulch, crop rotations and practices for enhanced crop competitiveness with a combination of pre- and post-emergence herbicides should be integrated to develop sustainable and effective weed management strategies under CA systems.

Constraints in adoption of CA

Conservation agriculture is not a panacea to solve all agricultural production constraints but offers potential solutions to break productivity barriers, and sustain natural resources and environmental health. Despite several benefits, the adoption of CA systems by farmers in developing countries is still in its infancy as they require a total paradigm shift from conventional agriculture with regard to crop management (Table 6). CA technologies are essentially herbicide-driven, machine- driven and knowledge-driven, and therefore require vastly-improved expertise and resources for adoption in large areas. For wider adoption of CA, there is an urgent need for researchers and farmers to change their mindset and explore these opportunities in a site- and situation-specific manner for local adaptation.

Table 6: Two sides of the conservation agriculture system

Payoffs	Trade-offs
• Timeliness of operations • Reduced soil erosion • Water conservation • Improved soil health • Reduced fuel and labour costs • Reduced sediment and fertilizer pollution in lakes and streams • Carbon sequestration • Climate smart production practices	• Mindset: transition from conventional farming to no-till farming is difficult • Relatively knowledge intensive • CA equipment not available locally and adds to cost for transport • Reliance on herbicides and their efficacy • Prevalence of weeds, disease and other pests may shift in unexpected ways • Reduced crop yield in initial year if not properly practiced • Need to refine nutrient and water management practices

Source: Huggins and Reganold (2008); Sharma *et al.* (2012)

Several factors including bio-physical, socio-economic and cultural limit the adoption of CA by resource-poor farmers. The current major barriers to the spread of CA systems are: (i) competing use of crop residues in rainfed areas, (ii) weed management strategies, particularly for perennial species, (iii) localized insect and disease infestation, and (iv) likelihood of lower crop productivity if site-specific component technologies are not adopted. In addition to these, there are several other factors restricting the adoption of CA technologies in India (Table 7).

Table 7: Nature of constraints towards adoption of CA technologies in India

Technical	• Non-availability of quality drill • Lack of regular monitoring of machines • Lack of training/ capacity building • Spare parts are not available locally • Lack of local manufacturers of machines
Extension	• Lack of extension support from state extension agencies • Lack of extension literature • Lack of attention by mass media • Lack of knowledge of extension agencies • Inadequate extension facility at disposal of input agencies • Lack of cooperation from fellow farmers
Financial	• Lack of credit facilities • Lack of money to buy new machines and inputs • No subsidy on machines • High cost of seed-drill

Source: Meena and Singh (2013)

Conclusions

CA is a complex suite of resource-efficient technologies. It is possible to achieve the same or even higher yields with CA compared with CT. Altering tillage practices changes the depth of weed seeds in the soil, which play a role in weed species shifts and affect the efficacy of control practices. Zero tillage systems cause a shift in weed flora, and may result in emergence of perennial weeds like *Cyperus rotundus, Cynodon dactylon* and *Sorghum halepense* in most crops; and others like *Rumex dentatus* and *Malva parviflora* in wheat. Restricting tillage also reduces weed control options and increases reliance on herbicides; consequently, evolvement of weed resistance to herbicides has become a serious and escalating problem. ZT along with residue has beneficial effects on soil moisture, temperature moderation and weed control. CA is a machine-, herbicide- and management-driven agriculture for its successful adoption. Integrated weed management involving chemical and non-chemical methods (residue, cover crops, varieties etc.) is essential for success of CA systems in the long-term.

Indian Council of Agricultural Research has launched a Consortia Research Platform on Conservation Agriculture in the XII Plan, which is a major step for developing, capacity building and adoption of these resource conserving technologies. Directorate of Weed Research at Jabalpur has also taken lead in developing and promoting CA-based technologies in diversified cropping systems in the black soil region of Central India. Besides converting our own research farm with CA-based technologies, we have also taken up the task of disseminating these technologies on the farmers' fields on a large scale. Needless to say that the technology is spreading like wild fire for growing wheat and chickpea in the winter season, and greengram in the summer season. The farmers after having some initial apprehensions are fully convinced with these technologies. There is a growing demand for suitable farm machinery like happy seeder, which can do sowing, place fertilizer and also work under residue conditions. Undoubtedly, this technology has the potential for revolutionizing wheat cultivation in Central India, for which, greater collaborative efforts are needed by different institutions, state departments, and other agencies concerned with agricultural development.

Future outlook

Weed management research is lacking under conditions of CA. Therefore, development of integrated weed, disease or pest control strategies under CA systems is of paramount importance. Efforts are needed to understand weed, disease and insect responses to ZT soil and microclimate conditions on a long-term basis. Research should be conducted on soil biological aspects and the rhizosphere environment under contrasting soils and crops with particular emphasis on optimizing fertilizer and irrigation management. Other areas of research includes machinery development for local farming systems, sowing into crop residues, understanding herbicide performance in crop residues with reduced tillage, changes in nutrient cycling and nitrogen demand, leaf and root diseases. More focus is required on the influence of crop residue and weed management components.

Herbicides cannot be eliminated from no-tillage systems; and therefore, crop management, degradation pathways, adsorption–desorption and transport processes of herbicides are important research areas. Further, over-reliance on herbicides in a CA system is a concern from an environmental point of view. A major research effort in this area should be towards developing economically-viable strategies to prevent and manage herbicide resistance. Inclusion of allelopathic crop cultivars for managing weeds in the CA systems could be a strategy to avoid development of herbicide resistance. Concerted efforts using advances in plant biotechnology will help to unveil the genetics of this trait. A breeding programme to transfer the allelopathic genes into modern cultivars to enhance their allelopathic activity for weed suppression may help to reduce over-reliance on herbicides in CA systems.

References

Chauhan BS and Abugho SB (2012) Interaction of rice residue and pre herbicides on emergence and biomass of four weed species. Weed Technology 26: 627–632.

Chauhan BS, Gill G and Preston C (2006) Tillage system effects on weed ecology, herbicide activity and persistence: a review. *A*ustralian Journal of Experimental Agriculture 46: 1557-1570.

Chauhan BS and Johnson DE (2010) Implications of narrow crop row spacing and delayed *Echinochloa colona* and *Echinochloa crusgalli* emergence for weed growth and crop yield loss in aerobic rice. Field Crops Research 117: 177-182.

Chauhan BS, Singh RG and Mahajan G (2012) Ecology and management of weeds under conservation agriculture: A review. Crop Protection 38: 57-65.

Chauhan DS, Sharma RK, Verma OS, Singh G and Tyagi BS (1998) Wheat: Research needs beyond 2000 AD, pp. 263-273. Proceedings of International Conference, 12-14 August 1997, Karnal, India,.

Chhokar RS and Malik RK (2002) Isoproturon-resistant little seed canary grass (*P. minor*) and its response to alternate herbicides. Weed Technology 16: 116-123.

Chhokar RS, Sharma RK, Jat GR, Pundir AK and Gathala MK (2007) Effect of tillage and herbicides on weeds and productivity of wheat under rice– wheat growing system. Crop Protection 26: 1689–1696.

Chopra NK and Chopra N (2010) Evaluation of tillage system and herbicides on wheat (*Triticum aestivum*) performance under rice (*Oryza sativa*)–wheat (*Triticum aestivum*) cropping system. Indian Journal of Agronomy 55: 304-307.

Chopra P and Angiras NN (2008a) Influence of tillage and weed control methods on weeds, yield and yield attributes of maize (*Zea mays* L.). Indian Journal of Weed Science 40: 47-50.

Chopra NK and Angiras NN (2008b) Effect of tillage and weed management on productivity and nutrient uptake of maize (*Zea mays*). Indian Journal of Agronomy 53: 66-69.

Crabtree W (1999) Stubble, harrows, ryegrass, soil residual herbicides and no-till sowing, pp. 35–37. In: Stewart V (Ed), Western Australian GRDC crop updates-weed update. Department of Agriculture and Food, Government of Western Australia, Perth, Australia.

Dubey RP (2008) Effect of weed control and nitrogen application rates on weed infestation and productivity in maize– cowpea intercropping system. Indian Journal of Weed Science 40: 155-158.

Farooq U, Sharif M and Erenstein O (2007) Adoption and impacts of zero-tillage in the rice-wheat zone of irrigated Punjab, Pakistan, pp. 9–18. CIMMYT and the Rice-Wheat Consortium for the Indo-Gangetic Plains, New Delhi.

Franke AC, Singh S, Mcroberts N, Nehra AS, Godara S, Malik RK and Marshall G (2007) *Phalaris minor* seedbank studies: longevity, seedling emergence and seed production as affected by tillage regime. Weed Research 47: 73–83.

Gangwar KS, Singh KK, Sharma SK and Tomar OK (2006) Alternative tillage and crop residue management in wheat after rice in sandy loam soils of Indo-Gangetic Plains. Soil and Tillage Research 88: 242–252.

Gathala MK, Ladha JK, Saharawat YS, Kumar V and Sharma PK (2011) Effect of tillage and crop establishment methods on physical properties of a medium-textured soil under a seven-year rice-wheat rotation. Soil Science Society of America Journal 75: 1851-1862.

Gupta R and SethA (2007) A review of resource conserving technologies for sustainable management of the rice–wheat cropping systems of the Indo-Gangetic Plains (IGP). Crop Protection 26: 436–447.

Harker KN, Clayton GW, Blackshaw RE, O'Donovan JT and Stevenson FC (2003) Seeding rate, herbicide timing and competitive hybrids contribute to integrated weed management in canola (*Brassica napus*). Canadian Journal of Plant Science 83: 433-440.

Huggins DR and Reganold JP (2008) No-till: the quiet revolution. *Scientific American*, July 2008, pp. 70-77.

Jat ML, Pal SS, Subbe Rao AVM and Sharma SK (2003) Improving resurce use efficiency in wheat through laser land leveling in an Ustochrept of Indo-Gangetic Plain. In: National Seminar on Developments in Soil Science, 68th Annual Convention of the India Society of Soil Science, November 4-8, CSAUAT, Kanpur, UP.

Jat RK, Banga RS and Yadav A (2013) Resource conservation techniques and pendimethalin for control of weeds in *durum* wheat cultivars. Indian Journal of Weed Science 45: 93-98.

Kang MS (2013) Happy seeder technology is well received by punjab farmers. http://bharatsandesh.com/bharat/index.php?option=com_content&task =view&id=16957&Itemid=27 dt. 02.12.2013

Kumar V, Singh S, Chhokar RS, Malik RK, Brainard DC, Singh M, Sharma PC, Kamboj BR, Mcdonald A and Ladha JK (2012) Conservation agriculture and weed management: experiences from rice– wheat cropping systems of the Indo-Gangetic Plains, pp. 82-83. Lead papers, 3rd International Agronomy Congress, November 26-30, 2012, New Delhi, India, Vol 1.

Ladha JK, Dawe D, Ventura TS, Singh U, Ventura W and Watanabe I (2000) Long-term effects of urea and green manure on rice yields and nitrogen balance. Soil Science Society of America Journal 64: 1993–2001.

Meena MS and Singh KM (2013) Conservation Agriculture: Innovations, Constraints and Strategies for Adoption (September 1, 2013). http://ssrn.com/abstract=2318985 dt. 01.12.2014

Mishra JS and Singh VP (2012) Tillage and weed control effects on productivity of a dry seeded rice– wheat system on a vertisol in central India. Soil and Tillage Research 123: 11–20.

Ram H, Kler DS, Singh Y and Kumar K (2010) Productivity of maize (*Zea mays*)–wheat (*Triticum aestivum*) system under different tillage and crop establishment practices. Indian Journal of Agronomy 55: 185-190.

Rathi JPS, Rathi PK and Singh OP (2007) Studies on integrated weed management technique in chickpea with mustard intercropping system. Plant Archives 7: 909-912.

Rickman JF (2002) Manual for land leveling, Technical Bulletin Series 5, p. 24. Rice– Wheat Consortium for the Indo-Gangetic Plains, New Delhi, India.

Sharma AR, Jat ML, Saharawat YS, Singh VP and Singh, R (2012) Conservation agriculture for improving productivity and resource-use efficiency: Prospects and research needs in Indian context. Indian Journal of Agronomy 57(3rd IAC Special Issue): 131-140.

Sharma AR, Singh R, Dhyani SK and Dube RK (2010) Effect of live mulching with annual legumes on performance of maize (*Zea mays*) and residual effect on following wheat (*Triticum aestivum*). Indian Journal of Agronomy 55: 177-184.

Shrestha A, Knezevic SZ, Roy RC, Ball-Coelho BR and Swanton CJ (2002) Effect of tillage, cover crop and crop rotation on the composition of weed flora in a sandy soil. Weed Research 42: 76-87.

Singh A, Kang JS, Kaur M and Goel A (2013) Farmer's participatory approach for the *in-situ* management of paddy straw with Happy Seeder and rotavator. International Journal of Agriculture Innovations and Research 2: 178-185.

Singh RK, Bohra JS, Srivastava VK and Singh RP (2008) Effect of diversification of rice–wheat system on weed dynamics in rice. Indian Journal of Weed Science 40: 128-131.

Singh S, Ladha JK, Gupta RK, Bhushan L, Rao AN, Sivaprasad B and Singh PP (2007) Evaluation of mulching, intercropping with *Sesbania* and herbicide use for weed management in dry-seeded rice (*Oryza sativa* L.). Crop Protection 26: 518-524.

Tuti MD and Das TK (2011) Carry-over effect of metribuzin applied to soybean (*Glycine max*) on weeds and wheat (*Triticum aestivum*) under zero and conventional tillage. Indian Journal of Agronomy 50: 121-126.

Van Der Weide RY, Bleeker PO and Lotz LAP (2002) Simple innovations to improve the effect of the false seed bed technique, pp. 3-4. In Proc. of the 5th Workshop of the EWRS Working Group on Physical and Cultural Weed Control, Pisa, Italy.

Varshney JG (2000) Final report of ICAR adhoc research scheme "Nutsedge management in field crops with particular emphasis on pulses", p. 163. Indian Institute of Pulse Research, Kanpur,

Wolf TM, Harrison SK, Hall FR and Cooper J (2000) Optimizing post-emergence herbicide deposition and efficacy through application variables in no-till systems. Weed Science 48: 761-68.

Yadav DS, Shukla RP and Kumar B (2005) Effect of zero tillage and nitrogen level on wheat (*Triticum aestivum*) after rice (*Oryza sativa*). Indian Journal of Agronomy 50: 52-53.

8

Scoping Agroforestry: Adaptation and Mitigation to Climate Change

Babita Bohra and Navin Sharma

Introduction

Agroforestry offers both adaptation and mitigation options to climate change especially for the poor rural communities in the developing world. It is now getting ample attention from different investors, scientists, policy makers and international bodies due to its multiple benefits to the environment; and remain an important component of sustainable land use in agricultural production. Agroforestry provides economic, environmental, and social security to the vulnerable communities who are dependent on agricultural production for their livelihoods, and on forest/ trees for their energy requirement. Agricultual production is vulnerable to climate change, and there is a need to cushion these communities to be less vulnerable to climate change, or natural hazards. The suitable, sustainable and smart method is agroforestry based agriculture which offer host of ecosystem services to cope the vagaries of climate change, particularly in rural areas.

Agroforestry is an integration of agriculture, trees and livestock sectors with multiple complimentary benefits to each other. The rising temperatures and changing pattern of rainfall affects trees and agricultural production as well as livestock health and production. In the current climatic scenario, extreme weather events is becoming more severe and frequent thereby increasing unpredictability in normal weather situation. Agroforestry systems can help farmers to make them less vulnerable towards the sudden changes in weather with multiple benefits (food and nutrition, fodder, timber, soil fertility, carbon sequestration, water conservation, soil moisture, reducing soil erosion, shelter, fuel and oil seeds for clean energy *etc.*) to their families and land management without disturbing the farm production. Therefore, from small-holder farmer's point of view we need to look for the adaptation-mitigation methods without changing the land use pattern. This paper defines how smart agroforestry systems to be an option for the vulnerable communities to adapt climate change, and act as a key component in climate change, mitigation measures. According to IPCC report (2007) there is no reason to believe that agroforestry cannot bring about mitigation of greenhouse gases but also produce a substantial set of co-benefits.

According to Cheikh (2014), agroforestry is one of the most startling land use systems across landscapes in different agroecological zones [2]. It has been practiced from ancient times by farmers in many parts of the world. Trees in the farm can give maximum benefits to farmers by providing a number of ecosystem services such as maintain soil nutrients through by nitrogen fixation, maintain soil fertility, reduce soil erosion from wind, heavy rain fall and floods, stabilizing sand dunes, adding organic fertilizer as leaf litter, and water conservation there from. Agroforestry as a discipline has come out of age, during the past few decades and indeed India is possibly the first country to recognise the significance of this area and came out with a separate policy in the coming year. Tree brings in the resilience to food production and also provide valuable ecosystem services and can support adaptation by: (1) providing alternate goods to local communities due to climate change; (2) trees in agricultural fields regulate water, improve soil, and microclimate and provide more resilient crop production; (3) also help in regulating water and protecting soils from erosion thereby reduce adverse climatic impacts; (4) forests protecting coastal areas from climate-related threats.

Impact of climate change on agriculture

Climate change is and will continue to impact the planet's ecosystems in innumerable and impulsive ways. As of now, the global temperature has increased to 0.8 °C compared to pre-industrial time as mentioned in IPCC report (2007). In the next 20-30 years, this rise will be even more pronounced, and the global temperatures are expected to rise by 2 °C[3]. More significantly, the temperatures are expected to further rise by 4 °C (with 40 per cent surety) and even up to 5 °C (10 per cent surety) by the year 2100 [4]. Increasing temperature will directly affect the agricultural production which will lead to food insecurity in the low income countries where agriculture is a main economic activity. The evidences are also showing that the world population is rapidly increasing year by year while agricultural land is shrinking and going to exist in the coming years (Fig. 1). The world population is expected to grow another 1 billion in just 12 years and according to UN forecasts a range from 8.3 billion to 10.9 billion people by 2050, with 9.6 billion as the mid-projection [5] which is an alarming situation. To feed the growing population without compromising the environment. is a big challenge in front of everyone.

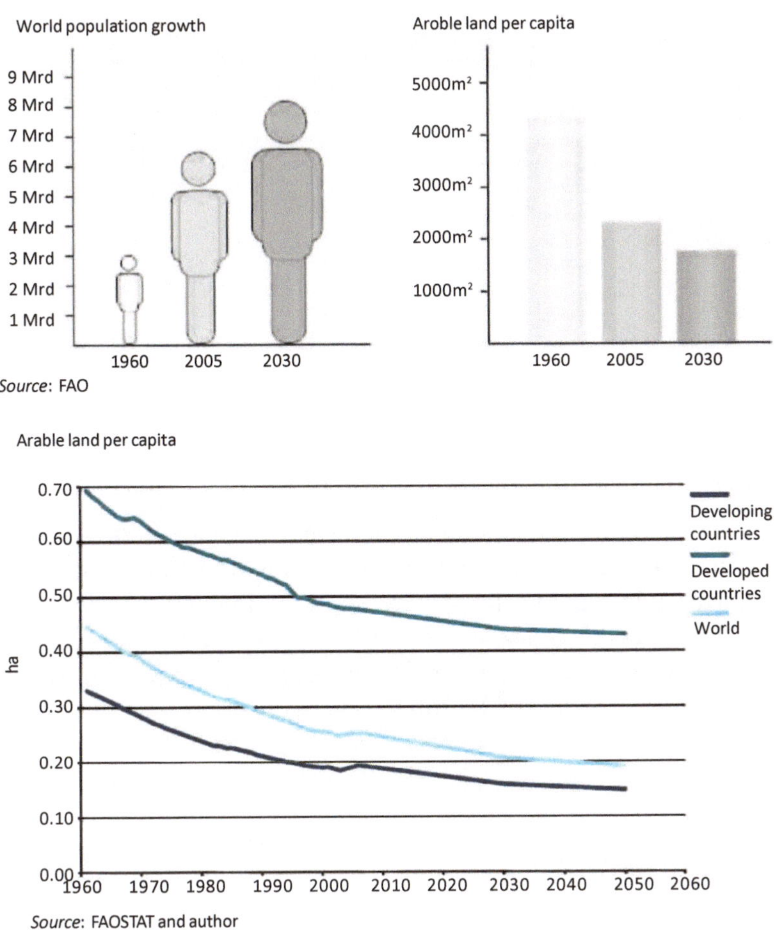

Fig. 1: World population growth rate and arable land per capita [6]

Agriculture is a main occupation of socio-economic development of the communities in the developing countries, and climate change is expected to adversely affect agricultural production in these countries [7]. The major operations of agricultural production are directly influenced by climate change as soil water, carbon and nitrogen cycles, crop growth and development, and increase rate of unwanted vegetation and diseases. The results of climate change impacts on agriculture production has been assessed by different scientists in the different years [7]. The results reflect that the climate change effect will be worst in future especially in the developing countries esp. on the agricultural dependent communities. Figure 2 is an example of significant yield loss due to climate change by IPCC AR5, 2013 [8] and figure 3 explains how human and earth systems are affecting by climate change through different drivers.

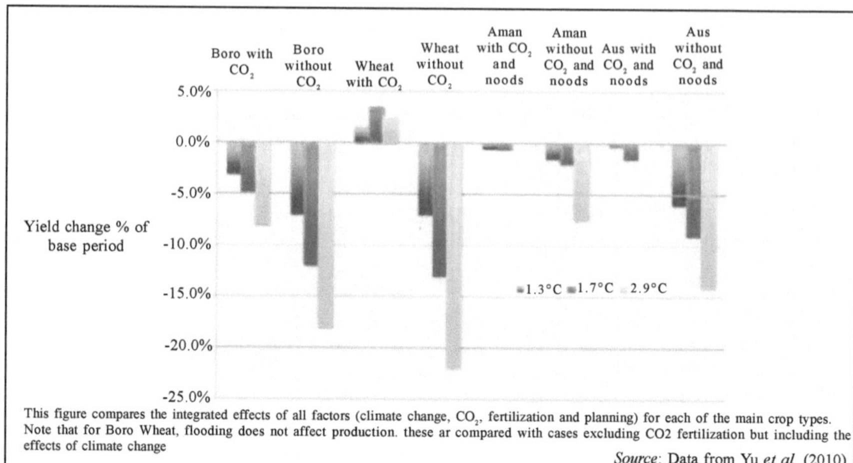

This figure compares the integrated effects of all factors (climate change, CO_2, fertilization and planning) for each of the main crop types. Note that for Boro Wheat, flooding does not affect production. these ar compared with cases excluding CO2 fertilization but including the effects of climate change

Source: Data from Yu *et al.* (2010)

Fig. 2: Causes and consequences of climate change on land use change and agriculture

IPCC (2007) mentioned that in high altitude areas crop production and pasture production increases due to of slight increase in temperature, but seasonally dry and low altitude regions slight increase in the temperature decreases the production; however this is a short term influence [9]. Table 1 presents some of the projected changes in climate and their potential impacts on agriculture production, summarised by IPCC, 2007 [3] [7].

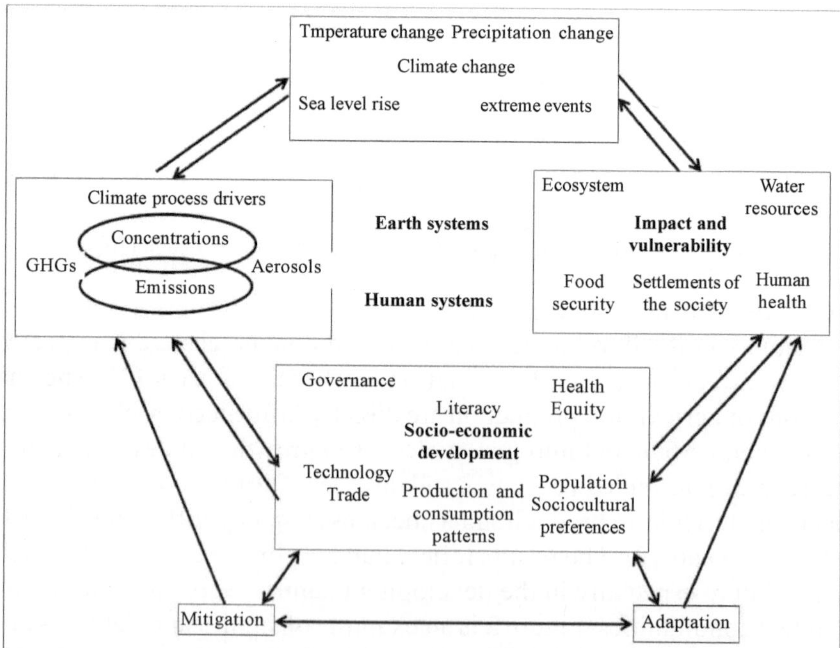

Fig. 3: Systematic framework of anthropogenic climate change drivers, impacts and responses [3]

Table 1: Impact of climate change on agriculture [7]

Likelihoods of occurrence	Phenomena and direction of change	Major projected impacts on agriculture
Likelihoods of occurrence	Phenomena and direction of change	Major projected impacts on agriculture
Warmer and fewer cold days and nights; warmer/more frequent hot days and nights over most land areas	Virtually certain (>99% chance)	Increased yields in colder environments decreased yields in warmer environments Increased insect outbreaks
Warm spells/heat waves: frequency increases over most land areas	Very likely (90-99% chance)	Reduced yields in warmer regions due to heat stress; wild fire danger increase
Heavy precipitation events: frequency increases over most areas	Very likely	Damage to crops; soil erosion, inability to cultivate land due to water logging of soils
Area affected by drought: increases	Likely (66-90% chance)	Land degradation, lower yields/crop damage and failure; increased livestock deaths; increased risk of wildfire
Intense tropical cyclone activity increases	Likely	Damage to crops; wind throw (uprooting) of trees; damage to coral reefs
Increased incidence of extreme high sea level (excludes tsunamis)	Likely	Salinisation of irrigation water, estuaries and freshwater systems

Food security and climate change

Around 974 million people suffering from chronic hunger in the year 1990, although this figure has fallen down to 842 million [10]. The challenges are bewildered by fact that almost half of the hungry people are from poor and rural communities of the developing world. Africa and Asia are home of 40% of the world's starved people. According to FAO (2012) that Africa number of hungry people grew double in recent times [11].

Steep growth in agricultural productivity is being required in reducing hunger and malnutrition in the developing world where population is totally depend on agricultural production for their livelihoods [10]. Increasing the farm productivity in the developing countries involving smallholders will be critical to reducing poverty and hunger.

As per one estimate, to feed 9.6 billion people by 2050, the food production needs to increase by 70 per cent with added burden on the land [12]. Most of the future population may prevail in the developing countries, and meeting sustainable food demand will be a big challenge. In the year 2008 there were food riots in several parts of the world due to of higher food prices. Decreased farm productivity would also make farmers less interested in agriculture. Hence, improving accessibility and availability of food for poor is important goal of agroforestry.

Different parts of the world will have different impacts of climate change on agriculture, depending on the current climatic and soil conditions, and the actions taken in that region on making suitable resources available to cope with climate change [13]. Moreover, every crop has an optimal thermal requirement to achieve maximum yield, any rise in temperature affects the productivity negatively. Even a relatively minor increase of 1 to 2 °C temperature during the growing period of a crop can significantly affect the yield of many food crops. Higher temperature affects crop yield by decreasing physiological processes of the plants. It is argued that the elevated carbon dioxide will add to photosynthesis and add to biomass; however, elevated temperature can offset this phenomenon. Scientists have now confirmed that a 1 °C rise in temperature above the optimum temperature can affect the yields by 10 per cent for many cereals [14]. According to Lobell & Asner (2003) 1-degree Celsius rise in temperature affects corn yields by 17 per cent [15]. Similarly, Kumar & Parikh (2005) concluded that a 2-degree rise in temperature above norms lowers the yield of wheat by 37 to 58 per cent [16].

With an increase in temperature above 1.5°C, the significant yield and crop production losses will occur in regions such as Sub-Saharan Africa, South East Asia and South Asia. South Asia region will be specially burdened with these changes as it will have a challenge to feed an ever growing population of 2.2 billion by 2050 with world's largest yet poorest population. The agriculture accounts for 18 per cent in GDP of this region and houses more than 50 per cent of people relying on agriculture and depended on it. Around 28 per cent of global rice cultivation takes place in this region and with increased temperature the estimated yield losses of -21 per cent is forecasted [17].

Agroforestry as an option

Agroforestry is defined as a land use system where trees are incorporated in the agricultural land scape to enhance diversity, ecosystem sustainability and provide additional livelihood options by diversification (Table 2).

Table 2: Multiple benefits of agroforestry systems

Agroforestry	Multiple benefits
	• Managing the heat stress both on crops and animals • Reversing land degradation through adoption of practices • Effective management of climate related risks through promotion of innovative and sustainable diversification • Efficient capture, storage and utilization of rainfall through adoption of appropriate soil and water conservation practices • Maintaining soil fertility and productivity by arresting nutrient mining and building or sustaining soil fertility • Guarding against pest and disease pressure • Enhancing resilience of communities by better targeting investments • Ensuring maintenance of food and nutritional security • Protection of women and other disadvantaged groups from adverse impacts of climate change • Ensuring maintenance of food and nutritional security

In addition, trees play a critical role in reducing vulnerability to uncertain and shifting climates. Trees can balance microclimates, modulate water flows, store carbon, provide habitat for plants and animals in protected areas and corridors, and provide food for people. It is now certain that if drastic actions are not taken to mitigate the climate change impacts, it will not be possible to feed the world in the future. While crop diversification, breeding for heat resistant varieties and agroforestry can contribute significantly to food production. A landscape approach of integration of trees in agriculture and livestock production system and increasing access to clean energy will play a vital role in boosting the crop productivity (Fig. 4). Agroforestry has tremendous potential to provide avenues to adapt to the climate change [22] [23]. It combines a long list of landscape management, including crop diversification, soil nutrient balance, soil health management, soil fertility management and if practiced properly it also regulates the water flow [24] [25] [26] [27].

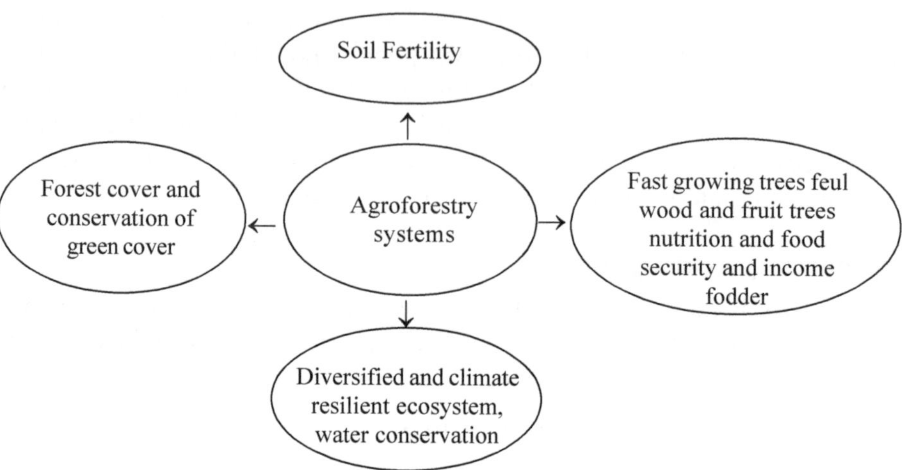

Fig. 4: Benefits of Agroforestry systems in agriculture

Presently there are interests from investors to invest on agroforestry because of its multiple benefits to the ecosystem [27] [28]. It is more profitable to farmers than agriculture or forestry for a particular area of land. A comprehensive analysis indicated economic viability with internal rate of return (IRR) ranging from 25 to 68 and B : C ratio of 1.01 to 4.17 for 24 agroforestry systems from different agroclimatic regions of India [29].

In general, agroforestry systems include livestock silvo-pastoral systems; home gardens; on-farm timber production, and biomass plantations within a wide diversity of biophysical conditions and socioecological characteristics [30]. With increased risk of food security caused by climate change, it seems to offer various on-farm adaptation needs, and fulfil many roles in mitigation pathways [2]. It is globally gaining focus from different stakeholders as a potential to address various on- farm adaptation needs. The systems are complex and have integration of different ecosystem components each of which responds to climate change [31]. It can be one of the options for future growth and development for the growing population especially in developing countries. It is a well-established method practiced from ancient times and fits well with the three pillars of sustainability e.g. economic, environmental and social. It is a mean to improve the overall growth of the communities and environmental sustainability. Identification of different models in different regions and ecologies can be a future option for sustainable agricultural development through smart agroforestry (Fig.5) There is thus great need for methods to project climate change impacts on agroforestry systems. Three main methods are available: (1) process-based models, (2) species distribution models and (3) climate analogue analysis (Fig. 5) [31].

Method modeling	Process-based modeling	Species distribution analysis	Climate analogue
Preparatory steps	Detailed characterization and quantification of all relevant processes	Collection of system occurrence data over entire distribution range	Identification of analogue locations
Analysis	Generalization of processes Run model for future climates cenarios	Characterization of environmental niche	Trials or observations at target and analogue sites
Output	System performance in future climate	Projection of niche onto future climates Habitat suitability for agroforestry system from observations	Infer climate change impacts Example of plausible future land use

Current Opinion in Environmental Sustainability

Fig. 5: Main approaches through different models of agroforestry to combat climate change

In an agroforestry system, trees are used in two ways as productive and protective. Productive uses are direct which can help farmers to enhance their livelihoods and protective uses can have environmental benefits. The potential agroforestry tree species for different agroforestry systems are given in the Table 3.

Adaptation strategies and mitigation measures through agroforestry

"Adaptation" refers to efforts by society or ecosystems to prepare for adjust to future climate change. These adjustments can be protective (i.e., guarding against negative impacts of climate change), or opportunistic (i.e., taking advantage of any beneficial effects of climate change)" [1]. Adaptation to climate change has been practiced by human being for centuries such as migration to new place, growing different crops etc. However, the current rate of climate change is unprecedented and will require a different approach than hitherto practiced. Although they are not the main contributor to GHG emissions, the poor rural communities get affected by climate change problems. These communities in the developing countries are directly involved in the agricultural production for their livelihoods and on forest/ trees for their nutrition and energy requirements. There is a need to reduce vulnerabilities and build resilience to these communities against the impacts of climate change. Ecosystems provide important services that can help people adapt to climatic variability and thus there is a need to give importance to this service.

Table 3: Agroforestry systems with the potential tree species [32] [29]

Agroforestry systems	Description	Potential tree species
Agroforestry systems	Description	Potential tree species
Commercial Agroforestry systems	Large scale production	Elaeis spp., Cocos nucifera, Ficus elastica, Eucalyptus kitsoniana, Populus ilicifolia, Casuarina glauca, Leucaena leucocephala, Tectona grandis, Salix spp.
Intermediate Agroforestry systems (partially commercial)	Between commercial and subsistence scale of production and management	Acacia nilotica, Albizia lebbeck, Azadirachta indica, Leucaena leucocephala, Gliricidia sepium, Sesbania grandiflora Gliricidia sepium, Sesbania grandiflora, Erythrina sp, Acacia sp. Madhuca indica, Prosopis cineraria, Morus alba, fruit trees and other mixed tree species
Subsistence Agroforestry systems (small holder)	The use of land is directed towards satisfying basic needs and is managed mostly by the owner and his family	Acacia nilotica, Acacia leucophloea ,Tamarindus indica, Azadirachta indica. Alnus nepalensis, Madhuca indica, Pogamia, fruit trees and other mixed species

Smart agroforestry models have multiplc benefits to the communities especially as a coping mechanism for the climate impacts on agriculture in terms of adaptation, which are again not separate by substantially contribute to climate change mitigation. There are two examples of successful agroforestry models recently from CGIAR [33]. For example, *Gliricidia sepium* in boundaries has a total biomass of 7 Mg km^{-1} per year (35 Mg C ha^{-1} per year) when pruned every 4 months and 9.5 Mg km^{-1} per year (50 Mg C ha^{-1} per year) when pruned every 6 months and monitored in a four years cycle. *Erythrina bertroana* has similar results (30 and 55 Mg C ha^{-1} per year, respectively) in the same study [34]. According to de Jong (1995) in Mexico, trees in the boundaries had stored 24–36 Mg C ha^{-1} during a cycle of 25–30 years [34]. Table 4 shows the potential of agroforestry systems for different regions globally [36]

Table 4: Carbon storage potential of agroforestry systems in different ecoregions of the world [36] [37] [38] [39]

Region	Ecoregion	System	Mg C ha^{-1}
Africa	Humid tropical high	Agrosilvicutural	29–53
South America	Humid tropical low	Agrosilvicutural	39–102a
	Dry lowlands		39–195
Southeast Asia	Humid tropical	Agrosilvicutural	12–228
	Dry lowlands		68–81
Australia	Humid tropical low	Silvopastoral	28–51
North America	Humid tropical high	Silvopastoral	133–154
	Humid tropical low	Silvopastoral	104–198
	Dry lowlands	Silvopastoral	90–175
Northern Asia	Humid tropical low	Silvopastoral	15–18

a Carbon storage values were standardised to 50-year rotation

Current agroforestry practices for adaptation or mitigation to climate change (Table 5) and its vulnerability of agriculture and impact of climate change (Fig. 6).

Table 5: Agroforestry practices for adaptation or mitigation to climate changes [27]

		Mitigation	
		Positive	Negative
Adaptation	Positive	Soil carbon sequestration, improved water holding capacities, use of manure instead, mixed agroforestry for commercial products, income diversification with trees, reduced nitrogen fertilizer, fire management	Dependence on biomass energy, overuse of ecosystem services, increased use of mineral fertilizers Poor management of nitrogen and manure, over extraction of non-timber products, timber extraction
	Negative	Integral protection of forest reserves, limited rights to agroforestry trees, forest plantation excluding harvest	Use of forest fires for pastoral and land management, tree exclusion in farming lands

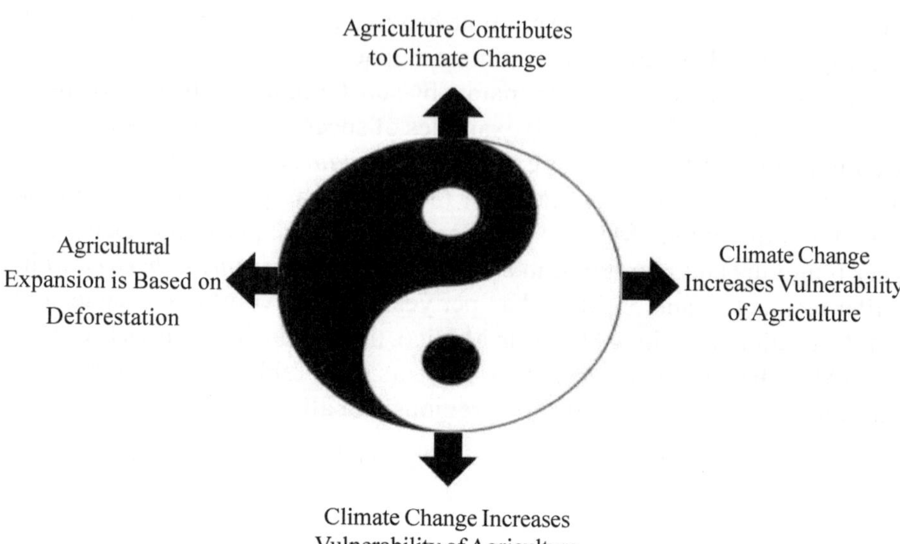

Fig. 6: Vulnerability of agriculture and climate change (by intensifying agriculture on existing land and protecting the remaining forests, we can eliminate emissions from land-use change)

Smart agroforestry system- multiple benefits

Trees planted along with agricultural crops have direct effects on vulnerable communities by improving ecosystem services, increasing agricultural production there by creating food and nutrition security. These benefits improve community capacities to cope the climatic hazards. Therefore in integration of trees crops and livestock called agroforestry are adaptation to climate change, and also climate smart agroforestry. Rural communities can adjust to climate change variability, moderating potential damages and helping them cope with adverse consequences and making them less vulnerable through adaptation [35].

The figure 7 shows how agroforestry can be used as an option for climate adaptation and mitigation tool to the vulnerable communities in the developing countries. Adaptation is the best way of mitigation that too from locally available resources generating livelihood option and increased agricultural and livestock production.

Table 6: Examples of planned adaptation by sector, key constraints and opportunities to implementation

Sector	Adaptation option/strategy	Underlying policy framework	Key constraints and opportunities to implementation
Water	Expanded rainwater harvesting; water storage and conservation techniques; water re-use; desalination; water-use and irrigation efficiency	National water policies and integrated water resources management; water-related hazards management	Financial, human resources and physical barriers; integrated water resources management; synergies with other sectors
Agriculture	Adjustment of planting dates and crop variety; crop relocation; improved land management, e.g. erosion control and soil protection through planting i.e. agroforestry	R&D policies; institutional reform; land tenure and land reform; training; capacity building; crop insurance; financial incentives, e.g. subsidies and tax credits	Technological and financial constraints; access to new varieties; markets; longer growing season in higher latitudes; tree revenues from 'new' products
Energy	Strengthening of overhead transmission and distribution infrastructure; underground cabling for utilities; energy efficiency; of renewable sources; reduced dependence on single sources of energy	National energy policies, regulations, and fiscal and financial incentives to encourage use of alternative sources; incorporating climate change in design standards	Access to viable alternatives; financial and technological barriers; acceptance of new technologies; stimulation of new use technologies; use of local resources

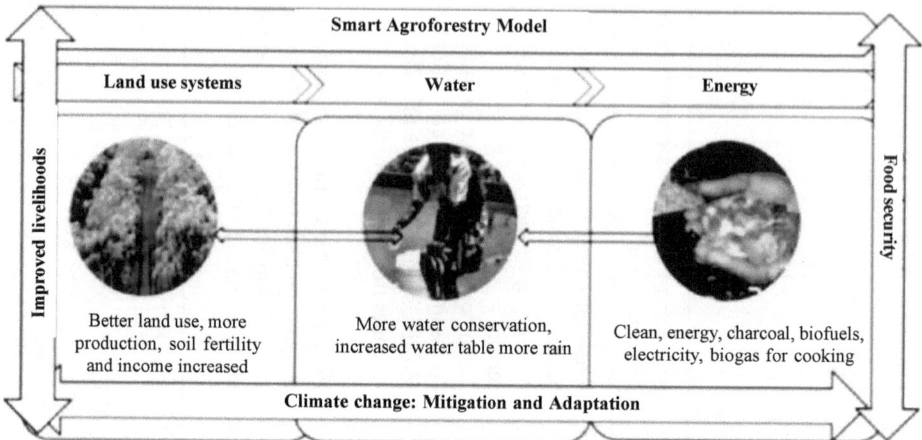

Fig. 7: Agroforestry as an adaptation method

India as an example

India has launched an agroforestry policy in the year 2014, and following map (Fig 8) showed area under agroforestry in different states of India [29]. The map indicate agroforestry is practiced in almost every part of India. It has a number of successful programme and agroforestry models for instance the model which got attention is BAIF's "Wadi Model" (agri-horti-silvi) which was implemented in Maharashtra, Rajasthan, Gujarat, Karnataka, Uttar Pradesh, Uttarakhand, Madhya Pradesh, Chhattisgarh, Bihar, Andhra Pradesh, and Jharkhand and covered 1.81 lakh families to establish 68,586 hectares [29].

Fig. 8: Area under agroforestry in different states of India: a map

Conclusion

Agroforestry can't be a single strategy for the climate adaptation and mitigation option; however it can be a smart solution to avert problems arising though climate change. Smart agroforestry systems have a potential to provide livelihood options to the poor households without compromising the crop production and land use pattern with a number of co- benefits. The greatest advantage is that the farmers can use their trees as insurance for their future during difficult times libe drought where crop can't survive and trees can play a congenial role by providing feed/ fodder and shelter for their livestock. By practicing agroforestry in modern/ sustainable way without losing its originality of diversification, it can be one of the best climate adaptation methods to the vulnerable communities. Agroforestry – a climate-smart agricultural practice - can help farmers and communities adapt to changes in climate through maintaining and enhancing environmental services and providing opportunities for diversification that build resilience and generate additional income. There are innumerable examples of traditional land use practices involving the growing trees with agricultural crops to augment food production and more importantly buffer farmers against the risks of crop failure by providing alternate sources of income. The incorporation of livestock and bio-energy also provide diversification opportunities to farmers. The higher production of agroforestry products for home consumption and market provides increased income and thus helps in alleviating poverty. Agroforestry interventions are uniquely placed to achieve multiple objectives of food and environmental security. Can also help to create rural employment and thus help open up more rural development prospects.

References

[1] UN, "EPA Region 10 Climate,change Adaptation,and implementation Plan," 2014.

[2] C. Mbow, P. Smith, D. Skole, L. Duguma and M. Bustamante, "Achieving mitigation and adaptation to climate chnage through sustainable agroforestry practices in Africa," Current Opinion in Environmental Sustainablity , vol. 6, pp. 8-14, 2014.

[3] IPCC, "Climate Change 2007: Sysnthesis report-A report of the intergovernmental panel to climate change".

[4] WorldBank, 2013. [Online]. Available: http://www.worldbank.org/.

[5] TMP, "http://www.millennium-project.org/millennium/Global_Challenges/chall-03.html," [Online]. Available: http://www.millennium-project.org/millennium/Global_Challenges/chall-03.html.

[6] B. Jelle, "The resources outlook: by how much do land, water and crop yields need to increase by 2050?," [Online]. Available: http://www.fao.org/docrep/014/i2280e/i2280e06.pdf.

[7] K. Rao, L. V. Verchot and J. Laarman, "Adaptation to Climate Change through Sustainable Management and Development of Agroforestry Systems," Sat e-journal , vol. 4, no. 1, 2007.

[8] [Online]. Available: http://www.worldagroforestry.org/downloads/Publications/PDFS/WP14143.pdf.

[9] D. 2. Maddison, "The perception of and adaptation to climate change in Africa, Development Research Group, Sustainable Rural and Urban Development Team, the World Bank, policy workign paper," 2007.

[10] FAO, "http://www.fao.org/news/story/en/item/161819/icode," 2013.

[11] FAO. [Online]. Available: http://www.fao.org/news/story/en/item/161819/icode/.

[12] A. Nikos and B. Jelle, "World Agriculture: Towards 2030/2050," 2012. [Online]. Available: http://www.fao.org/economic/esa/en/.

[13] J. Olesen, M. Trnkab, K. Kersebaumc, A. Skjelvågd, P. Seguine, Peltonen-Sainiof, P. Rossig, J. Kozyrah and F. Micalei, "Impacts and adaptation of European crop production systems to climate change," European Journal of Agronomy, vol. 34, pp. 96-112, 2011.

[14] N. Nguyen, "Global climate changes and rice food security," FAO, [Online]. Available: http://www.fao.org/forestry/15526-03ecb62366f779d1ed45287e698a44d2e.pdf.

[15] D. Lobell and G. Asner, "Climate and management contributions to recent trends in U.S. agricultural yields," Science, vol. 14, 2003.
[16] [Online]. Available: http://www.earth-policy.org/books/pb2/pb2ch4_ss3.
[17] T. W. Bank, 2013. [Online]. Available: http://www.worldbank.org/en/news/press-release/2013/06/19/warming-climate-to-hit-south-asia-hard-with-extreme-heat-floods-disease-world-bank-report-says.
[18] T. Thorlakson and H. Neufeldt, "Reducing subsistence farmers' vulnerablity to climate change: evaluating the potential contribution of agroforestry in western Kenya," Agriculture and Food Security , vol. 1, p. 13, 2012.
[19] A. Challinor, T. Wheeler, C. Garforth, P. Craufurd and A. Kssam, "Assessing the vulnerability of food crop systems in Africa to climate change," 2007.
[20] W. Bank, "World development report," World bank publications, Washington DC, 2008.
[21] R. Neupane and G. Thapa, "Impact of agroforestry intervention on soil fertility and farm income under the subsistence farming system of the middle hills, Nepal," Agric Ecosyst Environ, vol. 84, pp. 157-167, 2001.
[22] R. Rice, "Agricultural intensification within agroforestry: the case of coffee and wood products," Agric Ecosyst Environ, vol. 128, pp. 212-218, 2008.
[23] O. Oke and K. Odeblyi, "Traditional cocoa-based agroforestry and forest species conservation in Ondo State, Nigeria," Ecosyst Environ, vol. 122, pp. 305-311, 2007.
[24] C. Mbow, P. Smith, D. Skole, L. Duguma and M. Bustamante, "Achieving mitigatiing and adapatation to climate change through sutainable agroforestry practices in Africa," Current opinion in environmental sustainablity , vol. 6, pp. 8-14, 2014.
[25] Q. Nguyen, M. Hoang, I. Oborn and M. Noordwijk, "Multipurpose agroforestry as a climate change resiliency option for the farmers: as an example of local adaptation in vietnam," Climatic Change, vol. 84, pp. 241-257, 2013.
[26] S. Chavan, A. Keerthika, S. Dhayni, A. Handa, R. Newaj and K. Rajarajan, "National agroforestry policy in India: a low hanging fruit," Current Science , vol. 108, pp. 1827-1834, 2015.
[27] R. J. Zomer, A. Trabucco, R. Coe and F. Place, "Trees on Farm: Analisys of Global Extent and Geographical Patterns of Agroforestery," 2009.
[28] E. Luedeling, R. Kindt, N. I. Huth and K. Koenig, "Agroforestry systems in a changing climate — challenges in projecting future performance," Current opinion in environmental sustainability , vol. 6, pp. 1-7, 2014.
[29] TNAU, "Forestry technologies," [Online]. Available: http://agritech.tnau.ac.in/forestry/agroforestry_index.html.
[30] L. V. Verchot, M. V. Noordwijk, S. Kandji, T. Tomich, C. Ong, A. Albrecht, J. Mackensen, C. Bantilan, K. Anupama and C. Palm, "Climate change: linking adaptation and mitigation through agroforestry," Mitigation and adaptation strategies for global change , vol. 12, no. 5, pp. 901-918, 2007.
[31] P. A. Sanchez, "Linking climate change research with food security and poverty reduction in the tropics," Agriculture, Ecosystems & Environment, vol. 82, pp. 371-383, 2000.
[32] F. C. C. M. J. S. L. G. G. I. Romero, "Productividad de Gliricidia sepium y Erythrina bertroana en cercas vivas manejadas bajo tres frecuencias de poda en la zona atlantica de Costa Rica," Agroforestria, p. 4, 1991.
[33] B. deJong, G. Montoya-Gomez, K. Nelson, L. Soto-Pinto, J. Taylor and R. Tipper, Community forest management and carbon sequestration: a feasibility study from Chiapas, Mexico, vol. 20, 1995, pp. 409-416.
[34] A. Albrecht and S. Kandji, "Carbon sequestration in tropical agroforestry systems," Agriculture, Ecosystem and Environnment , vol. 99, no. 1-3, pp. 15-27, 2003.

[35] R. Dixon, K. Andrasko, F. Sussman, M. Lavinson, M. Trexler and T. Vinson, "Tropical forests: their past, present and potential future role in the terrestrial carbon budget," Water Air Soil Pollut., pp. 71-94, 1993.

[36] P. Schroeder, "Agroforestry systems: integrated land use to store and conserve carbon," Climate Res., pp. 53-60, 1993.

[37] J. Winjum, R. Dixon, Schroeder and PE, "stimating the global potential of forest and agroforest management practices to sequester carbon," Water Air Soil Pollut., vol. 64, pp. 213-228, 1992.

[38] L. Verchot, M. Noordwijk, T. T. Kandji S, O. C. Albrecht A, M. J, B. C, A. KV and P. C, "Climate change: linking adaptation and mitigation through agroforestry," Mitigation Adapt Strat Global Change, vol. 12, pp. 901-918, 2007.

[39] IPCC, "Climate Change 2001: impacts, adaptation and vulnerabilty. contribution of working group II the third assessment report of the IPCC," 2001.

[40] M. Altieri, "The ecological role of biodiversity in agroecosystems.," Agric Ecosyst Environ, vol. 74, pp. 19-31, 1999.

[41] F. Muchena, D. Onduru, G. Gachini and J. DA, "Turning the tides of soil degradation in Africa: capturing the reality and exploring opportunities," Land use Policy, vol. 22, pp. 23-31, 2005.

[42] P. Tschakert and G. Tappan, "The social context of carbon sequestration: considerations from a multi-scale environmental history of the old peanut basin of senegal," Arid Environ., pp. 535-564, 2004.

[43] J. Lehmann, I. Peter, C. Steglich, G. Gebauer, B. Huwe and W. Zech, "Below ground ineractions in dryland agroforestry. forestry," Forest Ecology and Management , vol. 111, pp. 157-169, 1998.

[44] J. Lott, C. Ong and B. CR, "understorey microclimate and crop performance in a Grevillea robusta-based agroforestry system in semi-arid Kenya," Agric Forest Meteor , vol. 149, pp. 1140-1151, 2009.

[45] E. Molua, "The economics of tropical agroforestry systems: the case of agroforestry farams in Cameroon," Forest Policy Econ, vol. 7, pp. 199-211, 2005.

[46] R. Rice, "agriculrural intensification within agroforestry: the case of coffee and wood products," Agric Ecosyst Environ, vol. 128, pp. 212-218, 2008.

9

Conservation Agriculture for Sustainable Intensification in Madhya Pradesh

Raj Gupta and Nikhil Kumar Singh

Conservation Agriculture (CA) is defined as a process of 'sustainable intensification of agriculture' for enhancing and sustaining production at low cost, without deteriorating soil health and flow of environmental services. The process of agricultural intensification increases food production and spares land for the other uses including eco-system services. In line with the vision of Norman Borlaug, agricultural intensification is also known as *'The Borlaug Effect'.* Frameworks such as 'Conservation Agriculture', 'Ecological Intensification' and 'Ever Green Revolution' share a common view of cropping systems as agro-ecosystems designed to make maximum use of fixed resources (land, light and temperature etc.) with optimum use of agri-inputs for attaining sustainable production levels (Cassman 1999)'. Conservation agriculture is also considered as an important component of the overall strategy for ensuring food security, poverty alleviation, health for all, rural development, enhancing productivity, improve environmental quality and preserve natural resources. With limited scope for further expansion of area under agriculture, production gains can be accomplished through intensification of agriculture, achieved by pursuing one or more strategies (with rationale use of natural resources as well as external production inputs) enumerated as under:

- Increasing yields per hectare (e.g. timely planting and with increased inputs (e.g. water and fertilizer nutrients),
- Increasing cropping intensity per unit of land (e.g. use of short season crop cultivars, relay and inter- cropping, and grow an additional crop through cropping system optimization), and
- Changing land use from low value crops/commodities to those that receive higher market prices.

In context of the state of Madhya Pradesh, however, large tracts of black soils (nearly 1.99 Mha per season) remain fallow each in the kharif and rabi seasons (Subbarao *et al.* 2001). Fallow land, if addressed appropriately offer a huge opportunity of reducing pressure on the other cultivated lands and also bring transformation of agriculture in the state.

Potential for agricultural intensification in Madhya Pradesh

The key issues for the state of Madhya Pradesh should be how to seed crops in the 'fallow land parcels' before the onset of the monsoon. In rabi season, it is crucial to sown crops in the receding soil moisture conditions in rainfed agriculture. The other issues that relate with intensification of agriculture includes

- Fewer aquifers formations, low water use efficiency and water shortage
- Cracking Vertisol - narrow soil-moisture working range, timeliness of operation crucial and useful mechanization for sustainability in intensive agriculture
- Practice of summer deep ploughing leading to high run off and extensive soil erosion and no cover crop.
- Fallow land ~ 5 million hectare in kharif and rabi seasons (ICRISAT, 2002)
- Large 'management' yield gaps - agronomic failures
- Residue burning and air pollution
- Open grazing, makes it difficult to enhance cropping intensity, many field remain fallow during summers
- Climate variability increasing – impact crop production, require capacity building however weak R-E-F linkages.

Large management yield gaps in crops exists such as in rice (27-40%), wheat (35-47%) and maize (50-65%) and dwindling productivity trends in soybean and pulses calls for identification of location specific strategic entry points and potential interventions to address the declining crop yields and bridging the yield gaps. The common practices of open grazing of cattle after harvest of the Rabi crops together with above stated issues are detrimental to options for sustainable intensifications. It appears that the situation can be significantly improved by making some adjustment in conservation agriculture by integrating traditional knowledge of the farmers with new information relevant to the specific agro-ecologies (rainfed and irrigated and assured rainfall regions). It is strongly believed that 'Tree-Crop-livestock' systems developed on conservation agriculture platforms in watershed approaches will make many grey areas -green.

Conservation agriculture not only help bridge yield gaps but also reduce the demand supply gaps of irrigation water which otherwise is likely to exacerbate in changing climate scenarios. It is worth mentioning here that conservation agriculture is still co-evolving with agents for change.

Agriculture production of essential commodities can be increased by adopting a two-pronged strategy to deal with gaps in production and productivity and reduce acreage of kharif and rabi fallow lands in the state. Several reports suggest that significant yield gaps exist in different geographic regions/districts in different crops (horizontal yield gaps). The strategic entry points and the action that one adopt under each entry point are listed in the fig 1. For specific locations, the strategic entry points and desirable intervention can belong to more than one set, such as in case for preventing the menace of soil erosion (mechanical, and soil crop management measures).

1. Water scarcity/ Rainwater mange	2 Soil Erosion	3. Nutrient Management	4. Terminal heat Imbalances
• Land leveling • Raised beds • DSR, avoid puddling • Micro-irrigation • Watershed & *in situ* water management • Water recharging • Mulching	**In-situ* moisture conservation • Watershed approach • Peripheral Bunding • Gully plugs • Grassed waterways • Mulching	• Conjunctive use of nutrient inputs • LCC/SPAD/GS • Super granules	• Cultivar choices • Seed priming • Water schedules • Residue manage.
5. Labor, Energy shortages, High production cost	6. Diversification and 'Fallows'	7. Herbicide resistance /Weed mgt.	8. Networks and Capacity Building
• Minimal tillage • Zero tillage • Raised bed •Double no-till system • New Machines	• Relay/Para cropping • Crop substltu tlon • Cotton-wheat) • S.Cane -wheat) • Rice Fallows	• Test new molecules • In tegra ted weed managementapproach • Crop rotation	• Public-Private partnerships • Trainings • Traveling seminars • SMS service

Fig. 1: Conceptual framework of strategic entry points and potential interventions for conservation agriculture

It is now widely recognized that yield gaps in cereal crops grown in many states result from agronomic failings, and that future yield increases depend heavily on this science-based resilient interventions for bridging management yield gaps (yield gains through 'better-bet' crop management).

The 'Borlaug effect' is all about producing more output from the same area of land while reducing the negative environmental impacts and at the same time increasing contributions to flow of environmental services (Pretty, 2008; Godfray *et al.*, 2010). Cropping systems such as rice-wheat face serious challenges due to their rising production and environmental costs. Thus, sustainable agricultural systems should be less vulnerable to shocks and stresses and have to be more profitable to the practitioners.

In our paper, we discuss evidences for science-based opportunities for sustainable intensifications (including viable diversification & cropping system optimization options) while addressing the challenges of agriculture (water shortages, natural resource degradation, emerging climatic variability) for national food security. The paper also provides an evidence for the need of reorientation of cropping systems and reveals that proper choice of cultivars and smart mechanization solutions, wheat can be planted in late October-early November to buffer year to year fluctuations due to climatic extremes/ terminal heat stresses (Fig. 2).

Results presented in the Fig.1 indicate that early planting of wheat, cause fewer productivity fluctuations than planted late in the season. Therefore, it appears to be a good strategy to plant wheat early for buffering the productivity variations year after year. Early planting of wheat in late October in the residual moisture of the preceding rice crop, saves water and increases the period for vegetative growth/biomass. This option requires weekly vernalized wheat cultivars such as HD 2967 and CSW18 are bred that would need some mild winters to flower and escape terminal heat. This strategy can save pre-plant irrigation and watering

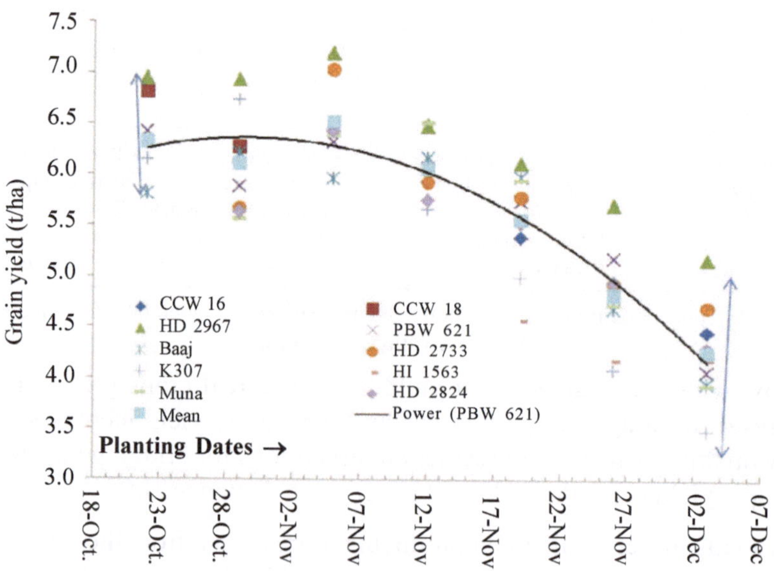

Fig. 2: Effect of planting dates on productivity of the wheat cultivars at Pusa, Bihar (2013-14)

to prevent terminal heat effects in many areas where wheat is grown after rice. With new machinery, it is now possible to plant wheat without tilling, spread the rice straws uniformly and avoid residue burning to sustain air quality and soil health and harness benefits accruing from surface mulching (weeds, water, and temperature effects).

Water Availability and Crop Production

Water availability holds key to the vexed issue of terminal heat stresses with terminal heat stresses grains become shrivelled, lose weight and lustre. The three fundamental ways to bridge demand- supply gap of water include (a) increase water supplies, (b) improve water productivity 'crop per drop', and (c) make economic choices- reduce water withdrawals, and make crop substitutions. In above context, higher research priorities for water productivity improvement should be accorded to i) areas with high poverty and low water productivity, ii) areas of physical water scarcity with intense competition for water e.g. in areas with declining water tables (eg. drought prone areas), and iii) areas with less water resource development (eg. assured rain and rainfed regions). The report of the World Water Resource Group (2011) published 'water availability-cost curve' implications for India. It suggests that lost cost measures such as land levelling, recently brings out that conservation agriculture practices save irrigation water, improve water productivity and also crop productivity. Measures such as laser assisted precision land levelling, peripheral field bunding; check-dams and gully plugs, raised-bed planting, mulching, zero-tillage, furrow/ skip-furrow and intermittent irrigation, sprinkler irrigation and surface seeding etc. save irrigation water and promote intensification. The mixed and intercropping systems in vogue can potentially improve water productivity and crop yields. The relative importance of the options for productivity improvement and trade off amongst various water saving options need to be studied in order to prioritise the 'better bet practices'.

Direct seed rice (DSR), nutrient-water interactions and germplasm development, drip / sprinkler irrigation, and enhanced fertiliser use, raised bed and furrow irrigation measures have the potential of closing 80% of the projected water gap (for India ~ 750 billion m^3 by 2030) in irrigated and rainfed ecologies in India. Additionally, low cost supply measures (e.g. last mile completion of water courses, repair and cleaning of canals etc.) have the potential of bridging the remaining 20% gap (as mentioned in the report on World Water Resource Group). Micro irrigation systems can be of significantly benefit the crop and water productivity in sloping as well as levelled field. Intermittent irrigation, skip furrow and permanent skip furrow irrigation practices can significantly improve water use efficiency particularly when combined with mulching.

Reducing acreage of fallow lands through conservation agriculture

Black soils occupy close to 60 Mha in the Central Plateau region and parts of Rajasthan. A number of land-surface management practices such as raised and sunken beds, ridge and furrow and broad-bed and-furrow (BBF) have been recommended for black soils to increase the opportunity time, infiltration of rain water and to reduce soil erosion through surface runoff (Kampen, 1982; Pathak et al. 1985; Singh et al .1999). In the vertisol, pre- monsoon dry seeding on BBF is a general recommendation. However, farmers are reluctant to practice pre-monsoon dry seeding recommendation for fear of seed loss through:

- Seed loss through picking by birds,
- Loss of seed viability due to extended exposure of seed to peak summer temperatures ranging between 38- 47 °C for more than 40 days, and
- Mortality of young seedlings in alternate wetting-drying cycles

As a consequence, farmers are forced to grow puddled / unpuddled rice after the start of the rainy season in low lying fields and large tract of land remain fallow in kharif in spite of> 1000mm rainfall. Geo-spatial multi-temporal analysis of the remotely sensed satellite data suggested that nearly 2.02 million ha (6.57% of the total area of the state) remains fallow in the state of M.P alone. This is in spite of a strong recommendation for dry sowing of crops before the onset of the rainy season (ICRISAT, 2006).

It is a common knowledge that condition of the black soils change very quickly from the stone dry soils with wide and deep cracks to extremely plastic slippery muddy soils during monsoon rains. Such situations disrupt preparatory tillage and sowing operations. Planting on the flat surface and/or even on the raised beds during the rainy season often results in poor germination in the wet black soils. This is likely due to swift changes in macro-porosity in the shrinking-swelling clay black soils due to change in soil water content. Above field capacity, changes in soil water content impacts low-tension oxygen diffusion processes more than the water content itself in black soils (Lin, et al., 1998), adversely affecting seed germination and emergence in the monsoon season.

The challenge for the tropical farmers is to "close the rainy season window" with crops to cover soils with residues for maintaining aggregate stability, modulate soil temperature, promote plant / root growth, sequester carbon and mop-up residual NO_3-N from the surface soil layers before it is leached out of the soil profile. Pre-monsoon dry seeding enables germination and emergence of seedlings before the start of the monsoon rains. Thus acreage of the current kharif fallow lands can be reduced with direct dry seeding following CA practices.

The practice of direct dry seeding was evaluated before thc onset of monsoons on the 541 acre BISA farms were initiated on April 28 and completed by June 16, 2014 (Fig. 3). Crops such as rice, maize, pearl-millet, sorghum, sorghum, pigeon-pea and cluster beans were planted using a Turbo-Happy seeders and or Multi-crop zero till planters attached with double disc openers in presence of 5-8 t/ha of residues of the previous wheat crop in flats and raised bed system.

Between April 28 to June 16, 2014, there were four rainfall events of (5-20 mm) interspersed with several windy and sunny days. It was observed that in first four rainfall events, seed placed under the surface mulch, imbibed some moisture but insufficient to trigger germination event. During the alternate wetting-drying cycles experienced by seeds, the content of the imbibed water never reached threshold moisture levels required to trigger /initiate seed germination, preparatory to emergence of the seedlings. For this reason, seeds germinated not observed even after 4 rainfall events except in some low-lying spots. After the first few showers and onset of monsoons (on June 14 & 17, 2014, when it rained 20 & 25mm respectively, additional water through sprinkler irrigation was supplied to facilitate and ensure successful germination events.

Germination counts of the dry seeded crops were recorded several weeks after the rain events also supported by sprinkler irrigation. Extended periods of seed exposure to high temperature and alternate wetting- drying cycles of the seed during pre-monsoon showers did not adversely affect the seed quality. Rice seedling population was around 30 hills/m2 and the maize population was around 10 plants / m^2. The germination counts of rice and maize are an average of 100 and 50 counted sites, respectively, spread over 150 acres. . Normal plant population/ m^2 in farmer fields generally vary between 25-35 seedlings in rice and 6-7 seedlings in maize. Results of the large field trials indicate that pre-monsoon direct dry seeding in zero-till black soil, is a safe recommendation in presence of surface mulches. It is because surface mulching mediates the genotype-temperature-moisture interactions in conservation agriculture to prevent any loss of seed or the viability of the seed. It appears that pre-monsoon direct dry seeding (PmDDS) practice recommended earlier has huge potential only if practiced and popularised through conservation agriculture. This technology has a major potential to reduce the acreage of kharif fallows, reduce soil erosion by providing vegetation cover, increase cropping intensity, rainwater use efficiency and livelihoods of the people in the Central Plateau region. In other areas, crop yields will benefit through direct dry seeding before the onset of monsoons through efficient rain water use, better crop stand becoming more competitive to rainy season weeds.

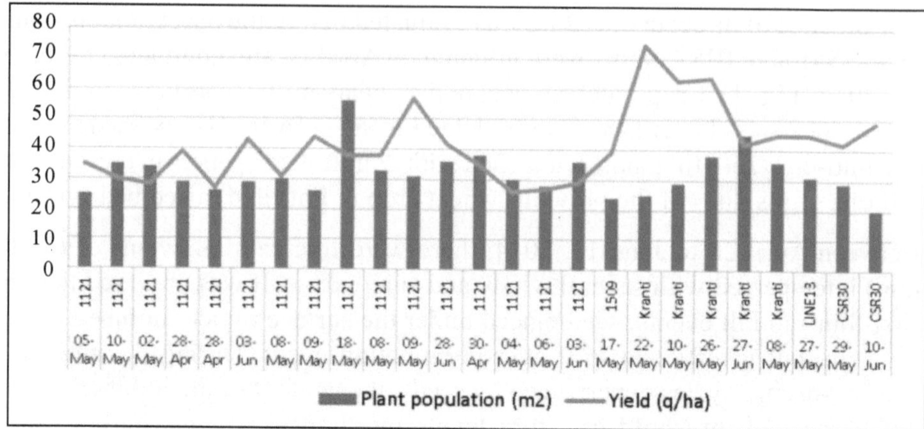

Fig. 3: Plant population and grain yield of Pre-monsoon dry seeded rice cultivar Pusa 1121, Kranti and CSR30) grown with sprinkler irrigation systems in a Vertisol at BISA Jabalpur. Numerals on the curves refer to plant population m-2 grain yield of (rice cultivar Kranti (K); CSR 30 and Pusa Basmati 1121 (without any alphabet notation).

Results collected from large sized field trials seeded in pre-monsoon season using sprinkler irrigation are presented in the figure 3 for plant population and crop productivity (cultivars Pusa 1121, Pusa 1509 and Kranti). Data show that pre-monsoon direct dry seeding of rice can maintain sufficient plant population for normal productivity of rice as observed with transplanted rice in the area with less water.

Reduce rabi fallows in low rainfall areas

In many rainfed regions, crops such as mustard, toria, and chickpea are grown. Due to paucity of irrigation facilities, these crops have to be planted in receding soil moisture contents. This requires that seed should be placed at deeper soil layers having sufficient moisture to enable germination. Deeper placement of seed often results in poor seedling emergence and result in sporadic crop stand. This problem can be overcome through alignment and adjustment of a triangular furrow opener, followed by the knife opener to place fine size seed below the sill of the furrow (below the furrow) in moist soil layers for proper crop stand and ultimately, yield.

Summary

Many research reports have pointed out that conservation agriculture can sequester carbon, save water and reduce the erosion hazards of rainwater, particularly when CA is adopted in watershed approaches. The technology used in CA is economically and ecologically sound and can play a vital role on both large and small land holdings. Conservation agriculture plays two major roles namely (i) it helps to counters the negative impact of climate change on agriculture and food production and at the same time it (ii) helps to conserve the biodiversity and ecological balance and slow down the pace of climate change. It's one of the most viable options for sustainable intensification of agriculture. It's only prudent that the methods of conservation agriculture find a place in policy frameworks and extension programs. Conservation agriculture provide options to the farmers for timely seeding of the crops (DSR, pre-monsoon direct seeding in broad bed furrow systems, controlled traffic plots etc.) there has a huge potential for reducing acreage of fallow lands during kharif and rabi seasons particularly in the vertisol belt of the country.

References

Cassman, K.G. 1999. Ecological Intensification of cereal production system: Yield Potential, soil quality, and precision agriculture. Proceedings of the National Academy of Sciences of the USA. 96: 5952-5959.

Godfrey, H.C.J. , J.R. Beddington, I.R. Crute, L. Haddad, D. Lawrence, J.F. Muir, J. Pretty, S. Robinson, S.M. Thomas, and C. Toulmein. 2010. Food Security: The Challenge of Feeding 9 Billion People. Science. 327, 812-817.

ICRISAT. 2002. ICRISAT Annual Report 2002 'Research for Impact'. Documentation. International Crops Research Institute for the Semi-Arid Tropics.

ICRISAT. 2006. ICRISAT Annual Report 2006 'Nurturing the seeds of success in the semi-arid tropics'. Documentation. International Crops Research Institute for the Semi-Arid Tropics. Kampen, J. 1982. An Approach to Improved Productivity on Deep Vertisols. Manual. International Crops Research Institute for the Semi-Arid Tropics, Patancheru, Andhra Pradesh, India.

Lin H.S; K. J. Mc Innes, L.P. Wilding; and C.T .Hallmark. 1998. Macro porosity and initial soil moisture effects on infiltration rates in Vertisols and Vertic intergrades. Soil Sci. 163: 2-8). Pathak, P., S.M. Miranda and S.A. El-Swaify. 1985. Improved rainfed farming for the semi-arid tropics; implications for soil and water conservation. In: 'Soil Erosion and Conservation'. (Eds. S.A. El-Swaify, W.C. Moldenhauer, and L. Andrew). Pp. 338-354. Soil Conservation Society of America, Ankeny, Iowa.

Pretty, J. 2008. Agricultural sustainability: concepts, principles and evidence. Philosophical Transactions of the Royal Society of London 363:447–466.

Singh, P., Alagarswamy, G., Hoogenboom, G., Pathak, P., Wani, S. P. and Virmani, S. M. 1999. Soybean-chickpea rotation on Vertic Inceptisols II. Long-term simulation of water balance and crop yields. Field Crops Research, 63 (3). pp. 225-236.

Subbarao, G.V., J.V.D.K. Kumar Rao, J. Kumar, 2001. Spatial distribution and quantification of rice-fallow in South Asia- potential for legumes, Res. Bulln: International Crops Research Institute for Semi-Arid Tropics (ICARSAT).

World Water Resources Group. 2011. Charting our water future. Economic Framework to Inform Decision Making. pp.12.

10

Response of Organic Agriculture to Climate Change

Manish Bhan, S.S. Tomar and K.K. Agrawal

Climate change is a statistical variation in properties of the climate system that includes changes in global temperature, precipitation, etc due to natural or human drivers over a long period of time. It may affect distribution and quality of natural resources thereby influencing livelihood security of the people (Status of Indian Agriculture, 2013). Recent climate change activities is the increase in earth's temperature by 0.74 °C between 1906 to 2005 (IPCC 2007) due to increase in anthropogenic emissions of green house gases such as carbon dioxide (CO_2), methane (CH_4), ozone (O_3), nitrous oxide (N_2O) and chlorofluorocarbon (CFCs). Increase in concentration of green house gases influence future climate as well as agriculture (Mall *et al*., 2006; IPCC, 2001; Aggarwal, 2003: Bhatia *et al*., 2004). For Indian region under South Asia, Intergovernmental Panel on Climate Change (IPCC), 2007 projected a rise in temperature from 0.5-1.2 °C by 2020, 0.88-3.16 °C by 2050, and 1.56 - 5.44 °C by 2080. This may anticipate greater instability in agricultural production (Aggarwal, 2008). Beside this, climate change also affects the dynamics and frequency of extreme events like tropical cyclones, associated storm surges, and rainfall events. Low lying regions are more vulnerable to climate change as increase in temperature melt down the glaciers causing floods, submergence, and degradation of these regions. Therefore, there is a need for a 'wake up call' at this stage to aware farmers about the climate change, extreme weather events like hailstorm, heat and cold waves, drought, flash and river floods and landslides, and its affect on crop and agriculture, and strategies to overcome it.

Challenges Due to Climate Change on Agriculture

The challenge is to enhance productivity in agriculture as it is declining due to loss in soil fertility, increase cost of production, and rainfed nature of agriculture in the state. As per the current climate change scenario, rainfed agriculture will be affected in the coming years due to shift in spatial and temporal distribution of rainfall; increase in intensity of rainfall and longer dry spells; gradual increase in temperature across all the seasons; frequent extreme weather events like frost, drought, flash floods, etc. Erratic and uneven distribution of rainfall poses a major constraint for achieving target levels of production in rainfed area. Due to higher occurrence of rainfall, farmers plant/sown crops without equipped irrigation practices. Change in rainfall pattern and its amount will definitely affect net sown area of those agro-climatic zones, which receives higher amount of rains during south-west monsoon. Beside this, increase in green house gases through burning of agricultural residues cause carbon dioxide emission; paddy cultivation causing methane emission also induced the shift in cropping pattern.

Greenhouse gases (GHGs) are responsible for global warming and climate change and it is increasing rapidly. The emission of GHGs into the atmosphere comes with industrialization, deforestation, shifting cultivation, and expansion of intensive agriculture. Agriculture is a major contributor to emissions of GHGs to an approximately 15 % of all GHGs emissions globally (IFOAM, 2004). In the current scenario, carbon dioxide (CO_2),Nitrous oxide (N_2O), methane (CH_4) concentration increasing rapidly with no signs of their reduction due to current practices, technologies, and population growth and life-style of human beings. One third of all CO_2 emissions comes from changes in land use (deforestation, shifting cultivation, and intensification of agriculture) whereas two-thirds of CH4 and majority of N_2O emitted through agricultural practices (Kotschi and Müller-Sämann, 2004).

Current observations of data sets of various weather parameters indicate a gradual increase in temperature across all the seasons with erratic and uneven spatial and temporal distribution of monsoon. Climate projections indicate a future increase in temperature by 2050s with uneven distribution of rainfall across the state, with perceptible decrease in rainfall during winter period, and almost no change in rainfall during monsoon with respect to current climate. These predicted changes could lead to spatial and temporal shift of cropping centres and decline in the productivity of crops. It is likely that the onset of south-west monsoon may shift from June to the first fortnight of July that could affect cropping sequence and sowing time. Also, increase in rate of evapotranspiration also increase with temperature, which will lead to depletion in moisture retention capacity of different soil types and may pose a threat to agriculture. On the other side, increase in rainfall leads to faster runoff causing higher soil erosion, which will deteriorate soil health and its fertility. The rise in temperature cause stored carbon to be released from the soil.

Certain areas have soils with low underlying storage capacity and low water availability for irrigation. Due to less irrigation, farmers used to plant one crop in a year. Under utilization of large tract of fallow land, or use for one season contribute to under realization of the potential productivity of crops. Beside this, monocropping also makes the crop susceptible to failure if there is a change in climatic conditions for optimal production. On the other hand, use of fertilizers coupled with chemicals for longer period affects soil fertility and productivity in high cropping intensity regions.

In the present agricultural scenario, without taking further risks, there is a need to look for solutions that are based on ecological and biological principles and have fewer environment costs. Therefore, the quest for climate-proof food systems is of interest to all (Scialabba and Müller-Lindenlauf, 2010). Organic agriculture is a holistic production management system, which promotes sustainable agriculture and enhances agro-ecosystem health. It ensures qualitative development of soil, water and environment on sustainable basis. It is widely accepted that organic agriculture contributes to the alleviation of poverty. The experimental evidences could prove that a holistic soil and crop management had similar productivity as compared to conventional agriculture and often higher yields in the regions of the world where the production environment is much fragile and tough (rainfed and hilly areas). In India, organic agriculture has made credible performance during past ten years. It is a combined effect of farmer's efforts, NGO's work, Government interventions, scientists' initiatives and market forces that has put Indian agriculture in a safe zone.

Definitions of Organic farming

As codified in the *Codex Alimentarius* from the FAO/WHO: 'organic agriculture is a holistic production management system that avoids use of synthetic fertilizers, pesticides and genetically modified organisms, minimizes pollution of air, soil and water, and optimizes the health and productivity of interdependent communities of plants, animals and people' (Codex Alimentarius Commission, 2001). In this farming, growers applied various crop management practices to optimize nutrients through symbiotic relationship while enhanced biodiversity to suppress pest influence, sustaining yield levels with improving soil fertility and productivity. This is accomplished by using, where possible, cultural, biological, and mechanical methods, as opposed to using synthetic materials, to fulfill any specific function within the system. An organic production system is designed to a) enhance biological diversity within the whole system; b) increase soil biological activity; c) maintain long-term soil fertility; d) recycle wastes of plant and animal origin in order to return nutrients to the land, thus minimizing the use of non-renewable resources; e) rely on renewable resources in locally organized agricultural systems; f) promote healthy use of soil, water, and air, as well as minimize all forms of pollution that may result from agricultural practices... (Codex Alimentarius 1999, 2).

On this note, organic agriculture may prove to be an alternating strategy to avert climate change with basic principles as mentioned by Kotschi and Müller-Sämann (2004) in organic agriculture with regard to mitigation of climate change, and it include to:

- encourage and enhance biological cycles within the farming system
- maintain and increase long-term fertility in soils
- use as far as possible , renewable resources in locally organized production systems
- minimize all forms of pollution

Beside this, organic agriculture give priority to the optimal (recycled and reuse) use of inputs with an aim to achieve maximum output.

India developed National Action Plan on Climate Change in 2008 with intent to enhance sectoral responses towards vulnerability caused by climate change as well as its mitigation and adaptation. Madhya Pradesh (M.P.) Organic Farming Policy, 2011 was drafted with a vision to provide sustainable livelihoods, conserve natural resources, and provide employment to the villagers. It is being implemented to reduce the impact of climate change thereby improving crop productivity. In M.P., State Action Plan on Climate Change (SAPCC) was established in 2009 within Environment Planning and Coordination Organization (EPCO) of the Housing and Environment Department. The major highlights of SAPCC include pro-active approach in identification of policy reforms, institutional arrangements, stakeholder participation, and estimation of regional climate change vulnerabilities.

Certain policy targets as mentioned in the website of State Knowledge Management Centre on Climate Change (SKMCCC), Bhopal (Available at www. skmccc.net) will be included as:

Long-term Targets: By achieving renewable environment through agriculture-ecosystem management, implementation of main schemes as soil carbon storage and manuring, enhancing soil health, reducing ground water pollution specially harmful chemicals and enhancing biodiversity.

Medium-term Targets: To standardize agricultural investment and enhance income on investment in present agriculture system. On the other hand, making agriculture a profitable business through enhancing the net income by adapting market based mechanisms.

Short-term Targets: Developing appropriate atmosphere for families dependent on small and marginal agricultural economy by developing agricultural technology, market security, developing capable economic human resources and institutions. Simultaneously, ensuring availability of required quality resources for producing organic products, developing renewable energy resources by utilizing biodegradable resources are also stressed upon.

Role of Organic Farming to Reduce Risk of Climate Change

The major strategies to avert climate change under organic farming is to apply diversification practices and increase in soil organic matter, which both could enhance resilience against extreme weather events. It not only reduces nutrient exploitation but also increases soil organic matter. Similarly, harvested produce under organic farming systems is less prone to extreme weather conditions, such as drought, flooding and water logging conditions.

Table 1: Direct and indirect reduction on agricultural greenhouse gas emissions arising from the principles of Organic Agriculture

	CO_2	CH_4	N_2O
1.Agricultural land use and management			
Permanent soil cover	+++	-	+
• Reduced soil tillage	+	-	+
• Restriction of fallows in (semi)arid regions	+	-	-
• Diversification of crop rotations incl. fodder production	++	-	+
• Restoring the productivity of degraded soils	++	+	-
• Agro-forestry	++	-	-
2.Use of manure and waste			
• Recycling of municipal waste and compost	++	-	+
• Biogas from slurry	-	++	-
3.Animal husbandry			
• Breeding and keeping for longevity	-	++	+
• Restriction of livestock density	-	+	+
• Reduction of fodder import	+	+	-
4.Management of fertilizers			
• Restriction of nutrient input (nutrient recycling)	++	-	++
• Leguminous plants	+	-	+
• Integration of plant and animal production	++	-	+
5.Change of consumer behavior			
• Consumption of regional products	+++	-	-
• Shift towards vegetarian products	+	++	-
++ high, + low, - no potential			

Source: Sauerbeck 2001; Cole *et al.* (1997) cited in FAO (2002); Kotschi and Müller-Sämann, 2004

In valuing single measures, an overview of the benefits of Organic Agriculture is given in Table 1. It comprises technologies that either reduce emissions or sequester carbon or do both. It also includes consumer behavior, which may contribute to emission reductions from agriculture. Other benefits of organic farming on farmland include:

- Organic farming is low-risk farming strategy with reduced input costs, therefore, lower risks with partial or total crop failure due to extreme weather events or change and variability (Eyhorn 2007). As such, it is a viable alternative for the farmers with higher farm incomes are possible due to lower input costs and higher sale prices. The coping capacity of the farms is increased and the risk of indebtedness is lowered. Risk management, risk reduction strategies, and economic diversification to build resilience are also prominent aspects of adaptation, as mentioned in the Bali Action Plan (UNFCCC).
- Comprises of diverse cropping systems , which increases the diversity of income sources and the flexibility to cope with adverse effects of climate change and vulnerability, such as changed rainfall patterns. This leads to higher economic and ecological stability through optimized ecological balance and risk spreading.
- Crops are well adapted to the local environment under organic farming systems. Local effects of climate variability cannot be foreseen in detail because, at local level, climate change models are not very accurate or even available. Adaptation thus may utilize measures that build on self-adaptive capacity, such as local crop-breeding. The systemic character (on farm breeding, *etc*.) in organic farming is especially adequate to provide such measures (Eyhorn 2007).
- Organic farming is an adaptation strategy that can be targeted at improving the livelihoods of rural populations and those parts of societies that are vulnerable to the adverse effects of climate and variability (Eyhorn 2007).

Potential of Organic Agriculture to Combat Climate Change

There is considerable ability in organic agriculture for mitigating climate change (IFOAM, 2004; Niggli, 2007; IFOAM, 2008; Goh,2011) by reducing GHGs emissions like nitrous oxide (N_2O), carbon dioxide (CO_2) and methane (CH4), adopting techniques to increase soil carbon sequestration, and enhance effects of organic farming practices which favors the above two processes (Goh, 2011).

There are both mitigation potential and adaptation strategies as described by Scialabba and Müller-Lindenlauf (2010) and other authors practiced under organic agricultural systems to reduce the concentration of GHGs gases along with addition of adaptative techniques into the cropping system in the wake of climate change.

Reduced Input Levels: Organic farming systems require fewer external inputs. This help in reducing the fossil fuels as the energy used for preparing nitrogenous fertilizers release about 0.4–0.6 Gigatons (Gt) CO_2 emissions (FAOSTAT, 2009).

Williams *et al.* (2006) estimated energy utilized under conventional wheat production in UK to be allocated as 56% to mineral fertilizers and by 11% to pesticides. Similarly, Pimental (2006) in the USA calculated 30–40% for fertilization and 9–11% for plant protection energy burden utilized under conventional systems for wheat and corn. These GHGs gases released for producing these inputs are avoided completely under organic agriculture however manual labor is a constraints and their availability is needed along with more fossil fuel energy for machinery due to the use of mechanical weed control (Scialabba and Müller-Lindenlauf, 2010).

In organic agriculture, reduced dependency on energy inputs reduces vulnerability to rising energy prices, and hence volatility of agricultural input prices. On the other note, no application of nitrogen fertilizer enhances nutrient use efficiency and therefore, reduces the risk of nitrous oxide emissions. Between 1960 and 2000, while agricultural productivity increased substantially with increased utilization of fertilizers, the global efficiency of nitrogen use for cereal production decreased from 80 to 30%, while the risk of nitrogen emissions increased (Erisman *et al.*, 2008). Reduce use of external inputs including synthetic fertilizers and chemicals reduce crop yield effectively however it varies among countries. In developed countries, yield may reduce from 0-20 % than conventional farming while yield improves or no losses in the developing countries (Badgley *et al.*, 2007), as these countries plant their crops in a smaller areas with more crop diversity in low-input systems by applying manures, composting, and diversification, particularly in arid and semi-arid regions. Similarly, nitrogen supply in paddy through nitrogen-fixing algae like Azolla proves a strategy for climate resilience (Augstburger *et al.*, 2002).

Landscape Management and Agroforestry: Landscape elements and their integration may be an effective mitigation strategy due to its multiple adaptation effects (Smith *et al.*, 2007). The elements include hedges and trees to reduce soil erosion with fewer losses of soil organic matter, thereby improving soil fertility. In addition, they provide habitats for predators and beneficial insects that keep pests under check, while hedges protect from winds (Zehnder *et al.*, 2007). The problem is that these elements hinder the process of mechanization. Similarly, agroforestry systems are recommended by IPCC as a mitigation strategy under organic agriculture as it holds the biggest potential of agricultural carbon sequestration especially in the tropical countries (Kotschi and Müller-Sämann, 2004).

Crop Residue Management: Organic agriculture restricts the ban of biomass burning as it accounts for 12 % of agricultural GHGs emissions with the release of CH_4 and N_2O gases. Additionally, carbon sequestered in the burned biomass is lost in the atmosphere (IFOAM, 2002). Similarly, ban on deforestation reduce

12% of global GHG emission and act as a useful strategy to mitigate climate change (Andrasko *et al.*, 2007).

Cropland Management: The major aim is to sustain soil fertility levels with reduce population of insect-pest and diseases through preventive, cultural, mechanical, and biological measures. It includes strategies and potential as described below:

Cropping system and diversification: Cropping systems in organic agriculture should be a holistic approach towards the management of water, land, crop, and other natural resources to alleviate the adverse impact of climate change. Cultivars developed under crop breeding programmes should be climate resilient (temperature- and drought-tolerant), high yielding, respond to high CO_2 concentration, and easily available in the wake of climate change (Chauhan *et al.*, 2014).

Diversification of crops in cropping systems is one of the major strategies in organic agriculture. Crops selected to adopt local soil and climate, different nutrient uptake, root depth, and their tolerance to local pests and diseases (Scialabba and Müller-Lindenlauf, 2010). In organic system, uniform crops (crops of same family) are not selected in the holistic cropping system to avoid nutrient uptake and pest problems. It is rather to opt for planting of robust traditional species to avoid pest problems however yield levels are low that create yield instability in these systems. Beside this, growing different assemblages of crops in time and space seeks to enhance agro-ecosystem resilience to external shocks such as extreme weather events or price variation, which are all at risk due to increase in the problems caused by climate change (Smith *et al.*, 2007).

Crop-based approaches: Adjustment in sowing dates is an effective adaptation strategy under climate change scenario (Chauhan *et al.*, 2014). This will help in minimizing the deleterious effect of weather thereby reducing yield instability. Beside this, other agronomic practices include bed planting of crops improves water-use efficiency, reduced waterlogging, better access for interrow cultivation, weed control, banding of fertilizers, better stand establishment, less crop lodging,, and reduced seeding rates (Bhardwaj *et al.*, 2009; Chauhan *et al.*, 2014).

Organic manuring: In permanent cropping system, soil organic carbon improves by use of organic manures added through plant residues, mixed cropping, legume based crop rotations, or agroforestry (Drinkwater *et al.*, 1998). On the opposite side, sole use of synthetic nitrogen fertilizer application increases oxidation of organic matter, thereby reducing organic carbon from the soil. Kotschi and Müller-Sämann (2004) mentioned key issues related to the application of organic fertilization as:

- To optimize quantity and application of organic manure. A close integration of crop with animal husbandry and systematic recycling of organic waste are basic elements.
- To improve organic waste processing techniques to obtain high quality manure. The composting of animal and plant residues has become a key technology to minimize losses in the humification process and to achieve a high proportion of the solid humus fraction (but this may lead to higher methane emissions).

Crop rotation: Diversified crop rotation with inclusion of legumes in a cropping system is a major characteristic of organic agriculture. A system comparison conducted by Haas and Köpke (1994) revealed that organically grown crops sequester less carbon in the upper biomass of crops but more in the root biomass. Similarly, sequestration potential was twice as high under organic agriculture than conventional systems (42:1 compared with 21:1 in conventional) (Table 2).

Table 2: Carbon sequestration by organic and conventional farming systems

	---------- t CO_2/ha -----------		
	Organic	Conventional	Difference
Cash crops			
Above ground biomass	3.76	4.95	-1.18
Root biomass	1.44	0.89	0.55
Catch crops*			
Above ground biomass	0.55	0.22	0.33
Root biomass	0.22	0.09	0.13
Weeds			
Above ground biomass	0.22	0.04	0.17
Root biomass	0.04	0.01	0.03
Gross output (sequestration)	6.23	6.19	0.04
Energy Input (emission)	0.15	0.29	-0.14
Net output (sequestration)	6.08	5.91	0.18
Carbon-sequestration efficiency	41.5	21.3	

* Catch crops (intercrops) are sown after the harvest of the main crop in order to capture the nutrients and to provide soil cover. They can also be sown into the main crop

Source: Haas and Köpke, 1994; Kotschi and Müller-Sämann , 2004

Conservation tillage: It embraces zero- and minimum tillage combined with the use of organic soil covers. The benefit is that it reduces production costs and improves soil fertility with significant effect on carbon sequestration however it is dependent on herbicides combined with the use of genetically modified crops both are not practiced under organic agriculture. Hence, it may be an option to combat climate change but not acceptable due to the herbicide technique involved in it.

Other agronomical management: Organic farming practice like crop rotation, planting cover crops, manuring, and application of organic amendments are the strategies to restore degraded soils and improve soil fertility, carbon sequestration, and yield levels (Blanco and Lal, 2008). These practices also suppresses plant-parasitic root knot nematode population, attracts predators and beneficial insects of major insect-pests, and lowers weed germination followed by poor establishment of emerging weeds, which proves favorable in organic crop production system (Bhan, 2010).

Mitigation Potential of Organic Agriculture

Scialabba and Müller-Lindenlauf (2010) described the mitigation potential of organic agriculture to reduce GHGs emissions and enhance carbon sequestration in the soil as mentioned in the Table 3. It also has a potential for building resilience of uncertainties due to climatic variability with adaptative approaches (Table 4) that may be helpful for future research prospects and developing public policies.

Mitigating N_2O Emission From Soils: N_2O emission contributes 38 % of agricultural GHGs emissions (Smith *et al.*, 2007) of which 1 % as direct N_2O emission from applied nitrogenous fertilizers. In other publications, 3.5 kg N_2O-N emitted per 100 kg nitrogen based fertilizers (Crutzen *et al.*, 2007). These emissions can be reduced in organic farming system by adding catch or cover crops that have the capacity to extract stored nitrogen from the soil that was not used by the previous crops. Incorporation of cover crops back into the soil used to keep and sustain nitrogen into the cropping system. However, study suggests higher N_2O emission after manure application compared to mineral fertilizer application due to higher oxygen consumption for decomposition of organic matter (Flessa and Beese, 2000). These can be reduced by improving soil aeration either by lowering bulk density, recue incorporation of legumes, and increase tillage practices (as no-tillage cause low aeration with more release of N_2O gas.

Carbon Sequestration For Mitigating Climate Change: Lal (2008) implies soil carbon sequestration is the transfer of atmospheric CO_2 into long-lived soil pools and storing it securely so that it may not be reemitted back instantly. It means increasing soil organic and inorganic carbon stocks through judicious land use with best recommended management practices. The global soil carbon pool of 2500 gigatons (Gt) include 1550 Gt of soil organic carbon and 950 Gt of soil inorganic carbon (Lal, 2004). Of which, approximately, global mitigation potential of organic agriculture accounts 3.5–4.8 Gt CO_2 through carbon sequestration (Niggli *et al.*, 2008). Organic farming increases carbon sequestration in the soil (Freibauer *et al.*, 2004) as depletion of soil organic carbon degrade soil quality, reduces biomass productivity, and adversely impact

Table 3: Mitigation potential of organic agriculture

Source of GHG	Share of total anthropogenic GHG emissions	Impacts of optimized organic management	Remark
Direct emissions from agriculture	10-12%		
N_2O from soils CH_4 from enteric fermentation	4.2%3.5%	Reduction opposed effects	Higher nitrogen use efficiency Increased by lower performance and lower energy concentration in the diet but reduced by lower replacement rate and multi-use breeds
Biomass burning Paddy rice	1.3%1.2%	Reduction Opposed effects	Burning avoided according to organic standards Increased by organic amendments but lowered by drainage and aquatic weeds
Manure handling	0.8%	Equal	Reduced methane emissions but no effect on N2O emissions
Direct emissions from forest clearing for agriculture	12%	Reduction	Clearing of primary ecosystems restricted
Indirect emission Mineral fertilizers Food chain	1%	Totally avoided (Reduction)	Prohibited use of mineral fertilizers Inherent energy saving but still inefficient distribution systems
Carbon sequestration Arable lands Grasslands		Enhanced Enhanced	Increased soil organic matterIncreased soil organic matter

Source: Scialabba and Müller-Lindenlauf (2010)

water quality, and this depletion increase the cause of global warming (Lal, 2004). Cultural practices like no-tillage improves carbon sequestration however it also increases N_2O emissions. Beside this, carbon stored through no-tillage is released by single ploughing due to its labile quality (Stockfisch *et al.*, 1999) that showed that removal of GHGs from the atmosphere through carbon sequestration is limited. Most of the soil-sequestrated carbon is stored as soil organic matter, which is higher in organically managed soil. Soil organic matter has high water-capturing capacity (strengthening resilience to drought and reduce risk of floods) enhances nutrient buffer capacity, and microbial activity (Lotter *et al.*, 2003).

Drought Tolerance Under Organic Systems

It was found that water capture in organic plots was twice as high as in conventional plots during torrential rains (Lotter *et al.*, 2003). Similarly, Pimentel *et al.* (2005) reported that the amount of water percolating through the top 36 cm was 15-20% greater in the organic systems compared to conventional systems. The organic soils held 816,000 litres per ha in the upper 15 cm of soil. This water reservoir was likely the reason for higher yields of corn and soybean in dry years. In India, most of the organic cotton farmers stated that the capacity of their soils to absorb and retain water was increased after conversion to organic management (Eyhorn *et al.*, 2009). Many farmers also said that they need less rounds of irrigation and the crops can sustain longer periods of drought. In Central America, farmers using organic and sustainable methods reported substantially lower economic losses, and 90% of the neighbours of the study farms indicated a desire to adopt their neighbours' methods after observing the environmental stress tolerance of the organic/sustainable farms (Holt-Gimenez, 2002). In the 21-year Rodale farming systems trial, in which two organic and a conventional crop rotation were compared, the organic crop systems performed significantly better in 4 out of 5 years of moderate drought. In the severe drought year of 1999, three out of the four crop comparisons resulted in significantly better yields in the organic systems than the conventional (Lotter *etal,* 2003).

Plant water uptake and ability to withstand drought are significantly improved by mycorrhizal associations (Syliva and Williams, 1992). Mycorrhizae have been shown to be more abundant in the roots of crops from organically managed systems relative to those of conventionally managed crops (Eason *et al.*, 1999; Mader *et al.*, 2000). This suggests both a physicochemical and biological basis for the increased drought tolerance of organic cropping systems (Lotter *et al.*, 2003).

Table 4: Adaptation potential of organic agriculture

Objectives	Means	Impacts
Objectives	Means	Impacts
Alternative to industrial production inputs (i.e., mineral fertilizers and agrochemicals) to decrease pollution	Improvement of natural resources processes and environmental services (e.g., soil formation, predation)	Reliance on local resources and independence from volatile prices of agricultural inputs (e.g., mineral fertilizers) that accompany fossil fuel hikes
In situ conservation and development of agro biodiversity	Farm diversification (e.g., polycropping, agroforestry and integrated crop/livestock) and use of local varieties and breeds	Risk splitting (e.g., pests and diseases), enhanced use of nutrient and energy flows, resilience to climate variability and savings on capital-intensive seeds and breeds
Landscaping	Creation of micro-habitats (e.g., hedges), permanent vegetative cover and wildlife corridors	Enhanced ecosystem balance (e.g., pest prevention), protection of wild biodiversity and better resistance to wind and heat waves
Soil fertility	Nutrient management (e.g., rotations, cover crops and manuring)	Increased yields, enhanced soil water retention/ drainage (better response to droughts and floods), decreased irrigation needs and avoided land degradation

Source: Scialabba and Müller-Lindenlauf (2010)

Methane Emission From Organically Planted Rice Cultivation

Methane (CH_4) emitted from paddy field accounts for 11 % of global agricultural GHG emissions, with organic amendments increases emissions while drainage reduces emissions (Neue *et al*., 1996; Yang and Chang, 2001). Organic systems add more organic amendments but these should be added during the time of drainage to avoid higher CH_4 emission (Cai and Xu, 2004). On the other side, aquatic weeds observed more in the paddy fields and have the tendency to reduce methane emissions (Inubushi *et al*., 2001). Further research is necessary for recommending the quantity of organic amendments and the levels of aquatic weeds in the paddy fields. System of Rice Intensification (SRI) also is a promising technique in paddy management in which field were kept un-flooded most of the time thereby reducing levels of CH_4 emissions (Dobermann, 2004).

Integrated Organic Livestock Management

Methane emissions from enteric fermentation accounts for 4-5 % of global anthropogenic GHGs emissions (IPCC, 2007). Feeding high energy feedstuff (e.g. grains and soya) can reduce emissions as methane emissions mainly derive from the digestion of fiber from roughage (Pelchen, 1996). However, it neglects the unique ability of ruminants to digest roughage, and also impact food security if crops given as feed rather than food to humans. Furthermore, high energy diets may lead to rumen acidification and secondary inflammations that leads to animal illness (Plaizier *et al*., 2009).

Basic enteric fermentation by cattle leads to CH_4 and N_2O emissions from manure that accounts for about 7 % of agricultural GHG emission (Smith *et al*., 2007). Liquid manure emits more CH_4 while N_2O emitted higher through solid manure systems. Methane emission can be reduced by fermenting the slurry in biogas plants, which not only helps in generating renewable energy but also in line with the organic standards (Smith *et al*., 2007). In addition, pastures are the feeding strategy for organic cattle. Therefore, organic livestock management is an option for maintaining grasslands having high carbon sequestration potential (Neill *et al*., 1997).

Beside this, successful organic farming include an integration of both plant and livestock production to optimize nutrient use and recycling. Manure comes through livestock system should be utilized as an impact in crop production, a mitigation strategy recommended by IPCC to reduce N_2O emission and leaching (Smith *et al*., 2007). A nutrient excess on the farm leads to high N_2O emission whereas manure application require more energy costs for transportation.

A case study showed reduction in CH_4 emission levels by 20 % through milk and beef production by keeping double-use breeds i.e. for milk and meat production. Double-use breeds are avoided in conventional systems due to more inputs with lower milk yield however less yield losses was observed in organic systems as less external inputs required in managing double-use breeds (Scialabba and Müller-Lindenlauf, 2010).

Conclusion

Climate change is a reality due to elevated carbon dioxide concentration with increase in temperature and humidity. Organic agriculture has the capacity to reduce GHGs emissions and also enhance soil carbon sequestration. However, more data is needed to confirm these facts by establishing long-term trials at different types of soils, climates, and practices.

Similarly, most of the research work has been carried out in the developed countries. There is a need for further investigation of mitigation potential in the developing countries residing in tropical and sub-tropical areas to unlock the potential of organic agriculture and its application to a larger scale. Finally, organic agriculture has a strong potential for building resilience in the face of uncertainties arise from climatic variability and these approaches should be focused in developing public policies and research. However, organic agriculture proved disadvantageous in competition with the other strategies. Unlike conservation agriculture that offer one technology, organic agriculture follows a site-specific farming-system approach with a whole set of technological changes. Monitoring and impact assessment with respect to carbon sequestration (or GHG avoidance) are somehow complicated and expensive. Overall, strategic research is needed followed by technological demonstration to enhance the resilience of crop improvement, natural resource management, livestock, fisheries, etc. Beside this, capacity building of scientists, SHGs, NGOs, and other stakeholders related to agricultural research may also help to avert climate change problems.

References

Aggarwal, P. K. 2003. Impact of climate change on Indian agriculture. J. Plant Biology 30(2): 189-198.

Aggarwal, P.K., 2008. Global climate change and Indian agriculture: impact, adaptation and mitigation. Indian J. Agric. Sci. 78, 911–919.

Andrasko, K., Benitez-Ponce, P., Boer, R., Dutschke, M., Elsiddig, E., Ford-Robertson, J., Frumhoff, P., Karjalainen, T., Krankina, O., Kurz, W., Matsumoto, M., Oyhantcabal, W., Ravindranath, N.H., Sanz Sanchez, M.J., and Zhang, X. 2007. Forestry. In B. Metz, O.R. Davidson, P.R. Bosch, R. Dave, L.A. Meyer, (eds). Climate Change 2007: Mitigation. Contribution of Working Group III to the Fourth Assessment Report of the Intergovernmental Panel on Climate Change. Cambridge University Press, Cambridge, UK.

Augstburger, F., Berger, J., Censkowsky, U., Heid, P., Milz, J., Streit, Ch., Panyakul, V., den Braber, K., and Naturland. 2002. Organic Farming in the Tropics and Subtropics. Exemplary Description of 20 Crops—Rice. Naturland e.V., with support of German Agency for Technical Cooperation. Gra¨felfing, Germany.

Badgley, C., Moghtader, J., Quintero, E., Zakem, E., Chappell, M.J., Avile´s-Va‘quez, K., Samulon, A., and Perfecto, I. 2007. Organic agriculture and the global food supply. Renewable Agriculture and Food Systems 22:86–108.

Bhan, M. 2010. Cropping system complexity for suppressing pests in organic vegetable production.Ph.D. Dissertation, University of Florida, Gainesville, FL.

Bhardwaj, V., Yadav, V., Chauhan, B.S. 2009. Effect of nitrogen application timings and varieties on growth and yield of wheat grown on raised beds. Arch. Agron. Soil. Sci. 56, 211–222.

Blanco, H. and Lal, R. 2008. Chapter 15: Restoration of eroded and degraded soils. In Principles of Soil Conservation and Management. Springer, Dordrecht, Netherlands.

Bhatia, V. S., Mall, R. K., Lal, M., Rathore, L. S. and Singh, R.2004. Mitigating climate change impact on Soybean productivity in India: A simulation study Agricultural and Forest Meteorology, 121 (1–2), pp. 113–125.

Cai, Z.C. and Xu, H. 2004. Options for mitigating CH4 emissions from rice fields in China. In Y. Hayashi (ed.).Material Circulation through Agro-Ecosystems in East Asia and Assessment of Its Environmental Impact. NIAES Series 5, Tsukuba, Japan. p. 45–55.

Chauhan, B.S., Prabhjyot-Kaur, Mahajan ,G., Randhawa,R. K., Singh,H., Kang, MS. 2014.Global Warming and its Possible Impact on Agriculture in India. In Donald L. Sparks, editor: Advances in Agronomy, Vol. 123, Burlington: Academic Press, 2014, pp. 65-121.

Codex Alimentarius Commission. 2001. Guidelines for the Production, Processing, Labelling and Marketing of Organically Produced Foods. First Revision. Joint Food and Agriculture Organisation (FAO) and World Health Organisation (WHO) Food Standards Program, Rome, Italy. Available at Web site http://www.codexalimentarius.net/download/standards/360/CXG_032e.pdf (verified 15 October 2009).

Cole CV, Duxbury J, Freney J, Heinemeyer O, Minami K, Mosier A, Paustian K, Rosenberg N, Sampson N, Sauer-beck D and Zhao Q. 1997. Global estimates of potential mitigation of greenhouse gas emissions by agriculture. Nutrient Cycling in Agroecosystems 49. 221-228.

Crutzen PJ, Mosier AR, Smith KA and Winiwarter W. 2007. N20 release from agro-biofuel production negates global warming reduction by replacing fossil fuels. Atmospheric Chemistry and Physics Discussions, 7:11191-11205.

Dobermann, A. 2004. A critical assessment of the system of rice intensification (SRI). Agricultural Systems 79:261–281.

Drinkwater, L.E., Wagoner, P. and M. Sarrantonio, M. 1998. “Legume-Based Cropping Systems Have Reduced Carbon and Nitrogen Losses,” Nature 396: 262–65.

Eason WR, Scullion J and Scott EP. 1999. Soil parameters and plant responses associated with arbuscular mycorrhizas from contrasting grassland management regimes. Agriculture Ecosystems and Environment, 73(3): 245-255.

Erisman, J.W., Sutton, M.A., Gallowaz, J., Klimont, Y., and Winiwarter, W. 2008. How a century of ammonia synthesis changed the world. Nature Geoscience 1:636–639.

Eyhorn, F., Ramakrishnan, R. and Mäder,P . 2007. The Viability of Cotton-Based Organic Farming Systems in India, International Journal of Agricultural Sustainability 5(1): 25–38.

Eyhorn F, Mader P and Ramakrishnan M. 2009. The impact of organic cotton farming on the livelihoods of smallholders: Evidence from the MaikaalbioRe project in central India. Organic Farming Newsletter, 5(1): 19-21.

FAO .2002. Organic Agriculture, environment and food security. Environment and Natural Resources Series 4, Rome, Italy.

FAOSTAT. 2009. FAO Statistical Database Domain on Fertilizers: ResourceSTAT-Fertilizers. Food and Agriculture Organisation of the United Nations (FAO) Rome, Italy. Available at Web site http://faostat.fao.org/site/575/default. aspx#anchor (accessed 7 October 2009).

Flessa, H. and Beese, F. 2000. Laboratory estimates of trace gas emissions following surface application and injection of cattle slurry. Journal of Environmental Quality 29:262– 268.

Freibauer, A., Rounsevell, M.D.A., Smith, P., and Verhagen, J. 2004. Carbon sequestration in the agricultural soils of Europe. Geoderma 122:1–23.

Goh KM. 2011. Greater mitigation of climate change by organic than conventional agriculture: A review. Biological Agriculture & Horticulture, 27(2): 205-229.

Haas G and Köpke U 1994. Vergleich der Klimarelevanz ökologischer und konventioneller Landbewirtschaftung. In: Enquete Kommission Schutz der Erdatmosphäre des Deutschen Bundestages (ed): Schutz der Grünen Erde, Klimaschutz durch umweltgerechte Landwirtschaft und Erhalt der Wälder. Bonn, Economica Verlag.

Holt-Gimenez E. 2002. Measuring farmers' agroecological resistance after Hurricane Mitch in Nicaragua: a case study in participatory, sustainable land management impact monitoring. Agriculture Ecosystems and Environment, 93:87¬105.

IFOAM. 2008. FAO Workshop on Organic Agriculture and Climate Change. Workshop at the 6* IFOAM Organic World Congress, Modena, Italy, June 18,2008.

Inubushi, K., Sugii, H., Nishino, S., and Nishino, E. 2001.Effect of aquatic weeds on methane emission from submerged paddy soil. American Journal of Botany 88(6): 975–979.

IPCC .2001. Climate Change 2001: The Scientific Basis. Contribution of Working Group I to the Third Assess-ment Report of the Intergovernmental Panel on Climate Change [Houghton, J.T.,Y. Ding, D.J. Griggs, M. Noguer, P.J. van der Linden, X. Dai, K. Maskell, and C.A. Johnson (eds.)]. Cambridge University Press, Cambridge, United Kingdom and New York, NY, USA, 881 pp.

IPCC. 2007. Contribution of Working Group 1 to the Fourth Assesment Report of the Inter Governmenttal Panel on Climate Change. pp 996. Solomom S, Qin D, Manning M, Chen Z, Marquis M, Averyt KB, Tignor M and Miller HC (Eds).Cambridge University Press, Cambridge, UK and Newyork, USA.

Kotschi, J., and Müller-Sämann, K. 2004. The Role of Organic Agriculture in Mitigating Climate Change: A Scoping Study. IFOAM, Bonn, Germany. Available at http://www.ifoam.org/press/positions/Climate_study_green_house-gasses.html. Accessed February 12, 2009.

Lal, R., 2004. Soil carbon sequestration impacts on global climate change and food security. Science 304, 1623–1627.

Lotter, D. 2003. Out of the Ashes of the Coffee Crash, Costa Rica Organic is Born. The New Farm, Rodale Institute, Rodale, PA, USA. Available at Web site: http://www.newfarm.org (verified 15 October 2009).

Mader P, Edenhofer S, Boiler T, Wiemken A and Niggli U. 2000. Arbuscular mycorrhizae in a long-term field trial comparing low-input (organic, biological) and high-input (conventional) farming systems in a crop rotation. Biology and Fertility of Soils, 31 (2): 150-156.

Mall, R.K., Singh, R, Gupta, A., Srinivasan, G. and Rathore, L. S. 2006. Impact of climate change on Indian agriculture: A review Climate change: 78:445-478.

Neill, Ch., Melillo, J.M., Steudler, P.A., Cerri, C.C., de Moraes, J.F.L., Piccolo, M.C., and Brito, M. 1997. Soil carbon and nitrogen stocks following forest clearing for pasture in the south western Brazilian Amazon. Ecological Applications 7(4):1216–1225.

Neue, H.U., Wassmann, R., Lantin, R.S., Alberto, M.C.R.,Aduna, J.B., and Javellana, A.M. 1996. Factors affecting methane emission from rice fields. Atmospheric Environment 30:1751–1754.

Niggli, U., Fliessbach, A., Hepperly,P. and Scialabba, N. 2008. "Low Greenhouse Gas Agriculture: Mitigation and Adaptation Potential of Sustainable Farming Systems." Rome: FAO. (Similar to Niggli *et al.* 2007)

Plaizier, J.C., Krause, D.O., Gozho, G.N., and McBride, B.W. 2009. Subacute ruminal acidosis in dairy cows: The physiological causes, incidence and consequences. The Veterinary Journal 176:21–31.

Pelchen, A. 1996. Dynamik von Methanemissionen landwirtschaftlicher Nutztiere unter dem Einfluss verschiedener Fütterungssysteme—eine Modellbetrachtung zum Treibhauseffekt. PhD thesis, Humbolt-University, Berlin, Germany.

Pimentel, D., Hepperly, P., Hanson, J., Douds, D., and Seidel, R. 2005. Environmental, energetic and economic comparison of organic and conventional farming systems. Bioscience 55: 573- 582.

Pimentel, D. 2006. Impacts of Organic Farming on the Efficiency of Energy Use in Agriculture. The Organic Center, Cornell University, Ithaca, NY.

Sauerbeck D . 2001. CO2 Emissionen und Kohlenstoff-Senken im Landbau – Möglichkeiten und Grenzen ihrer Beeinflussung. In: Rosenkrantz D, Bachmann G, König W and Einsele G (eds.): Bodenschutz Vol 2, 4130, Berlin, 31 S.

Scialabba, N.E. and Müller-Lindenlauf, M. 2010. Organic agriculture and climate change. Renewable Agriculture and Food Systems: 25(2); 158–169.

Smith, P., Martino, D., Cai, Z., Gwary, D., Janzen, H.,Kumar, P., McCarl, B., Ogle, S., O'Mara, F., Rice, C., Scholes, B., and Sirotenko, O. 2007. Agriculture. In B. Metz, O.R. Davidson, P.R. Bosch, R. Dave, and L.A. Meyer (eds). Climate Change 2007: Mitigation. Contribution of Working Group III to the Fourth Assessment Report of the Intergovernmental Panel on Climate Change. Cambridge University Press, Cambridge, UK.

Status of Indian Agriculture. 2013. Directorate of Economics and Statistics. Department of Agriculture and Cooperation. Ministry of Agriculture, Government of India, New Delhi. Stockfisch, N., Forstreuter, T., and Ehlers, W. 1999.Ploughing effects on soil organic matter after twenty years of conservation tillage in Lower Saxony, Germany. Soil and Tillage Research 52:91–101.

Syliva DM and Williams SE. 1992. Vesicular-arbuscular mycorrhizae and environmental stress. In: Mycorrhizae in Sustainable Agriculture (GJ Bethenfalvay and RG Linderman, Eds.), Proceedings of a symposium, 31 October 1991. American Society of Agronomy, Crop Science Society of America, and Soil Science Society of America, Madison Wisconsin, pp 101-124.

Williams, A.G., Audsley, E., and Sandars, D.L. 2006. Determining the Environmental Burdens and Resource Use in the Production of Agricultural and Horticultural Commodities. Main Report. Defra Research Project IS0205. Cranfield University, Bedford, and the Department for Environment, Food and Rural Affairs (Defra) of the United Kingdom Government, London. Available at Web site http:// www.silsoe.cranfield.ac.uk, and www.defra.gov.uk (verified 15 October 2009).

Yang, S.S. and Chang, H.L. 2001. Effect of green manure amendment and flooding on methane emission from paddy fields. Chemosphere—Global Change Science 3:41-49.

Zehnder, G., Gurr, G.M., Kühne, St., Wade, M.R., Wratten, St. D., and Wyss, E. 2007. Arthropod pest management in organic crops. Annual Review of Entomology 52:57-80.

GREENHOUSE GAS EMISSIONS AND INDIAN AGRICULTURE

11

Greenhouse Gas Emissions in Rice and its Mitigation Strategies

P. Bhattacharyya, K.S. Roy, Ch.K. Swain and A.K. Nayak

Introduction

Greenhouse gases (GHGs) are the atmospheric trace gases which absorb and emit infra-red radiation and thereby trapping the heat radiated from the earth surface. The commonly known agriculturally important GHGs are carbon dioxide (CO_2), methane (CH_4) and nitrous oxide (N_2O). The major anthropogenic sources of CO_2 emission in atmosphere are biosphere, burning of fossil fuels and land use changes in tropical forests. The atmospheric concentration of CO_2 has increased from 280 parts per million by volume (ppmv) in 1750 to 394 ppmv in 2012 and is currently increasing at the rate of 1.9 ppmv year-1 (NOAA, 2012). Atmospheric CH_4 concentration has increased from about 715 to 1826 parts per billion by volume (ppbv) in 2012 over the same period and is increasing at the rate of 7 ppbv year-1 (IPCC, 2007; EPA, 2013). Similarly, the atmospheric concentration of N_2O has increased from about 270 ppbv in 1750 to 325 ppbv and is increasing at the rate of 0.8 ppbv year^{-1} (IPCC, 2007; EPA, 2013).

Agriculture is considered as one of the major anthropogenic sources of atmospheric GHGs (Lal, 2000). Carbon dioxide is the most important anthropogenic GHG and is primarily originated from industrial activities, deforestation, burning of fossil fuels, land use changes and microbial decomposition of soil organic matter. On the other hand, the increase in CH_4 concentration is predominantly due to agriculture and fossil fuel use while the increase in N_2O is mostly due to agriculture.

The variation of CO_2 concentrations in terrestrial ecosystem is determined by the balance between sources and sinks (Vinogradova *et al.*, 2007). Photosynthesis and respiration processes have to be linked with fossil fuel combustion for energy purposes and land-use changes (Buchmann, 2000; Artuso *et al.*, 2009) for terrestrial CO_2 budgeting. Methane has relatively higher infra-red absorption and radiation capacity than CO_2 and contributes greater extent in atmospheric chemistry and global warming (IPCC, 2007). Methane is presently the second most important GHG accounting for 15-20% of the anthropogenic radiative forcing. The major sources of CH_4 emission are the rice paddies, ruminants, landfills, natural wetlands and sediments (Zhu *et al.*, 2007). Tropospheric CH_4 has increased as a result of human activities related to agriculture, natural gas distribution and landfills. Although, the tropospheric CH_4 is increasing continuously, the rate of increase of CH_4 emission has declined during in past two decades (IPCC, 2007). Flooding of rice fields creates anaerobic soil conditions which are conducive for production of CH_4 (Neue, 1993; Tseng *et al.*, 2010). Nitrous oxide is generated by the microbial transformation of nitrogen in soils and manures, and is often enhanced where available nitrogen exceeds plant requirements, especially under wet conditions (Oenema *et al.*, 2005). The rice paddies act as sources of major N_2O emission upon nitrogenous fertilizer (e.g. urea) application. Nitrogenous fertilizer appears to be the single most important factor controlling N_2O emission from flooded rice fields.

GHGs emissions in rice

Annual global estimation of CH4 emission from flooded rice fields accounted for 7.08 Tg based on the biomass (Sinha, 1995). Wassmann *et al.* (2000) estimated 99 Kg CH_4 ha season from deepwater rice fields in Thailand. In -1^{-1} India the mean CH_4 emission from rice fields ranged between 3.5-4.2 Tg $yr^{-1}4$ (Parashar *et al.*, 1996). Bhatia *et al.* (2004) estimated 4.7 Tg yr^{-1} CH emission from the Indian paddy fields with the highest emission of 1.379 Tg yr^{-1} from the irrigated rice fields. On annual basis the net ecosystem methane exchange (NEME) was +262.62 kg CH_4 ha in tropical flooded rice ecology in Eastern-1 India (Bhattacharyya *et al.*, 2014).

Agriculture related activities account for around 90% of the total N2O emissions in India (Garg *et al.*, 2001). Parashar *et al.* (1998) estimated the total N_2O emission from Indian paddy and wheat fields were 199-279 Gg per annum. Sharma *et al.* (1995) estimated N_2O emissions from irrigated and upland paddy fields of India at 4-210 and 2-10 Gg yr^{-1}, respectively. Nitrous oxide emission from Indian agricultural field was estimated to be 0.08 Tg annually (Bhatia *et al.*, 2004; Pathak *et al.*, 2005). The irrigated rice-wheat system is a significant source of N O, emitting around 15 kg N O-N ha^{-1} yr^{-1} (Aulakh *et al.*, 2001). Chao *et al.* (2000) estimated around 0.67 Mg N O-N yr^{-1} from the paddy fields of Taiwan.

The net ecosystem CO_2 exchange (NEE) was predominantly negative (CO_2 sink) in daytime and positive (CO_2 source) in night hours during the cropping season as evident from the study from CRRI. Detailed studies conducted over the seasonal, annual and inter-annual scales revealed that the low land submerged rice ecology behaved as net carbon sink (Bhattacharyya *et al.*, 2014). The total CO_2-C budget integrated over the entire cropping period showed that the net ecosystem CO_2 exchange (NEE) in flooded rice fields (-258 g C m^{-2}) was about three times higher than that of aerobic rice fields (-85 g C m^{-2}) in IRRI, Philippines (Alberto *et al.*, 2009).

Quantification of CH_4, CO_2 and N_2O emission from rice

1. CH_4 emission from Indian rice field = 4.7 Tg yr^{-1}
2. Indian rice contributed 13.1% of total global CH_4 emission from agriculture
3. CH_4 emission from rainfed rice field in India = 40-70 kg ha^{-1} yr^{-1}
4. CH_4 emission from irrigated rice-rice field in India = 128-180 kg ha^{-1} yr^{-1}
5. CH_4 emission from irrigated field = 1.38 Tg yr^{-1}
6. In India irrigated rice contributed 30% and rainfed rice 70% of total CH_4 emission
7. Global N_2O emission from rice = 3.6 Tg $year^{-1}$
8. N_2O emission from Indian rice field = 0.05 Tg $year^{-1}$
9. Indian rice contributed 1.4% of total global N_2O emission from rice
10. Global CO_2 emission = 10 Pg $year^{-1}$
11. CO_2 emission from India= 16-48 Tg year-1

Technologies for GHGs monitoring and measurement

The GHG emissions from different major production systems should be monitored continuous and quantified precisely for budgeting. Manual as well as automatic and sensors based technologies are available for monitoring of GHG emission from rice fields. Conventionally, sampling of CH_4, CO_2 and N_2O gas emissions from rice is done with the help of static manual close chamber (photo 1& 2) and automatic chamber method. After gas collection the subsequent analysis of CH4, CO_2 and N_2O gas is done by gas chromatography (photo 3). Soil and plant canopy respiration chambers (infrared gas analyzer (IRGA)) are also used for *in situ* CO_2 measurement. Chamber methods integrate over small areas, and can alter local environmental conditions. Though the general errors caused by chambers are related to perturbations of the natural conditions at the sampling site, modifications of the microclimate, pressure-induced gas flows in open chambers and inhibiting effects of concentration build-up in closed chambers; but this methodology is widely followed for GHG flux study from agro-ecologies with little bit of site-and-component specific modifications. More recent and advanced techniques of GHGs emission monitoring and quantification are done by Eddy covariance techniques (Photo 4). This provides precise real time continuous GHGs fluxes on ecosystem basis.

Photo 1&2. Manual Chambers for greenhouse gas collection from rice field.

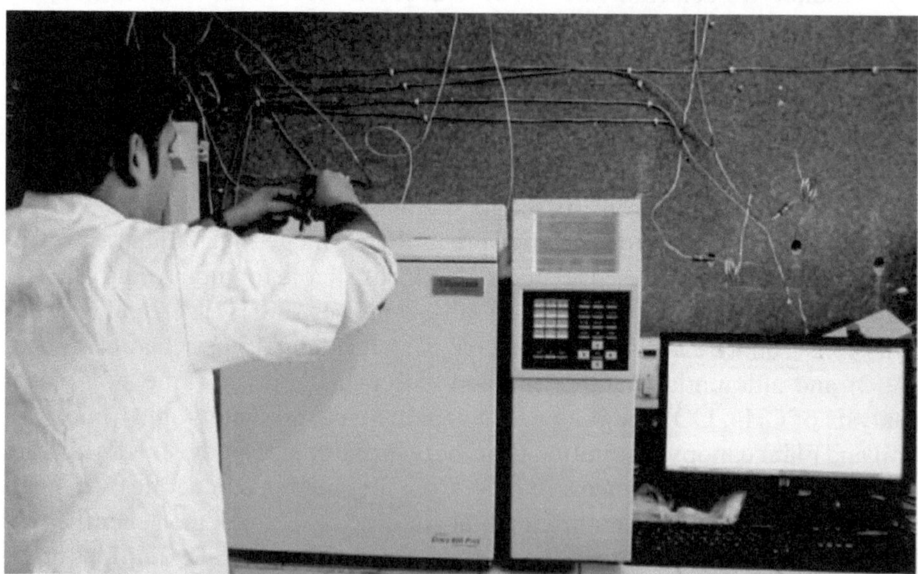

Photo 3: Analysis of greenhouse gases in Gas Chromatography

Photo 4: Analysis of greenhouse gases by Eddy covariance system

Principles of mitigation to reduce and avoid emissions

The principle of mitigation for reducing GHGs emissions from agriculture directly or indirectly depends on the mechanism of production and transformation of GHGs in soil and their emission to the atmosphere. Depending upon the detailed mechanisms and factors involved there in the process, one could be able to successfully formulate the mitigation technologies for different crop production systems grown under varied agro-ecologies.

Mitigation of carbon dioxide production from rice field

Thus, the principle of mitigate CO_2 emission from crop field to atmosphere should be focused on:

- Development of climate resilient crop varieties either through conventional breeding or by means of modern biotechnological tools, which would have the potential to harness more CO_2 in presence of ample sunlight or even under dim photosynthetically active radiation, and ability to transfer it into stable forms of C in the soil system.
- Manipulation for slow down the decomposition rate and process of the SOM. Controlling the soil heterotrophic respiration by means of management practices or by introduction of new amendments is the essence of mitigation principle.

The key principle to mitigate CH_4 emission from crop field to atmosphere should be focused on:

- To regulate readily decomposable and labile forms of C sources (from native soil nutrients, addition of external C sources like manure, rhizodeposits etc.) in the soils which could be used as substrate by methanogens for CH_4 production.
- Continuous and prolonged submergence in rice fields creates typical anaerobic condition in soil resulting into lowering down soil redox potential (Eh) below -200 mV, a situation favourable and conducive for methane production by the methanogens. So, periodical drying of soil in rice fields would help to create a situation where the redox potential rarely dips below -200 mV in absence of continuous submergence.
- Management practices should address the issues like supply of electron acceptors (Fe, S etc) in sufficient amount in such a way that Eh of the soil remain above -200mV.
- Development or introduction of rice cultivar with less CH_4 transport potential.
- Mid-season drainage substantially reduces CH_4 emissions by about 30-50% as compared to continuous flooding or water logging.

Mitigation of nitrous oxide production from rice field

The adaptation and mitigation options for reducing N_2O emissions that are widely adopted are: matching N supply with crop demand, use of controlled release fertilizers, placement and source of fertilizer, use of nitrification inhibitors, water and tillage management, biochar application etc.

- Application of nitrogen in splits as per crop requirement is an important strategy to improve N use efficiency, minimization of N loss and regulation of N_2O emission from the crop field.
- Site-specific nutrient management approach that includes site-specific quantitative knowledge of crop nutrient requirements, indigenous nutrient supply and the recovery efficiency of applied fertilizer nitrogen ensures about 30-40% increase in nitrogen use efficiency, has the potential of reducing N_2O-N emission from crop field.
- The N_2O losses from polyolefin-coated ammonium nitrate with uncoated ammonium sulphate are found to 3 to 7-fold less reduction in the emission of N_2O from arable soil.

- Addition of nitrification inhibitors such as nitrapyrin, 2-chloro-6-(trichloromethyl)- pyridine, dicyandiamide and wax-coated calcium carbide to soil after fertilizer application significantly reduces fertiliser induced loss of N_2O.
- Tillage practices like zero tillage or minimum tillage, bed planting, modifies N_2O emission through their impact on compaction, drainage, and aeration status of soil.

Some case studies of GHGs mitigation in rice Farming system approach

The emission of CH_4 and N_2O in a rice-fish farming system was evaluated. The four fish species namely, *Cirrhinus mrigala* H. (mrigal), *Labeo rohita* H. (rohu) , *Cyprinus carpio* L., (common carp) and *Catla catla* H. (catla) were introduced along with rice (*Oryza sativa* L.) cv. Varshadhan, in a rainfed shallow (0-25cm water depth) low land (Fig. 1). On seasonal basis, the emission of CH4 was significantly higher by 26% under rice-fish co-culture compared to rice alone. The maximum emission of CH4 (36% higher) was found under rice + common carp farming (Table 1). On contrary, the emission of N_2O was significantly lower by 9% under rice-fish compared to rice alone. Although the global warming potential (GWP) was highest (29.5% more) in rice + common carp, the CH_4 and N_2O emission from this treatment per unit of total yield (rice + fish) were at par and significantly lower, respectively, compared to rice alone. Further, it provided significantly higher yield (rice+fish) and economic benefit over rice and other rice + fish species combinations. The integration of common carp with rice in rainfed shallow lowland ecology is thus, one of the viable options for livelihood of the small and marginal farmers besides, higher carbon (C) credit compliance. However, Indian major carps, mrigal rohu and catla in order, are potentially less harmful in terms of GHGs emission in shallow rice fields besides, higher profitability and carbon credit compliance compared to rice farming alone.

The cultivation of fish along with rice in rainfed shallow lowland resulted in an increased emission of CH_4 and decreased release of N_2O irrespective of the fish species. However, considering the economic benefits of growing fish in rice fields, rice-fish farming with common carp can be an important crop management system, as evident from the additional fish output, an important source of protein to the marginal farmers and higher net profit that underwrites the increased carbon credit compliance of a rice-fish ecosystem due to larger cumulative GWP. Culture of the Indian major carps (catla, mrigal and rohu) in lowland rice fields are also alternate profitable farming system over rice alone with additional advantage of comparatively less cumulative GWP (Bhattacharyya *et al.*, 2013a).

Table 1: Comparison of GWP and other associated parameters among four rice-fish treatments including rice alone system

Treatment	GWP(kg ha^{-1})	CH_4 (kg) t^{-1} rice + fish yield	N_2O-N (kg) t^{-1} rice + fish yield	C-creditcompliance 1(Rs.)	Net profit (Rs.)
R	3126a	26.7a	0.22c	6,375a	29,274a
RF-A	3470b	26.1a	0.16ab	7,089b	41,259c
RF-B	3730b	31.5b	0.18b	7,599c	37,893b
RF-C	4051c	27.8a	0.15a	8,262e	52,326e
RF-D	3873c	31.3b	0.18b	7,905d	43,656d

Here, R= Rice only; RF-A= Rice + Fish (mrigal); RF-B= Rice + Fish (rohu); RF-C= Rice + Fish (common carp); RF-D= Rice + Fish (catla). Means followed by common letters are not significantly ($p<0.05$) different by Tukey HSD test.

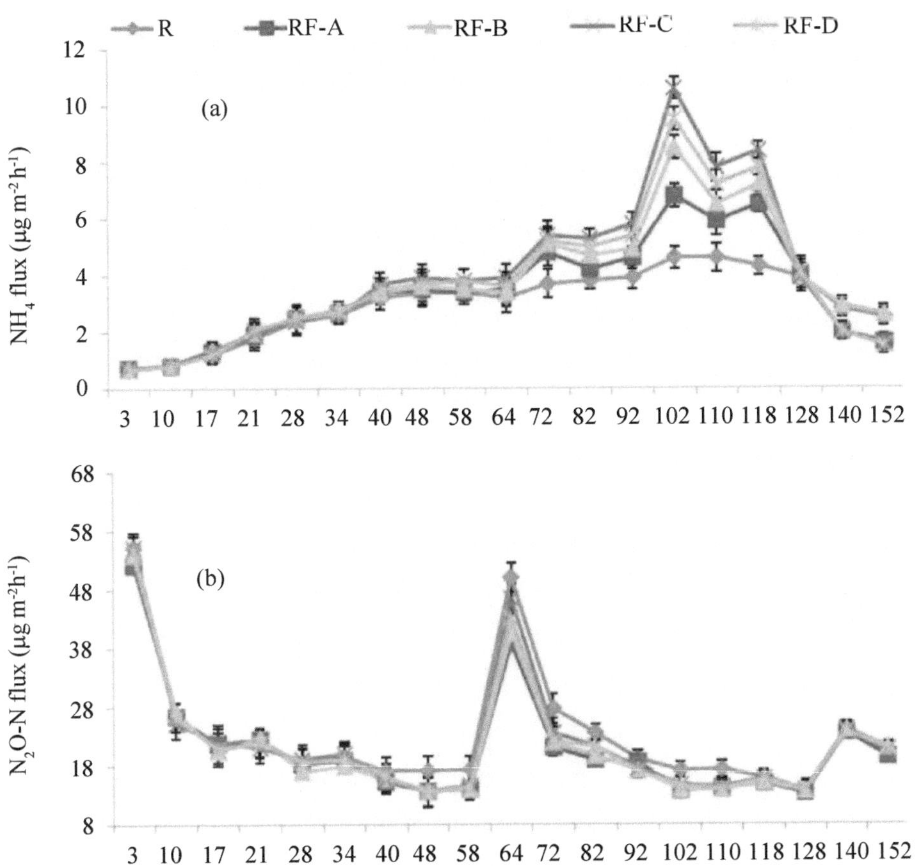

Fig. 1 (a): Methane emission flux (mg CH_4 m^{-2} h^{-1}) and (b) nitrous oxide (N_2O-N) emission flux (µg N_2O-N m^{-2} h^{-1}) under rice and rice-fish farming system. Means of three replicate values plotted, bars indicate the standard deviations. Here, R= Rice only; RF-A= Rice + Fish (mrigal); RF-B= Rice + Fish (rohu); RF-C= Rice + Fish (common carp); RF-D= Rice + Fish (catla)

Adoption of Resource Conservation Technologies (RCTs)

A resource conservation technology (RCT) for soil-nutrient management in low land rice ecology by residue incorporation practice was modified by applying rice straw along with urea (in 1:1 nitrogen (N) basis) that could conserve 30 kg N ha^{-1} yr^{-1}, sequester 0.35 t C ha^{-1} $year^{-1}$and reduce 9% GHGs emission with a sustainable yield of 5.57 t ha^{-1} in low land rice ecology (Table 2). This technique has the potential to save 1800 MJ ha^{-1} yr^{-1} energy based on fertilizer N conversion factor basis. The total soil C increased by 29% under rice straw + urea compared to application of urea alone. The technique is also useful to cut down the GHGs emissions (Bhattacharyya *et al.*, 2012).

Table 2: Greenhouse gas emissions on seasonal basis, GWP, grain yield and carbon storage after the application of inorganic fertilizers and organic manure to flooded soil planted with rice (cv. *Gayatri*)

Treatment	CH_4 emission Kg ha^{-1}	CO_2-C emission Kg ha^{-1}	N_2O-N emission Kg ha^{-1}	GWPKg CO_2 ha^{-1}	Yield Mg ha^{-1}	Carbon storage Mg C ha^{-1} yr^{-1}
Control	69.7a	1100.3a	0.23a	5862a	3.53a	-0.41a
Urea	92.6b	1447.7b	1.00d	8084b	5.13b	0.07b
Rice straw + urea	115.4c	1680.6c	0.84c	9418c	5.57b	0.35c
Rice straw + green manure	122.7c	1858.5d	0.72b	10188d	5.30b	0.22c

Balanced nutrient management

The impact of long term (39 years) integrated nutrient management (INM) (5 t ha^{-1} FYM + NPK) techniques for low land rice ecology revealed that continuous application of chemical fertilizers along with FYM results into higher soil quality index (SQI, 3.78) as compared to control (1.46) and enhanced sustainable yield index (SYI, from 0.39 in control to 0.70 in NPK+FYM). The annual GHGs emission of the system in terms of CH_4, N_2O and CO_2-C were 177.6, 1.82, 1407 kg ha^{-1} yr^{-1}, respectively in long term NPK+FYM application (Table 3). Though the GHGs emissions and GWP were higher under the combined application of FYM+NPK, emissions per unit grain yield were moderate under this treatment. Therefore, the combined application of FYM+NPK is a viable option in managing soil fertility, moderating GHG emissions and sustaining rice yield in tropical flooded soils (Bhattacharyya *et al*., 2013b; Shahid *et al*., 2013).

Conclusions

Greenhouse gas emissions can be mitigated through adoption of environment friendly agricultural operations. The increase in the storage capacity of carbon in agricultural soils through judicious land-use and appropriate nutrient and energy management practices can mitigate the climatic changes in future. The carbon dioxide emission depends on the soil and water management, cropping sequence, planting methods and tillage. Adoption of no tillage or minimum tillage rather than conventional tillage and mulching could reduce CO_2 emissions from soils, retains soil moisture and, thereby, helps to sequester more carbon in soils. Whereas, CH_4 emissions in case of rice could be mitigated by effective water management, use of situation specific rice cultivars, efficient management of iron and sulphur, manipulation of planting method, herbicide usage and effective land-configuration. Nitrous oxide emission primarily depends on source (urea, DAP) and method of nitrogenous fertilizer application in the field. Its emission could be reduced by real tine N management according to crop demand, water and tillage management and use slow release nitrogenous fertilizer. The strategies of reduction of these GHGs emissions, either singly or in combination largely depend upon the situations and ecologies and management practices, which must be standardized for curbing global warming.

Table 3: Annual greenhouse gas emission from flooded rice fields after 39 years of fertilizer and manure application

Treatment	CH_4 emission(Kg ha^{-1})	CO_2-C emission(Kg ha^{-1})	N_2O-N emission(Kg ha^{-1})
Control	93.3[a]	833.3[a]	0.58[a]
Nitrogen	107.3[b]	1034.0[b]	1.42[c]
FYM	128.8[c]	1106.3[c]	1.15[b]
FYM + nitrogen	160.2[d]	1271.6[d]	1.51[d]
NPK	124.1[c]	1210.1[cd]	1.46[c]
FYM + NPK	177.6[e]	1407.0[e]	1.82[e]

References

Alberto, M.C.R., Wassmann, R., Hirano, T., Miyata, A., Kumar, A., Padre, A., Amante, M. (2009). CO2/heat fluxes in rice fields: Comparative assessment of flooded and non-flooded fields in the Philippines. Agricultural and Forest Meteorology. 149: 1737-1750.

Aulakh, M.S., Khera, T.S., Doran, J.W., Bronson, K. (2001). Denitrification, N2O, and CO2 fluxes in rice-wheat cropping systems as affected by crop residues, fertilizer N and legume green manure. Biology and Fertility of Soils. 34: 375-389.

Bhatia, A., Pathak, H., Aggarwal, P.K., 2004. Inventory of methane and nitrous oxide emissions from agriculture soils of India and their global warming potential. Current Science 87, 317-324.

Bhattacharyya, P., Roy, K.S., Neogi, S., Adhya, T.K., Rao, K.S., Manna, M.C. (2012). Effects of rice straw and nitrogen fertilization on greenhouse gas emissions and carbon storage in tropical flooded soil planted with rice. Soil and Tillage Research. 124:119-130.

Bhattacharyya. P., Sinhababu, D.P., Roy, K.S., Dash, P.K., Sahu, P.K., Dandapat, R., Neogi, S., Mohanty, S. (2013a). Effect of fish species on methane and nitrous oxide emission in relation to soil C, N pools and enzymatic activities in rainfed shallow lowland rice-fish farming system. Agriculture, Ecosystems and Environment. 176:53-62.

Bhattacharyya, P., Nayak, A. K., Mohanty, S., Tripathi, R., Shahid, Md., Kumar, A., Raja, R., Panda, B. B., Roy, K. S., Neogi, S., Dash, P. K., Shukla, A. K., Rao, K.S. (2013b). Greenhouse gas emission in relation to labile soil C, N pools and functional microbial diversity as influenced by 39 years long-term fertilizer management in tropical rice. Soil and Tillage Research. 129:93-105.

Bhattacharyya, P., Neogi, S., Roy, K.S., Dash, P.K., Nayak, A.K., Mohapatra, T., 2014. Tropical low land rice ecosystem is a net carbon sink. Agriculture, Ecosystems and Environment 189: 127-135.

Buchmann, N. (2000). Biotic and abiotic factors controlling soil respiration rates in *Picea abies* strands. Soil Biology and Biochemistry 32: 1625-1635.

Chao, C.C., Young, C.C., Wang, Y.P., Chao, W.L. (2000). Daily and seasonal nitrous oxide fluxes in soils from hardwood forest and different agroecosystems of Taiwan. Chemosphere Global Change Scienc. 2:77-84.

Environmental Protection Agency (EPA). (2013). Climate Change Indicators in the United States: Atmospheric Concentrations of Greenhouse Gases - www.epa.gov/climatechange/indicators - Updated August 2013.

Garg, A., Bhattacharya, S., Shukla, P.R., Dadwal, V.K. (2001). Regional and sectoral assessment of greenhouse gas emission in India. Atmospheric Environment. 35: 2679-2695.

Intergovernmental Panel on Climate Change (IPCC). (2001). Climate Change: The Scientific Basis. Cambridge University Press, Cambridge, U.K.

IPCC [Intergovernmental Panel on Climate Change], 2007. Agriculture. In: Metz, B., Davidson, O.R., Bosch, P.R., (Eds.), Climate Change 2007: Mitigation. Contribution of Working Group III to the Fourth Assessment Report of the Intergovernmental Panel on Climate Change. Cambridge University Press, Cambridge, United Kingdom/New York, NY, USA.

Lal, R. (2000). World cropland soils as a source or sink for atmospheric carbon. Advances in Agronomy. 71:145-191.

Neue, H. (1993). Methane emission from rice fields: Wetland rice fields may make a major contribution to global warming. BioScience. 43: 466-73.

NOAA / ESRL. (2012). Available at http://www.esrl.noaa.gov/gmd/ccgg/trends Oenema, O., Wrage, N., Velthof, G.L., van Groenigen, J.W., Dolfing, J., Kuikman, P.J. (2005). Trends in global nitrous oxide emissions from animal production systems. Nutrient Cycling in Agroecosystems. 72: 51-65.

Parashar, D.C., Kulashrestha, U.C., Sharma, C. (1998). Anthropogenic emissions of NOx, NH_3 and N_2O in India. Nutrient Cycling in Agroecosystems. 52: 255-259.

Parashar, D.C., Mitra, A.P., Gupta, P.K., Rai, J., Sharma, R.C., Singh, N., Kaul, S., Lal, G., Chaudhay, A., Ray, H.S., Das, S.N., Parida, K.M., Rao, S.B., Kanungo, S.P., Ramasami, T., Nair, B.U., Swamy, M., Singh, G., Gupta, S.K., Singh, A.R., Saikia, B.K., Barua, A.K.S., Pathak, M.G., Iyar, C.P.S., Gopalkrishnan, M., Sane, P.V., Singh, S.N., Banerjee, R., Sethunathan, N., Adhya, T.K., Rao, V.R., Palit, P., Saha, A.K., Purkait, N.N., Chaturvedi, G.S., Sen, S.P., Sen, M, Sarkar, B., Banik, A., Subbaraya, B.H., Lal, S., Venkataramani, S., Sinha, S.K. (1996). Methane budget from paddy fields in India. Chemosphere. 33: 737-757.

Pathak, H., Li, C., Wassmann, R. (2005). Greenhouse gas emissions from Indian rice fields: calibration and upscaling using the DNDC model. Biogeosciences. 1: 1-11.

Sharma, C., Gupta, P.K., Parashar, D.C. (1995). Nitrous oxide estimates from paddy fields and forests in India. Indian Journal of Radio Space Physics. 24:311–313.

Sinha, S.K. (1995). Global methane emission from rice paddies: Excellent methodology but poor extrapolation. Current Science. 68:643-646.

Tseng, H.K., Tsai, L.J., Alagesan, A., Tsuang, J.B., Yao, H.M., Kuo, H.P. (2010). Determination of methane and carbon dioxide fluxes during the rice maturity period in Taiwan by combining profile and eddy covariance measurements. Agricultural and Forest Meteorology 150: 852-859.

Vinogradova, A., Fedorova, E., Belikov, I., Ginzburg, A., Elansky, N., Skorokhod, A. (2007). Temporal variations in carbon dioxide and methane concentrations under urban conditions. Izvestiya, Atmospheric and Oceanic Physics 43: 599-611.

Wassmann, R., Neue, H.U., Lantin, R.S., Makarim, K., Chareonsilp, N., Buendia, L.V., Rennenberg, H. (2000). Characterization of methane emissions from rice fields in Asia. II. Differences among irrigated, rainfed, and deepwater rice. Nutrient Cycling in Agroecosystems. 58: 13-22.

Zhu, R., Liu, Y., Sun, L., Xu, H. (2007). Methane emissions from two tundra wetlands in eastern Antarctica. Atmospheric Environment. 41: 4711-4722.

NEW TECHNOLOGIES IN RELATION TO CLIMATE CHANGE

12

Low Carbon Technologies in Indian Agriculture

H. Pathak

Global warming is the most prominent environmental issue across the world. It is caused by the increased concentration of greenhouse gases (GHGs) in the atmosphere and leads to a phenomenon widely known as 'greenhouse effect'. Amongst various sources of greenhouse effect, agriculture is considered a major contributor primarily through the emission of greenhouse gases such as methane, nitrous oxide and carbon dioxide.

Table 1: Greenhouse gas emissions from Indian agriculture during 2007

Source	CH_4	N_2O	CO_2 eq.
		Million ton	
Enteric fermentation	10.10	-	212.09
Manure management	0.12	-	2.44
Rice cultivation	3.37	-	84.24
Agricultural soil	-	0.22	64.7
Crop residue burning	0.25	0.01	8.21
Total	13.84	0.23	371.68

Source: INCCA (2010)

According to a report of Indian Network for Climate Change Assessment, the net emission of GHGs from India was 1728 million tons (Mt) of CO_2 eg (Table 1). Thc main sectors contributing to this emission are energy, industry, agriculture and waste. With a total emission of 334 Mt CO_2 eq., the major sources in the agricultural sector are enteric fermentation (63.4%), rice cultivation (20.9%), agricultural soils (13.0%), manure management (2.4%) and on-field burning of crop residues (2.0%). The crop production sector (rice cultivation, soil and field burning of crop residues), thus contributes 35.9% to the total emissions from agriculture. It is therefore, pertinent to develop technologies to reduce emission of GHGs from agriculture. This will not only mitigate climate change but also reduce consumption of costly inputs by enhancing their use efficiency and increase farmers' income by producing more with less of inputs.

Greenhouse gas emission from agriculture

Methane

Agricultural soils contribute to the greenhouse effect primarily through the emission and consumption of GHGs such as methane, nitrous oxide and carbon dioxide. Methane is produced in soil during microbial decomposition of organic matter under anaerobic conditions. Rice fields submerged with water, therefore, are potential source of methane. Continuous submergence, higher organic C content and use of organic manure in puddled soil enhance methane emission. Burning of crop residues also contributes to the global methane budget. Enteric fermentation in ruminant is another major source of methane.

In India, field and laboratory experiments are being conducted since early-1990s in several institutes including Indian Agricultural Research Institute (IARI), New Delhi; National Physical Laboratory (NPL), New Delhi; Central Rice Research Institute (CRRI), Cuttack, Orissa, to (a) measure methane emission from rice ecosystems, (b) evaluate the effect of irrigation and fertilizer management on methane emission, (c) assess the influence of organic amendment on methane emission, (d) measure the methane emission potential of different soils of India, and (e) develop an inventory of methane emission from Indian agriculture using indigenous, site-specific emission coefficients. These studies helped in rationalizing the methane emission estimates from Indian rice fields (Fig. 1).

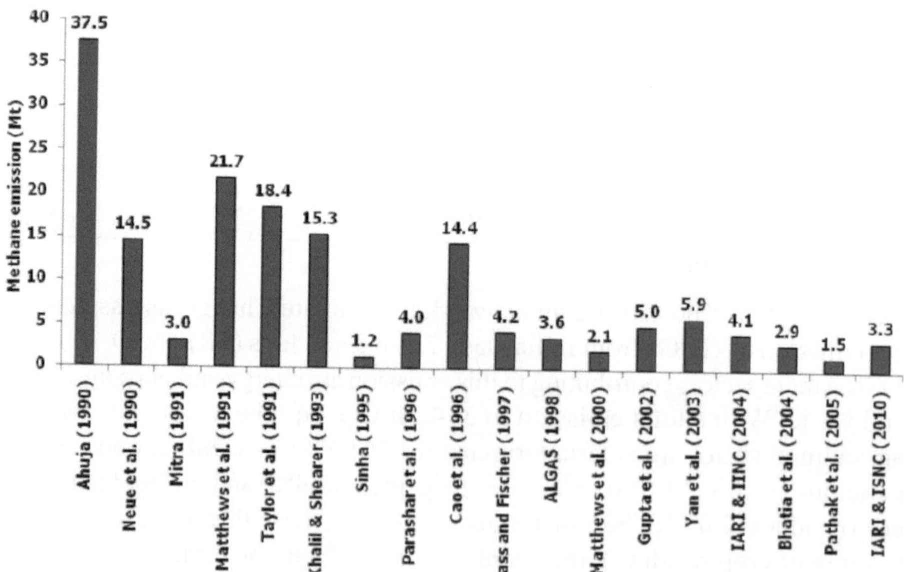

Fig. 1: Estimates of methane emission from Indian rice fields by various researchers over the years
(*Source:* Pathak *et al.* 2010)

Nitrous oxide

Nitrous oxide is produced naturally in soils through the processes of nitrification and denitrification. Nitrification is the aerobic microbial oxidation of ammonium to nitrate, and denitrification is the anaerobic microbial reduction of nitrate to nitrogen gas (N_2). Nitrous oxide is a gaseous intermediate in the reaction sequence of denitrification and a by-product of nitrification that leaks from microbial cells into the soil and ultimately into the atmosphere. One of the main controlling factors in this reaction is the availability of inorganic nitrogen (N) in the soil. Nitrogenous fertilizer is a source of N_2O in fertilized soils, whereas the indigenous N contributes to the release of this GHG in unfertilized soil. Soil water content and the availability of carbon enhance the production of N_2O provided a suitable nitrate source is available. Generally, an increase in N_2O emission is observed following irrigation and precipitation. Burning of crop residues also contributes to the global N_2O budget.

During the past one decade several experiments have been conducted in India to (a) measure nitrous oxide emission from soil, (b) evaluate the effect of crop management on nitrous oxide emission, (c) develop an inventory of nitrous oxide emission from agricultural soils of India, (d) evaluate the mitigation strategies, and (e) develop a simulation model for estimation of nitrous oxide emission.

Carbon dioxide

The main source of carbon dioxide from agriculture is through soil management such as tillage which triggers carbon dioxide emission through biological decomposition of soil organic matter. Tillage breaks soil aggregates, increases oxygen supply and exposes surface area of organic material promoting the decomposition of organic matter. Fuel use for various agricultural operations and burning of crop residues are other sources of carbon dioxide emission. An off-site source is the production of carbon dioxidefor manufacturing fertilizers and pesticides. Since agricultural soils act both as a source and a sink for carbon dioxide, the net flux is very small.

Inventory of GHGs emission from Indian agriculture

The research on GHGs emission from Indian agriculture started in 1990s when, based on very limited measurements done elsewhere, it was reported that Indian rice fields emit 37.5 Mt methane per year. With sustained and systematic indigenous research the methane emission estimates have been rationalized (Fig. 1). The current estimates show that Indian rice fields covering an area of 43.86 million ha (Mha) emit 3.37 Mt of methane per year. The nitrous oxide emission from Indian agricultural soils is 0.14 Mt. Several attempts have been made to estimate CH_4 emission from Indian rice fields. However, only a few studies have attempted to calculate detailed regional CH_4 emissions using simulation modeling. The emission estimates vary widely with the methodology adopted and assumptions made on the importance of different factors affecting CH_4 emission.

Several attempts have been made to estimate N_2O emissions from Indian soils. Recently, an inventory of GHG emissions from Indian agriculture for the year 2007 has been prepared (INCCA, 2010). The emission sources accounted for are enteric fermentation in livestock, manure management, rice cultivation, agricultural soils and burning of crop residues.

According to INCCA, the agricultural sector emitted 371.7 Mt of CO_2 eq. comprising 13.84 Mt of CH_4 and 0.227 Mt of N_2O. Enteric fermentation constituted 61% of the total CO_2 eq. emissions from this sector and 20% of the emissions were from rice cultivation. Agricultural soils emitted 16% of the total CO_2 eq. emission from agriculture (INCCA, 2010). The remaining 3% of the emissions are attributed to livestock manure management and burning of crop residues in field.

Because of increasing emission of nitrous oxide, the total global warming potential (GWP) (methane × 25 + nitrous oxide × 298) of Indian agriculture per unit area (kg CO_2 eq. ha^{-1}) is increasing. However, GWP per unit of produce (kg CO_2 eq. t^{-1}) is decreasing (Fig. 2). Similarly, GHGs intensity per unit agricultural gross domestic product (Ag-GDP) has also declined over the years. This decline has been due to increase in agricultural production of the country through adoption of high-yielding crop varieties and better crop management practices without increase in area under agriculture. Though application of N-fertilizer has increased the agricultural GWP, the increase in yield, and thereby agricultural GDP (because of higher N-use) has been greater. The analysis has shown that there is enough scope of improving agricultural production in the country without off-setting the GHGs emission. Moreover, agriculture has the potential to mitigate GHGs cost-effectively through changes in agricultural technologies and management practices. Mitigation of GHGs emission from agriculture can be achieved by sequestering C in soil and reducing methane and nitrous oxide emissions from soil through change in land-use management. Changes in the management of irrigation, fertilizer use, and soils can reduce emission of both nitrous oxide and methane. Such options are not only important for global warming mitigation but also for improving soil fertility and sustainable agriculture.

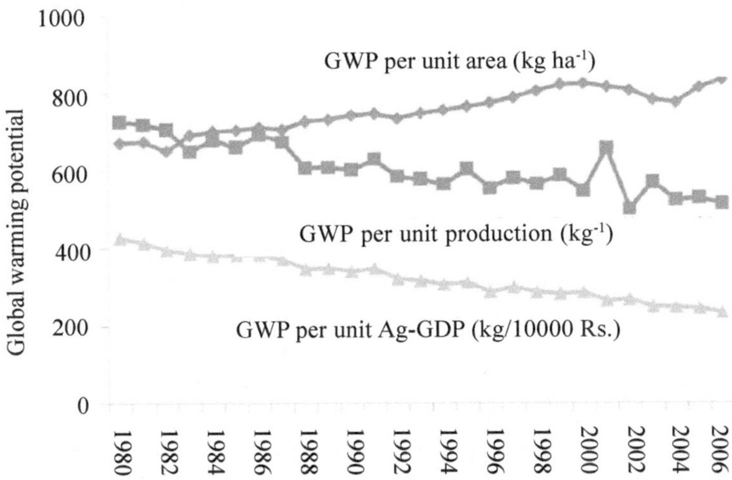

Fig. 2: Trend in intensity of greenhouse gas emission from Indian agriculture. *Note:* GWP, Global warming potential; Ag-GDP, Agricultural gross domestic product (*Source:* Pathak *et al.* unpublished)

Low carbon technologies in agriculture

Scientific agriculture can help in mitigating GHGs emission. The following strategies have been recommended for mitigating methane emission from rice cultivation.

- altering water management, particularly promoting intermittent irrigation and mid-season drainage;
- improving organic matter management by promoting aerobic degradation through composting or incorporating it into soil during off-season drained period;
- use of rice cultivars with few unproductive tillers, high root oxidative activity and high harvest index; and
- application of fermented manure like biogas slurry in place of unfermented farmyard manure (Pathak *et al.*, 2010). A single mid-season drainage may reduce seasonal methane emission. This emission could be reduced further by intermittent irrigation, yielding a 30% reduction as compared to mid-season drainage.

Emission of N_2O can be reduced by following management practices that improve N-use efficiency including using slow or controlled release of fertilizer or nitrification inhibitors which retard the microbial processes leading to N_2O formation. The most efficient management practices to reduce nitrous oxide emission are:

- site-specific nutrient management, and
- use of nitrification inhibitors such as nitrapyrin and dicyandiamide.

There are some plant-derived organics such as neem oil, neem cake and karanja seed extract which can also act as nitrification inhibitors. Nitrification inhibitors reduce N_2O emission directly by reducing nitrification, and indirectly by reducing the availability of NO_3 for denitrification.

Twenty technologies have been analysed for their potential to mitigate GHGs emission in rice in the upper and lower Indo-Gangetic Plains (IGP) and their economic feasibilities have been assessed (Pathak *et al.*, 2012). During crop production under conventional management practices, global warming potential (GWP) of rice cultivation was 3957 kg CO_2 ha in the upper-IGP and 2934 kg CO_2 ha in the lower-IGP. Compared to the current practices of farmers, 15 -1 technologies in the upper-IGP and 14 technologies in the lower-IGP have the potential to reduce the GWP. In the upper-IGP, only seven technologies, viz. sprinkler irrigation, direct seeded rice, use of nitrification inhibitor, use of urea super granules, use of leaf colour chart, site-specific nutrient management and crop diversification have depicted ability to reduce GWP without any additional cost (Fig. 3). In the lower-IGP, use of nitrification inhibitor, use of leaf colour chart, site-specific nutrient management and crop diversification have shown reduction in GWP with no additional cost.

Ten technologies have been assessed in wheat for GHGs mitigation potential and economics. The GWP in the upper-IGP is 1808 kg CO_2 ha^{-1}, whereas in the lower-IGP it is 1280 kg CO_2 ha^{-1}. Among the various technologies zero tillage, integrated nutrient management, organic farming, use of nitrification inhibitor and site-specific nutrient management have proved to be beneficial in terms of GWP reduction and profit enhancement in the upper-IGP. In the lower-IGP, zero tillage, integrated nutrient management, use of nitrification inhibitor and site-specific nutrient management technologies have been found both GHG-friendly and economically feasible (Fig. 3). Integrated nutrient management has caused 10% increase in income and 109% reduction in GWP in the upper-IGP. In the lower-IGP, zero tillage could cause 9% increase in net return and 105% reduction in GWP as compared to the conventional practices of farmers.

The major benefits of low carbon agricultural technologies are savings in irrigation water, labour and energy; reduction in GHGs emission; better water- and nutrient-use efficiency; provision of tolerance to moisture and heat stresses; improvement in soil health; and increase in income. There are some constraints also; these include high initial cost, infrastructure for installation and maintenance, knowledge-intensiveness and technical capability, high production cost, risk in rainfed areas, weed problem, yield loss, inadequate market facility, lack of

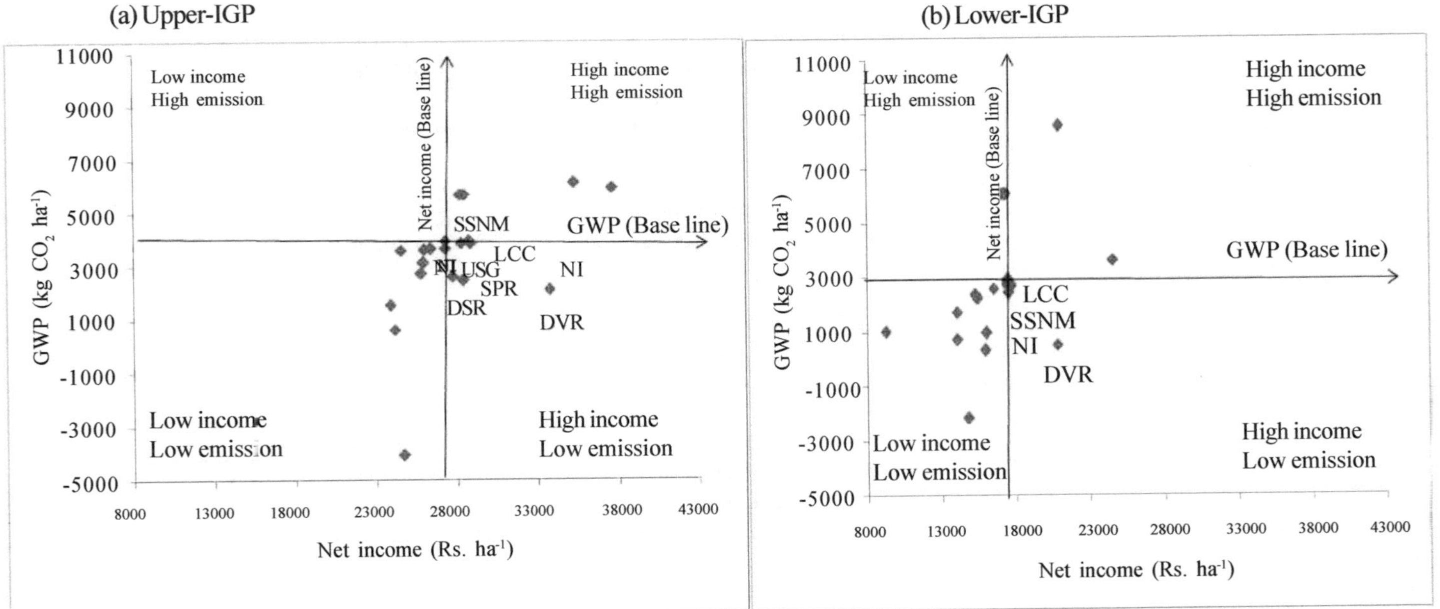

Fig. 3: Marginal abatement cost for reducing global warming potential from the baseline technology of continuously flooded and farmers' practice in rice in (a) upper and (b) lower Indo-Gangetic Plains

Note: DVR, Diversification; LCC, GWP, Global warming potential; Leaf colour chart; NI, Nitrification inhibitor; SPR, Sprinkler; SSNM, Site-specific nutrient management; USG, Urea super granule. Source: Pathak and Aggrawal , 2012

Table 2: Benefits, challenges and required interventions for promoting low-carbon and cost-effective mitigation technologies in rice and wheat in the Indo-Gangetic Plains

Technology	Major benefits[a]	Key challenges	Required interventions
Rice			
Alternate wetting and drying/Mid-season drainage	• Saving in irrigation water (25-30%) • Saving in labor (10-15%) and energy • Reduction in methane emission (70-75%)	• Weed problem • Yield loss (5-10%) • Difficulties and risk in rainfed areas • Increase in N_2O emission (5-10%)	• Development of effective, low cost, environment friendly herbicides • Accurate weather forecasting • Development of Irrigation facilities • Provision of incentives for saving water • Introduction of carbon credit for mitigation
Drip and sprinkler irrigation	• Saving in irrigation water (25-30%) • Saving in energy • Higher water- and nutrient-use efficiencies	• High initial cost • Need of infrastructure for installation and maintenance • Highly technical	• Introduction of subsidy for installation • Organization of trainings for skill development • Generation of public awareness • Introduction of carbon credit for mitigation
Nitrification inhibitor/urea super granules	• Reduction in losses and increase in N-use efficiency	• High production cost • Technical application process • Non-availability in the market	• Motivation of industries to manufacture these products • Provision of incentives to farmers • Generation of public awareness • Introduction of carbon credit for mitigation
Site-specific N management/ use of LCC	• Reduction in losses and increase in N-use efficiency	• Lack of awareness • Soil testing facility inadequate	• Establishment of more soil testing laboratories • Provision of incentives to farmers • Generation of public awareness • Introduction of carbon credit for mitigation
Direct seeding of rice	• Saving in irrigation water (25-30%) • Saving in labor (10-15%) and energy • Reduction in methane emission (70-75%)	• Knowledge intensive • Weed problem • Yield loss (5-10%) • Increase in N_2O emission (5-10%)	• Development of effective, low cost, environment-friendly herbicides • Training to farmers • Provision of incentive for saving of water • Introduction of carbon credit for mitigation

Contd.

Technology	Major benefits[a]	Key challenges	Required interventions
	• Increase in stress tolerance		
Crop diversification (rice to maize)	• Saving in irrigation water (50-60%) • Saving in labor (20-25%) and energy • Reduction in methane emission (90-100%)	• Minimum support price not available • Inadequate market • Lack of awareness • Limited post-harvest facilities	• Provision of minimum support price • Development of market infra-structure • Awareness generation • Development of post-harvest facilities
Wheat			
Zero tillage	• Saving in fuel (70-90 L diesel ha^{-1}) • Saving in water	• Knowledge intensive • Non-availability and maintenance of quality zero-till drill Lack of awareness	• Refinement of technology • Skill development for operation and maintenance of drill • Awareness generation
Integrated nutrient management/organic farming	• Improvement in soil health • Increase in N-use efficiency • Reduction in GWP • Increase in income	• Non-availability of manure • Poor quality of manure • High transportation cost	• Promotion of agri-wastes for compost • Promotion of biogas for energy and manure • Introduction of carbon credit for biogas
Nitrification inhibitor/Site-specific N management	• Reduction in losses and increase in N-use efficiency	• High production cost • Technical application process • Non-availability in the market	• Motivation of industries to manufacture these products • Provision of incentives to farmers • Generation of public awareness • Introduction of carbon credit for mitigation

Note: Figures within the parenthesis are compared to conventional practices
Source: Pathak and Aggarwal (2012)

awareness and limited post-harvest facilities. An example of promoting low carbon techniques in rice-wheat system in IGP as suggested in Table 2.

The policy interventions required to overcome the constraints are development of irrigation facilities, incentives for saving of water, carbon credits for mitigation, subsidy and incentive for installation of resource conserving infrastructure, trainings to farmers for skill development, public awareness generation, development of effective, low-cost, environment-friendly herbicides, accurate weather forecasting, development of post-harvest facilities and refining of technologies to make them simple, cheap and effective.

Several options are available for mitigating GHGs emission in agriculture. Policies and incentives would have to be developed to encourage farmers for adopting these mitigation options to harness benefits of improved soil health and better water- and energy-use efficiencies.

References

INCCA (Indian Network for Climate Change Assessment) (2010) Assessment of the Greenhouse Gas Emission: 2007. The Ministry of Environment & Forests, Govt. of India, New Delhi.

IPCC (Inter-governmental Panel on Climate Change) (2007) The Physical Science Basis. In: Solomon, S, Qin D, Manning M, Chen Z, Marquis M, Averyt KB, Tignor M, Miller HL (eds.) Climate Change 2007: Contribution of Working Group I to the Fourth Assessment Report of the IPCC. Cambridge University Press, Cambridge, UK.

Pathak H, Bhatia A, Jain N and Aggarwal PK (2010) Greenhouse gas emission and mitigation in Indian agriculture – A review. In: ING Bulletins on Regional Assessment of Reactive Nitrogen, Bulletin No. 19, Ed: Bijay Singh. SCON-ING, New Delhi.

Pathak H and Aggarwal PK (Eds.) (2012) Low Carbon Technologies for Agriculture: A Study on Rice and Wheat Systems in the Indo-Gangetic Plains. Indian Agricultural Research Institute, 76 p.

13

Weather Index Based Insurance for Risk Management in Agriculture

P. Vijaya Kumar and V. U. M. Rao

Introduction

In many developing countries including India, agricultural sector is of vital importance to the economy and welfare of the inhabitants. Agriculture contributes to 14% of the Indian GDP and supports the livelihood of two-thirds of the population. Despite tremendous technological developments, vagaries of weather continue to plague the agriculture production. Any shortfall in agricultural production will have its cascading effect on the economy of the country. In recent times, climate change and variability are confounding the troubles already faced by Indian agriculture and the farming community. Loss in crop production and farm income due to unexpected weather hazards is beyond the carrying capacity of resource poor farmers of the country. With the growing commercialization of agriculture, the magnitude of crop loss due to unfavourable weather hazards is increasing leading to suicides of farmers. The State and National governments on their part are coming to rescue of farming community by implementing various relief measures like reduction or suspension of land revenue taxes, loan waiving, relief from calamity relief fund etc. Though these measures are helpful, farmers cannot expect them as a right and they put lot of burden on the country's exchequer.

As agricultural yields are highly variable due to vagaries of weather, crops have to be covered under agriculture insurance for compensating yield losses and reduce the poverty of farming community. In India, since 1972 agricultural insurance has evolved as an adaptation strategy to mitigate / manage weather related risks in agriculture. However, experiences with traditional agricultural insurance schemes have shown that this approach is not often suitable in developing countries. In recent years, weather index insurance contracts in agriculture have emerged as an alternative to traditional agricultural (or crop) insurance.

Risks in agriculture

Though risks in agriculture are numerous, weather risks, biological risks and price risks are the most predominant ones in agriculture (Hess *et al.*, 2002; Bryla *et al.*, 2003 and Skees *et al.*, 2005). Agricultural risks can range from independent (like localized hail storms) to highly correlated or covariant (example: Price risks or widespread drought). Weather risks are broadly of two types viz. sudden, unforeseen events (thunder storms) and cumulative events occurring over extended period (example: drought). The adverse impacts of either of these two types vary according to the crop, crop variety and time of occurrence of the event. Some of the weather risks in agriculture are: Drought, excess rainfall or floods, high temperature, low temperature events like frost and freeze, high winds and hail storms etc.

Weather risk management

Farmers in developing countries have always been exposed to weather risks, and for a long time have developed a variety of weather risk management (WRM) techniques for reducing, mitigating and coping with weather risks (Dercon, 2002). Traditional risk management covers actions taken both before (ex-ante) and after (ex-post) the risky event occurs (Siegel and Alwang, 1999). Mitigation, coping and transfer are major strategies in agricultural risk management. Detailed description of these risk management strategies are:

Mitigation: It is the act of reduction of severity of adverse impacts of hazards and related disasters. Crop and livestock diversification, adoption of better agronomic practices, soil drainage, irrigation, use of resistant varieties, improved early warning systems and crop calendars etc are some of the risk mitigation options.

Coping: It refers to the improvement of resilience to withstand and manage risk events, through ex-ante (before the risk event occurs) preparation and making use of informal and formal mechanisms in order to sustain production and livelihoods after the risk event. Government assistance to farmers and debt restructuring etc are some of the coping measures.

Transfer: It refers to the transfer of potential financial consequences of some risk from one particular party to another willing party, at some cost. Insurance is the best-known form of risk transfer.

Agricultural insurance

Agricultural insurance is a means of protecting the farmers against financial losses suffered due to unforeseen and non-preventable weather risks (AIC, 2008). It is a financial mechanism in which loss of crop yields suffered by farmers in a particular area are borne by farmers in other areas (as premiums), so that burden of loss can be distributed among a large number of farmers. In agricultural insurance the reserves of premiums accumulated in good years are used to pay the indemnities in bad years.

Evolution of Agriculture Insurance in India

In India, crop insurance programme as a pilot started during 1972 on H-4 cotton in Gujarat and was extended later to few other crops and states. Pilot Crop Insurance Scheme (PCIS) was launched by the Government of India on 'area approach' in 13 states of the country in 1979-80 and continued till 1984. Comprehensive Crop Insurance Scheme (CCIS) was in operation from 1985-1999 and was implemented on "Homogeneous area approach" in 16 states and 2 UTs of the country. In 1999, CCIS was replaced with a broad based scheme called National Agriculture Insurance Scheme (NAIS). The NAIS is presently administrated by Agriculture Insurance Company of India Limited, which provides coverage to approximately 35 different types of crops during *Kharif* and *Rabi* seasons. The government announced a pilot on improved version named Modified NAIS (MNAIS) w. e. f. *Rabi* season 2010-11 for experimentation in 50 districts. The new version has, to a large extent, taken care of the drawbacks in the existing NAIS. From November 2013, National Crop Insurance Programme (NCIP) / Rashtriya Fasal Bima Karyakram (RFBK) was formulated by merging Modified National Agriculture Insurance Scheme (MNAIS), Weather Based Crop Insurance Scheme (WBCIS) and Pilot Coconut Palm Insurance Scheme (CPIS).

Crop insurance products

Crop insurance products can broadly be classified into two major groups: Indemnity-based insurance and Index insurance.

I. Indemnity-Based Crop Insurance

There are two main products in the Indemnity - based crop insurance. They are:

(i) ***Damage-based indemnity insurance* (or *named peril crop insurance*):** Damage-based indemnity insurance is the crop insurance in which the insurance claim is calculated by measuring the percentage damage in the field soon after the occurrence of the damage. The sum insured may be based on production costs or on the expected revenue. Where damage cannot be measured accurately and immediately after the loss, the assessment may be deferred until the end of the crop season. Damage-based indemnity insurance is best known for hail, frost, heat wave, deficit or excess rainfall.

(ii) ***Yield-based crop insurance* (or *Multiple Peril Crop Insurance, MPCI*):** Yield based crop insurance is cover-age in which insured crop yield is established as a percentage of the farmer's historical average yield. The insured yield is mostly between 50 to 70 percent of the average farm yield. If the realized yield is less than the insured yield, an indemnity (the amount payable by the insurer to the insured) is paid equal to the difference between the actual yield and the insured yield, multiplied by a pre-agreed value. Yield-based crop insurance protects against multiple perils, meaning that it covers many different causes of yield loss (it is generally difficult to determine the exact cause of yield loss).

II. Index-Based Crop Insurance

Index based insurance products for agriculture represents an attractive alternative for managing weather risks. Currently there are two types of index insurance products viz., Area yield index insurance and weather index insurance (WII), as distinguished by Skees (2003)

(i) ***Area yield index insurance*:** In Area yield index insurance, the indemnity is based on the realized average yield of an area such as a block or district, not the actual yield of the insured farmer. The insured yield is established as a percentage of the average yield for the area of the whole village or group of villages. An indemnity is paid if the realized yield for the area is less than the insured yield regardless of the actual yield on a policyholder's farm. This type of index insurance requires historical yield data of the area.

(ii) ***Weather Index Insurance (WII)*:** In WII, the indemnity is based on realization of a specific weather parameter measured over a pre-specified period of time at a particular weather station. The insurance was structured to protect against realization of either very high or very low indices that are expected to cause crop losses. For example, the insurance can be structured to protect against either too much or too little rainfall. An indemnity is paid whenever the realized value of the index exceeds a pre-speci-fied threshold (when protecting against excess rainfall) or when the index is less than the threshold (when protecting against deficit rainfall). The indemnity is calculated based on a pre-agreed sum insured per unit of index.

Skees (2003) argues that weather based indices are usually preferred to yield based indices, as in developing countries the quality of historical weather data is better than quality of yield data, and weather events, especially deficit or excess rainfall, are the major sources of crop losses in many regions.

Main features of Weather index-based Insurance

The essential feature of WII is that the insurance contract responds to an objective parameter (e.g. rainfall or temperature) at a referred weather station during an agreed time period. The parameters of the contract are set so as to correlate, as accurately as possible, with the loss of yield of a specific crop suffered by the policyholder. All policyholders within a defined area receive payouts based on the same contract and measurement at the same weather station, eliminating the need for in-field assessment of yield loss.

In order for the underlying index to be a sound proxy for loss, it has to be based upon an objective measure (for example, rainfall, wind speed, temperature) that exhibits a strong correlation with the variable of interest (in this case, crop yield). Additionally, weather variable that can form an index must satisfy the properties: i) Observable and easily measured, ii) Objective, iii) Transparent, iv) Independently verifiable, v) Reported in a timely manner, vi) Consistent over time and vii) Experienced over a wide area.

Important features of a WII contract are:

- A specific meteorological station is named as the *reference weather station*.
- A strike or *trigger* weather measurement is set (e.g. cumulative rainfall), at which the contract starts to pay out.
- A *lump sum* or an *incremental payment* is made (e.g. Rupees per mm of rainfall above or below the trigger).
- A *limit or exit* of the measured parameter is set (e.g. cumulative rainfall), at which a maximum payment is made.
- The *period of insurance* is stated in the contract and coincides with the crop growth period; it may be divided into *phases* (Maximum three), with each phase having its own triggers, increment and limit.

Advantages and disadvantages of weather index based insurance

Though the development and application of weather index insurance (WII) is still in its early stages, the advantages and disadvantages of WII were well-documented (World Bank, 2005; USAID, 2006; IFAD and WFP 2010). Some of the relative merits and demerits of the WII are presented below (IFAD, 2010).

Advantages of WII

In comparison to the traditional damage based agricultural insurance, WII has advantages on the following aspects:

Transparency: An Index Insurance contract usually allows the policyholder to have direct access to the information on which the payouts will be calculated. Hence, WII is more transparent to the clienteles.

No on-farm loss adjustment: This is the primary advantage of index insurance, as on-farm loss adjustment is quite complex and costly.

Lack of adverse selection: "Adverse selection" occurs when potential insured parties have hidden information about their risk exposure that is not available to the insurer, who then becomes more likely to erroneously assess the risk of the insured. As a result, traditional insurance encourages high-risk producers to insure, while risk and premium are calculated on the average producer. However, in Weather Index based insurance all the insured farmers within the defined area have the same insurance payout conditions, regardless of their specific risk exposure (USAID, 2006). Hence, insurers and clients benefit from reduced adverse selection.

Lack of moral hazard: "Moral hazard" refers to a phenomenon that the insured person's optimal decision may change as a result of purchasing the insurance, because the insurance contract reduces the loss associated with the insured event. Such changes in behaviour will normally increase the probability of the insured event occurring or increased severity of loss (Ashan *et al.*, 1982). In WII, individual farmers who are trying to influence claims will not get any benefit and all farmers in the defined area are treated equally for payment of pay outs or indemnity.

Addresses correlated risks: Index insurance works best, where there are correlated and widespread risks like drought. In traditional insurance, perils such as drought are challenging to insure.

Low operational and transaction costs: Index insurance requires limited individual underwriting (Client assessment). It can be distributed, or sold out and claims can be settled, at relatively lower cost than in traditional insurance.

Rapid payout: Measurement of meteorological data from a reference weather station, with no field loss adjustment, allows for providing rapid payouts to the clients in WII than in traditional agricultural insurance.

Despite the merits of weather index based insurance mentioned above, acceptance of WII by farmers is slow due to the following disadvantages:

Disadvantages of WII

Basis risk: Basis risk is a key constraint in WII and it has suppressed the growth of WII. Basis risk is the difference between the actual crop yield at farm unit level and yield projected by the weather index. As a result of basis risk, a farmer experiencing yield loss, may not receive a payout, and a payout may be triggered without any loss being experienced. Index insurance works best where losses are homogeneous in the defined area and highly correlated with the indexed peril. Diaz Nieto *et al.* (2006) mentioned three important types of basis risk, which are mentioned as under:

- ***Spatial or geographical basis risk:*** A geographical basis risk represents the risk that result from the difference between weather patterns at reference weather stations and the locations of farmers. Local variations in the weather (e.g. rainfall) within the area surrounding a reference weather station are responsible for geographical basis risk.

- ***Temporal basis risk:*** It results from inter-annual variations in crop phases, and the insurance phases are not temporally aligned with the intended crop growth stage.

- ***Product or crop specific basis risk:*** It means that the sensitivity to weather events varies across crop types due to different crop characteristics. Crop losses can be caused by many factors including weather. Where there is no clear-cut and strong relationship between yield loss and the indexed weather peril, risk can be high. WII is most likely to work, for rainfed crops and at severe levels of the risk event, when losses may be more widespread and homogeneous.

Limited perils: WII normally covers only one, or sometimes two, weather perils. Although this reduces the cost of operation compared to the multi peril crop insurance (MPCI), the product may not provide broader and enough coverage to more number of weather risks that are affecting the crop loss.

Replication or Scalability: The triggers, limits and increments of a specific product that were worked out for a reference weather station will not be valid for another reference weather station and they need to be adjusted to reflect the weather parameters of that new weather station. Different product designs are required for different crops (or at least generic crop types).

Requirement of Technical expertise: WII requires considerable technical work in its implementation and sustaining. Technical capacity and expertise in agro-meteorology are required, particularly during the initial design phase for new products, and also in operationalising the products.

Lack of weather data: WII depends on the availability of quality weather data, which drastically vary from country to country. In developing countries, the shortage of historical and real-time weather data is often a major hurdle for design of WII products.

Collection of weather data for weather index based insurance

The data used for constructing the weather indices should satisfy quality requirements, including:

- Reliable and trustworthy ongoing daily collection and reporting procedures;
- Periodic checks and quality control;
- An independent source of data for verification (e.g. surrounding weather stations).

The general criteria for weather data requirements for WII applications (ISMEA, 2006) are specified as follows:

- At least 20 years of historical weather data;
- Limited missing values and out-of-range values (preferably less than 3 per cent missing data from the entire dataset);
- Availability of a nearby weather station for fall-back verification purposes;
- Consistency of observation techniques – either manual or automated;
- Limited changes in instrumentation/orientation/configuration;
- Integrity of weather-data recording procedure;
- Little potential for measurement tampering.

Beyond the quality of data, definition of the boundaries of the area(s) covered by the weather station(s) is critical, so that WII contracts can be written for specific areas tied to a specific weather station. A general rule of thumb is to consider a specific WII contract marketable within a 20-km radius of the weather station; but in many cases the applicable area is smaller. The more the terrain varies, the more the acceptable distance from a station decreases.

Modalities must be defined for weather data collection and dissemination during the contract coverage period. Insurance and reinsurance industries tend to require the use of automated weather stations and availability of fallback verification measurements from nearby stations with comparable weather patterns. Manual measurement of weather variables (e.g. through manual rain gauges) is usually not regarded as sufficiently reliable and secure. As a result, low-cost, automatic weather stations are being implemented in some WII initiatives. Even if manual observation of weather risks are taken, all weather parameters should be taken at the same time for controlling hampering of weather data, as various weather variables are linked by specific relationship.

Collection of agricultural data

Agricultural information is the second most important component of the contract design equation of WII. The data on productivity (yield), and description of the agricultural production practices carried out in the areas is also necessary. Unfortunately, the availability of quality yield-data series of adequate length and at the appropriate spatial level is a major problem. However, lack of quality yield data does not pose as serious a problem as lack of good weather data, since it is still possible to find alternative approaches to estimate yield data. One possibility is to simulate synthetic yield-data series through crop growth simulation models. Information such as crop varieties adopted, planting dates, management practices, related costs, risk profiles, historical recollection of the impact of the peril, and the most sensitive phases in crop life are essential in designing a meaningful WII contract.

Technical issues concerned to the design of weather index

Diaz Nieto *et al.* (2006) mentioned the following technical requirements for the index insurance scheme to have a positive impact on farmers' livelihood.

- The weather index should be easily understood and well defined;
- It should take account of crop sensitivity at different growth stages;
- It should take account of the relationship between soil texture and rainfall effectiveness;
- It should define a protocol that reflects the actual planting date as closely as possible;
- It should ensure that the insured pays the price of spatial variation in risk;
- It should enable accurate estimation of the probability of the risk event.
- Reliability of the institution providing the weather data;
- Transparency and absence of corruption;
- Adaptation of the product to farmers' needs;
- Communication and training for farmers and field staff.

Design and validation of weather index based insurance products

The objective of the contract design is to define a structure that effectively captures the relationship between the weather variable and the potential crop loss and to select the index that is most effective in providing payouts when losses are experienced, eliminating basis risk as far as possible. The set of possible index combinations is unlimited, and numerous structures have been developed in the relatively short history of WII. Weather index-based contracts can be classified according to many different parameters (Table 1).

Table 1: Examples of parameters used for WII contracts

Contract parameter	Options
Triggering values of weather variable	Cumulative Average Maximum Minimum
Period covered by index	Entire life cycle of crop Fractions of crop life cycle
Number of phases into which covered period is divided	Usually 1 to 3 phases
Start of coverage period	Fixed Dynamic
Payout structure	Incremental Lump sum (single value payout)

One of the most commonly adopted structures is that of a continuous payout triggered and limited by a cumulative measure of the weather variable (e.g. rainfall) for each of the different crop growth stages, is discussed as follows.

Payout parameters in a WII contract – an example

Using the drought coverage case represented in Figure 1 as an example, the parameters that characterize an incremental payout structure can be defined as follows:

- ***Trigger:*** Threshold above or below which payouts are due. In this example, payments are due when the calculated value of the index is below the trigger level (300 mm).
- ***Limit:*** Threshold above or below which no additional incremental payout will be applied. In this example, the maximum payout is paid, if the calculated value of the index is equal to or below the exit threshold (100 mm).
- ***Tick:*** Incremental payout value per unit deviation increase from the trigger. With a maximum payout (the insured sum) of 12000, a trigger of 300 mm and an exit of 100 mm, the monetary value of each deficit mm of rainfall below the trigger is: 12000 / (300 mm-100 mm) or 60 per mm

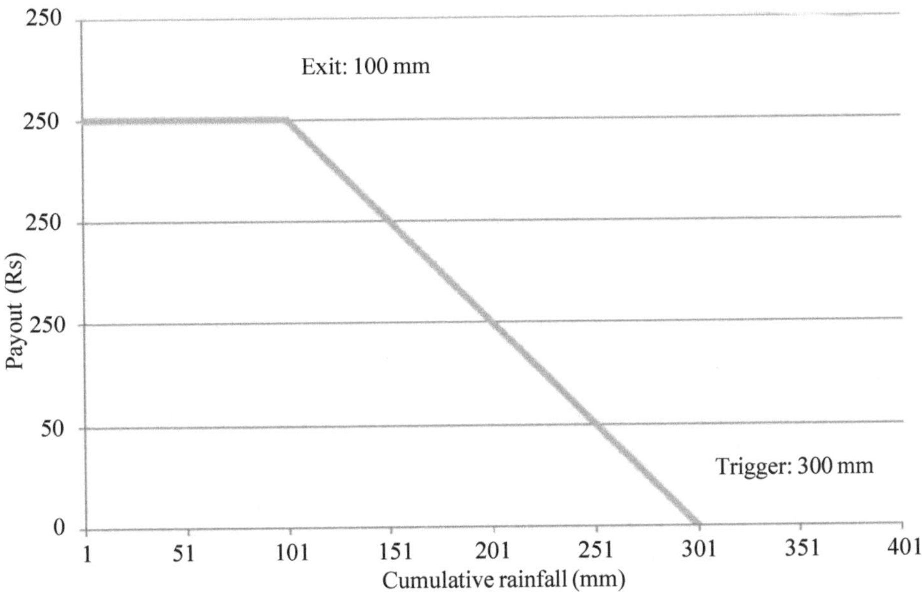

Fig. 1: Payout structure of a WII drought contract

Contract design is probably the most challenging part of developing a pilot programme and local insurance companies, usually have no expertise to carry out the design. WII pilot developed with the involvement of specialists of research organization is encouraged.

Weather index insurance: Indian experience

An impressive repository of historical weather data, high dependency of country's agricultural production on rains and huge pool of scientific resources place India in the forefront of piloting of weather index insurance.

Pilot Weather Risk based Crop Insurance

The first pilot on weather index insurance in India and also in the developing world was carried out in 2003 by ICICI Lombard. This was followed by pilots on weather risk index-based insurance by Agriculture Insurance Company of India (AIC) and IFFCO-Tokio, both during 2004. Building on the existing weather risk insurance products, the Government asked AIC in 2007 to design the Weather risk-Based Crop Insurance Scheme (WBCIS) as a pilot. An example of the product for multi-phase deficit rainfall and consecutive dry days is presented in Table 2.

Table 2: Illustration of WBCIS in *Kharif* groundnut at Mahaboobnagar, Telangana during 2004

	Crop: Groundnut	Season: *Kharif*		
		PHASE-I	PHASE - II	PHASE – III
1A. Rainfall Volume	PERIOD	21st June to 15th July	16th July to 15th Aug	16th Aug to 30th Sept
	TRIGGER I (<)	80 mm	160 mm	80 mm
	TRIGGER II (<)	40 mm	80 mm	40 mm
	EXIT	20	30	20
	RATE I (Rs./ mm)	25	25	25
	RATE II (Rs./ mm)	75	60	75
	Max. Payout (Rs.)	2500	5000	2500
	TOTAL PAYOUT (Rs.)	10000		
1 B. Rainfall Distribution (Consecutive Dry Days)	PERIOD	1st July to 31st August		
	TRIGGER DAYS (>=)	20	25	30
	PAYOUT (Rs.)	1500	3000	5000
		TOTAL PAYOUT (Rs.)5000		
Rainfall of less than	2.5mm in a day shall not be considered as a rainy day; and multiple payouts considered			
	Max. Payout (Rs.)	15000		

CRIDA's Contribution to WII

AIC has signed a MOU for generating weather indices for three crops viz., wheat, groundnut and cotton with CRIDA, having a network project All India Coordinated Research Project on Agrometeorology (AICRPAM) generating crop and weather data for more than 20 years in important crops at 25 of cooperating centres and having huge inter disciplinary scientific staff strength. Scientists of CRIDA identified critical phenological stages for temperature in wheat and for rainfall in groundnut and cotton. Thresholds of average maximum and minimum temperature in critical phenological stages of wheat were worked out in different varieties of wheat at eight cooperating centres for generating weather indices (Venkateswarlu *et al.* 2013). Likewise, thresholds of cumulative rainfall in critical stages of groundnut at four centres and cotton at three centres were worked out to serve as indices at the respective centres (Rao *et al.* 2013). Thresholds of weather parameters in wheat and groundnut are illustrated in Tables 3 and 4. These thresholds serve as triggers and temperature above these limits cause yield reduction in wheat at respective centres. Likewise, rainfall below these limits will cause reduction in yield of groundnut at respective centres / districts.

Table 3: Thresholds of temperature at critical stages for obtaining optimum wheat yield at different locations

Centre	Maximum temperature (°C)	Minimum temperature (°C)	Stage
Kanpur	25.6 - 27.5	9.9 - 11.3	Milk
Faizabad	32.0	14.0	Dough
Anand	26.9 - 28.1	9.9 - 11.0	Milk
Ranichauri	13.8 - 16.3	2.9 - 5.3	Jointing to Anthesis
Raipur	29.7 - 31.7	15.1 - 15.8	Milk
Ludhiana	20.0 - 31.3	6.4 - 15.4	Booting to Maturity
Udaipur	28.4	10.8	Dough
Ranchi	25.1 - 27.2	8.7 - 9.8	Milk

Table 4: Thresholds of Rainfall in critical stages for obtaining optimum yield of groundnut at Anantapur, Anantapur district, Bangalore and Anand

Station/District	Variety	Rainfall (mm)	Critical Stage
Anantapur	TMV-2	191.4	Pod initiation to Maturity
	Robut 33-1	224.3	Pod initiation to Maturity
Anantapur District	-	171.0	Pod initiation to Maturity
Bangalore	DH 3-30	118.7	Pod initiation to pod filling
	Robut 33-1	108.7	Pod initiation to pod filling
	TMV-2	155.7	Pod initiation to pod formation
	JL -24	138.4	Pod initiation to pod formation
	K-134	122.3	Pod initiation to pod formation
Anand	Robut 33-1	318.8	First seed to harvest
	GG-2	469.0	First seed to harvest
	Gaug-10	174.3	First seed to harvest

Other indices like cumulative dry days, water requirement satisfaction indices in groundnut and cotton were worked out. Efforts are on to develop weather indices in some more crops and horticulture systems.

Key Policy Directions

For making crop insurance more effective and meaningful, the following policy directions were suggested.

i. **Agricultural Risk Protection Act:** In countries like United states of America, Canada, Spain, Japan, Philippines, etc. crop insurance is being used as an integral part of agriculture risk management. In these countries separate statute is in force for facilitating smooth implemention of agricultural insurance. Many European Union Countries also formulated their own specific guidelines delineating the role of agriculture insurance vis-à-vis ad-hoc disaster relief. In India also, there is a need to bring out agriculture risk protection act for making agriculture insurance more viable and robust.

ii. **Catastrophic Insurance and Universal Coverage of Farmers:** Despite the existence of crop insurance programme for past 25 years in the country, only 20% of the farmers or cropped area could be brought under insurance net. There are some crops viz., Vegetables, fruits etc, in which insurance products are not yet available. To protect the non-borrowing farmers and farmers cultivating crops not having insurance products, from extreme financial distress, Government can introduce " Catastrophe Protection" or "Non Insured crop Loss Assistance" for farmers. Such protection can serve as an effective conduit for channelizing calamity and disaster relief funds from central and state Governments to crop insurance.

iii. **Defining the Role of Agriculture Insurance:** The role of agriculture insurance is to be defined clearly vis-à-vis other government programmes. Government has to channelize some portion of government funds allocated to other programmes of the agriculture sector to agricultural insurance for avoiding duplication of different government programmes and maximizing the utility of government funds. It would be an important policy initiative if eligibility or access to government subsidies in agriculture is linked to participation in agriculture insurance.

iv. **Financial Literacy:** Though agriculture insurance is in operation for last 25 years, not only farmers but also the policy makers and government functionaries are not much aware of it. Capacity building and financial literacy of farmers could help the farmers in understanding the merits of agriculture insurance, which in turn will help in increasing the adoption rate of agriculture insurance.

v. **Appropriate Mix of Insurance Models:** Various index insurance products viz., yield index and weather index products are not substitute to each other but are complimentary in nature. A combination of these indices like 'double trigger' or "multiple trigger' insurance products would bridge the 'gap' in indemnity and lower the basis risk.

vi. **Weather Station Density:** Poor density of reference weather stations was found out to be one of the reasons for basis risk in weather index based insurance. Though, India Meteorological Department (IMD) is putting efforts to increase the network of weather stations, it would not be adequate to meet the growing demands of weather insurance industry. For weather insurance to be effective and realistic, a rain gauge at every 5 km radius and weather station at every 10 km radius have to be established; the requirement amounts to 35,000 rain gauges and 10,000 weather stations over the country. Keeping in view the huge network of weather monitoring, both public and private weather stations need to be placed on a common "weather grid". To meet the weather data requirements, public-private partnership is to be encouraged.

vii. **Technology:** Lack of sufficient weather and crop data is a major problem in agriculture insurance or weather index based insurance. New or existing technologies are to be used for generating weather and yield data of sufficient years for developing insurance products. For example, it is possible to create 'virtual weather station' network at 1 km grid level using Terrestrial observation and prediction systems (TOPS) with capabilities to provide medium range weather forecast. Similarly, final resolution satellite imagery can be used to estimate yields of different crops and cropping systems required for crop insurance.

viii. **Incentives for Sustainable Agricultural Practices:** While subsidies are a must in agriculture insurance, it would be equally important to build risk management stipulations and incentives in the programmes for sustainable agricultural practices like integrated pest management, low green house gas releasing crops etc.

Conclusions

Weather risk is assuming importance in agriculture due to climate change. Agriculture insurance has been identified as one of the risk management strategies for adapting to the climate change. In contrast to the traditional crop insurance scheme, weather index based insurance is gaining prominence because of its transparency, low operational costs and fast pay out mechanism. Yet, there are major constraints associated with weather index products that need to be successfully addressed. Foremost among the constraints is high basis risk. There is an urgent need to bring down basis risk arising from insufficient network and spread of weather stations besides improving relationships between the weather triggers and yield loss. Finally, there should be an in-depth research (on a continuous basis) of the associated weather risks for various crops grown in the country.

References

Agriculture Insurance Company of India Limited. Available at www.aicofindia.org accessed 2006 to 2008.

Ashan, Syed, M, Ali, A.A.G. and Kurian, N.J. 1982. Towards a theory of agricultural crop insurance. Am. J. Agr. Econ 64(3): 520-529.

Bryla, E, Dana, J, Hess, U. and Varangis, P. 2003. Risk Management: Pricing, Insurance, Guarantees. The Use of Price and Weather Risk Management Instruments, prepared for the conference Paving the Way forward for Rural Finance, held in Washington D.C., June 2003.

Dercon S. 2002. Income Risk, Coping Strategies and Safety Nets, World Institute for Development Economics Research Discussion Paper 2002/22, Oxford.

Diaz Nieto J, Cook S, Lundy M, Fisher M, Sanchez D, Guevara E. 2006. A System of Drought Insurance for Poverty Alleviation in Rural Areas. Consultative Group on International Agricultural Research Working Paper. World Bank: Washington, DCHess, U, Richter, K. and Stoppa, A. 2002. Weather Risk Management for Agriculture and Agri- Business in Developing Countries, Climate Risk and the Weather Market Publication series, issue July 2002.

IFAD. 2010. Decision tools for rural finance, Rome. www.ifad.org/ruralfinance/dt/ index.htm.

IFAD and WFP. 2010. Potential for scale and sustainability in weather index insurance for agriculture and rural livelihoods, Rome. www.ifad.org/ruralfinance/pub/weather.pdf.

ISMEA. 2006. Risk management in agriculture for natural hazards. Rome: Istituto di Servizi per il Mercato Agricolo Alimentare.

Rao V.U.M, Bapuji Rao, B, Rajkumar Dhakar, Vijaya Kumar, P. And Rao, A.V.M.S. 2013. National Initiative on Climate Resilient Agriculture-AICRPAM Component, Annual Report 2012-2013, CRIDA, Hyderabad.

Seigel, P. and Alwang, J. 1999. An Asset Based Approach to Social Risk Management: A Conceptual Framework, World Bank Social Protection Discussion Paper 9926, Washington, DC.

Skees, J.R. 2003. Risk Management Challenges in Rural Financial Markets: Blending Risk Management Innovations with Rural Finance. Paper prepared for presentation at Paving the Way Forward for Rural Finance: An International Conference on Best Practices, June 2-4, 2003, Washington DC.

Skees, J.R., Barnett, B. and Hartell, J. 2005. Innovations in Government Responses to Catastrophic Risk Sharing for Agriculture in Developing Countries. Paper prepared for the workshop Innovations in Agricultural Production Risk Management in Central America: Challenges and Opportunities to Reach the Rural Poor, May 9-12, 2005, Antigua, Guatemala.

USAID. 2006. Index insurance for weather risk in lower-income countries, (Eds. J Skees, A Goes, C Sullivan, R Carpenter, M Miranda and B Barnett), GlobalAgRisk Inc., Lexington, Kentucky, USA. Washington, DC: United States Agency for International Development, www.microlinks.org/ev_en.php? ID=14239_201&ID2=DO_TOPIC.

Venkateswarlu, B, Maheswari, M, Srinivasa Rao, M, Rao, V.U.M, Srinivasa Rao, Ch, Reddy, K.S, Ramana, D.B.V, Rama Rao, C.A, Vijay Kumar, P, Dixit, S. and Sikka, A.K. 2013. National Initiative on Climate Resilient Agriculture (NICRA), Research Highlights (2012-13), Central Research Institute for Dryland Agriculture, Hyderabad,111 p.

World Bank. 2005. Managing agricultural production risk: Innovations in developing countries. Washington, DC: Agriculture and Rural Development Department (ARD), World Bank. Department (ARD), World Bank.

14

Agromet Advisory Services and Its Importance in Climate Change Scenarios

K.K. Singh

In recent past, extreme weather events are causing great concern in different sectors contributing to Indian economy. India Meteorological Department is doing yeomen's service by providing advance information including monitoring of the extreme events longs with proper advisories to the farming community by using state of art instruments & technology through efficient delivering mechanism of the information and ultimately help the farmers from incurring great loss. In many cases, the most severe impacts are felt when several extreme events occur together. The major extreme weather events related to agriculture are Tropical cyclones, floods, heavy rain and landslides, cold waves, fog, snow storms, Heat waves, Hailstorms, thunderstorms and dust storms, drought.

The challenges facing agriculture in India are ever increasing. First, agriculture is highly dependent on weather and subject to its variability. Second, the possible impacts of climatic variability and climate change leading to extreme weather events pose major challenges. Finally, the sustainability of intensive agriculture using current technologies is being questioned within the context of the global climate change. Long –term changes and increasing frequency of extreme weather events are likely to have adverse impacts on the agricultural sectors.

Weather and climatic information plays a major role before and during the cropping season and if provided in advance can be helpful in inspiring the farmer to organize and activate their own resources in order to reap the benefits. The Agro-meteorological Advisory Service (AAS) rendered by India Meteorological Department (IMD), Ministry of Earth Sciences (MoES) is a mechanism to apply relevant meteorological information to help the farmer make the most efficient use of natural resources, with the aim of improving agricultural production; both in quantity and quality. It becomes more and more important to supply climatological information blended with seasonal climate forecasts before the start of the cropping season in order to adapt the agricultural system to increased weather variability. Subsequent to this, short and medium range weather forecast based Agro-meteorological advisories become vital to stabilize their yields through management of agro-climatic resources as well as other inputs such as irrigation, fertilizer and pesticides. The main emphasis of the existing AAS system is to collect and organize climate/weather, soil and crop information, and to amalgamate them with weather forecast to assist farmers in taking management decisions. This has helped to develop and apply operational tools to manage weather related uncertainties in changing climate through agro-meteorological applications for efficient agriculture in rapidly changing environments.

Usefulness of Weather Forecasts for Agriculture

Climate-based strategic agronomic planning

For optimal productivity at a given location, crops and cropping practices must be such that their cardinal phased weather requirements match the temporal march of the relevant weather elements, and at the same time endemic periods of pests, diseases and hazardous weather are avoided. In such strategic planning of crops and cropping patterns, short period climatic data, both routine and processed (such as initial and conditional probabilities), have a vital role to play.

Occurrences of erratic and adverse weather

Agronomic strategies have to be devised to cope with the effects of erratic and adverse weather on agricultural production. For example, delay in the start of crop season can be countered by using short duration varieties of crops and thicker sowings and the effects of frosts can be prevented by resorting to irrigation or lighting trash fires. Medium range weather forecasts with a validity period that enable farmers to organize and carry out appropriate cultural operations to cope with are clearly useful. The following kind of forecast is mainly useful in operational Agrometeorology Short and Medium Range Forecast for agriculture is useful for the following activities

- Preparatory activities, including land preparation and preparation of plant material
- Planting or seeding/sowing; Management of crops, fruit trees and vines
- Application of fertilizer, irrigation; thinning, topping, weeding; pest and disease control; Management of grazing systems
- Harvesting, on-farm post-harvest processing Transport of produce; Livestock production activities (for dairy enterprises, beef systems, lamb and other livestock systems).

Weather and Climate Observation System

India Meteorological Department (IMD) has progressively expanded its infrastructure for meteorological observations on real time basis and to provide current and past observations for operational and research use. Surface observatories are located one in each district so as to meet the requirements of agricultural, transport and other sectors. IMD's present Upper air observational network comprises of radiosonde and pilot balloon, RS/RW observatories, Doppler Weather Radar etc. spread all over the country. Also a Doppler Weather Radar (DWR) provides velocity and spectrum width data (in addition to reflectivity). This capability permits the forecaster to view weather events like thunderstorms, hailstorm, heavy rainfall etc. in greater detail. The Doppler radar shows promise in depicting wind field patterns associated with cyclones, which can be used in assessment of wind damage potential and determination of the associated storm surge affecting the meteorological, hydrological and aviation products which are very useful in estimating the storm's centre, fixing its position and predicting its future path. In Numerical Weather Prediction models the Doppler data in digital form can directly be assimilated, which results in better rainfall estimation, better prediction of the storm track and point of landfall for tropical cyclones.

IMD is observing weather with satellite and radar. At present IMD is receiving and processing meteorological data from two Indian satellites Kalpana-1 and INSAT-3A. Kalpana-1 and INSAT-3A both have three channel Very High Resolution Radiometer (VHRR) for imaging the Earth in Visible (0.55-0.75 µm), Infra-Red (10.5- 12.5 µm) and Water vapour (5.7-7.1µm) channels having resolution of 2×2 kms in visible and 8×8 kms in WV and IR channels. In addition the INSAT-3A has a three channel Charge Coupled Device payload for imaging the earth in Visible (0.62-0.69µm), Near Infra Red (0.77-0.86µm) and Short Wave Infra Red (1.55-1.77µm) bands of Spectrum. All the received data from the satellite is processed and are archived in National Satellite Data Centre (NSDC), New Delhi.

The payload of INSAT-3D adds a new dimension to weather monitoring through its atmospheric sounding system, which provides vertical profiles of temperature, humidity and integrated ozone from surface to top of the atmosphere, and is expected for the next seven years making a difference for the weather forecasting and disaster warning systems for the country. This has two kinds of payloads 'Imager' and 'Sounder'. The six channel imager consists of broad visible (VIS), short-wave infrared (SWIR), Middle infrared (MIR), water vapour (WV) and two split-thermal infrared (TIR) bands. The optical bands (VIS & SWIR) have spatial resolution of 1 km. The MIR and TIR bands have 4 km spatial resolution. The WV band has 8 km spatial resolution. The 19 channel sounder will be able to probe the vertical structure of the atmosphere at regular time interval in a day. The Imager data will be able to provide satellite meteorological (sat met) products such as rainfall, land surface temperature (LST), land surface albedo, incident solar radiation, cloudiness, upper troposphere humidity (UTH), outgoing long wave radiation (OLR). The combination of all these products from 3D, improved weather forecasts, vegetation index product from INSAT 3A CCD and in situ data are valuable ingredients to generate real-time value-added information for enhanced operational agro-met advisory services in the country. The long-term datasets from INSAT suite will be able to aid in digital agro- climatic characterization in Fig.1.

IMD is maintaining a network of agro- meteorological observatories in farm environment. Beside this network, there are evapotranspiration (ET) stations, pan evaporation stations, soil moisture (SM) observatories, dew fall and radiation observations. Agro- meteorological Automatic Weather Stations (Agro-AWS) have been installed at 127 Agrometeorological Field Units (AMFUs). The data are being received in AWS Laboratory, Pune regularly. IMD has drawn plans to upgrade and strengthen its observation network in the coming years. Nationwide network of Automatic Weather Stations and Doppler Weather Radar network will also increase.

IMD has long time series of various climate data in its archive. National Climate Center (NCC) at Pune generates many climate data products for smaller spatial and temporal scales for the user community.

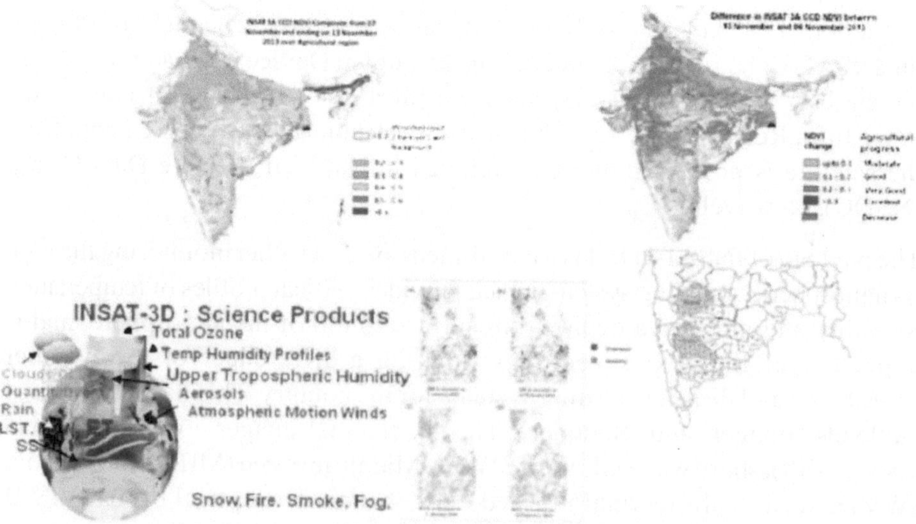

Fig. 1: INSAT 3A CCD NDVI, Agromet products

These data products include (i) Daily gridded (10 × 10) rainfall and temperature data, (ii) Daily gridded (0.50 × 0.5°) rainfall data, (iii) District wise normal for various surface parameters, (iv) Marine climate summaries for Indian Ocean region. IMD operationally prepares gridded rainfall data on 0.50 × 0.50 spatial resolution by processing spatially continuous satellite derived rainfall (TRMM) and weather observation from existing weather network (IMD Gauge). These data help in monitoring strength of monsoon, studying oscillation of different periodicity, extended range prediction and to fine tune the dynamical models. The Centre will very soon bring out daily rainfall data on 0.250× 0.250 spatial scale. Similarly daily gridded temperature data would also be brought out on smaller spatial scale.

Weather Forecasting System

India experiences large spectra of weather events having spatial scale of less than 1 km to more than 1000 km and temporal range of less than an hour to more than a week. Different parts of the country experience different kinds of weather conditions such as Winter season (Jan- Feb) is characterized by Western Disturbances, Cold Wave, Fog; Pre-Monsoon (Mar-May) by Cyclonic Disturbances, Heat Wave, Thunder Storms, Squalls, Hail Storm, Tornado; Monsoon (Jun-Sep) by Southwest Monsoon Circulation, Monsoon Disturbances; and Post-Monsoon (Oct- Dec) by Northeast Monsoon, Cyclonic Disturbances. Meteorological Department in the country was one of first scientific departments which started scientific research on weather systems as early as in 18th century. Indian monsoon is a field of active research all over the world as it is a unique weather event which covers several oceans and continents but depicts its characteristic phase over Indian sub-continent. The monsoon rains are crucial for agricultural production in the country as most parts of the country depend on this rain for day-to-day agricultural operations.

Prediction of weather systems in different spatial and temporal scale over the Indian region therefore assumes considerable importance. Several organizations are engaged in developing techniques for prediction of these systems. In the recent past, IMD has made enormous improvement in the accuracy and lead time of forecasts for various usages including AAS.

IMD has made Global Forecast System (GFS T574/L64, based on NCEP) operational in December 2009, implemented on IBM based High Power Computing Systems (HPCS). This incorporates Global Statistical Interpolation (GSI) scheme as the global data assimilation for the forecast up to 7 days. In horizontal, it resolves 574 waves (» 22 Km in the tropics) in spectral triangular truncation. The model has 64 vertical levels (hybrid; sigma and pressure). Currently, it runs twice in a day (00 UTC and 12 UTC). In addition to this, the meso-scale forecast system WRF (ARW) with 3DVAR data assimilation is being operated daily twice, at 27 km, 9 km and 3 km horizontal resolutions for the forecast up to 3 days using initial and boundary conditions from the IMD GFS-574/L64 (horizontal resolution over the tropics ~ 22 km). At ten other regional centres of IMD, very high resolution meso-scale models (WRF at 3 km resolution) are made operational. NWP based objective forecast products are prepared to support cyclone warning service. Doppler weather and meso-scale dynamical model based now-cast system was made operational for the national Capital of Delhi in Fig. 2. A full range of NWP products at different spatial and temporal scale are routinely made available on the IMD web site www.imd.gov.in. Next generation model will target more accurate ways to initialize forecasts by improving data assimilation techniques, more accurate parameterizations, and improved estimates of uncertainty in the modelling system and in forecasts. Coupling of the atmosphere to the underlying surface – land and ocean – is important and also critical for forecast lead times extending into fortnightly and monthly timescales. To adequately resolve the transient systems responsible for rainfall during monsoon and their movement, it is important to use models with high resolutions both in the horizontal and vertical. By 2020, plan is to run global models of about 10 km resolution with about 100 levels in vertical (high-resolution GFS such as T1148-L64 (16 km) to T1500-L128 (13 km), and regional models with about one km resolution. The physical parameterizations need to be improved/optimized for higher resolution models. As the model grid size approaches the horizontal scales of moist deep convection, special attention will be given to the parameterization of moist processes. Cloud resolving models are required to be embedded in the global and meso-scale models as clouds are the major cause of uncertainty in present day models. All current and future global NWP system will have to address the uncertainty inherent in the prediction over all temporal and spatial scales. The challenge is to increase the number of ensemble members which will further improve its value for all applications.

Recent studies suggest that an interactive upper-ocean in a high resolution coupled general circulation model significantly enhance the prediction skill of weather in medium range. Experiments at ECMWF and other centers have started showing the benefits of coupling in improving the skill in medium and 2 weeks to 1 month monsoon forecasts. Therefore ocean model with very high vertical resolution (1 m) in the upper ocean has to be coupled with a higher frequency (1 hour) of interaction with atmosphere model. Ocean data assimilation will be required to initialize the ocean model properly. Similarly a land surface model with its assimilation has to be also coupled to form the full 'Coupled Earth System Model'.

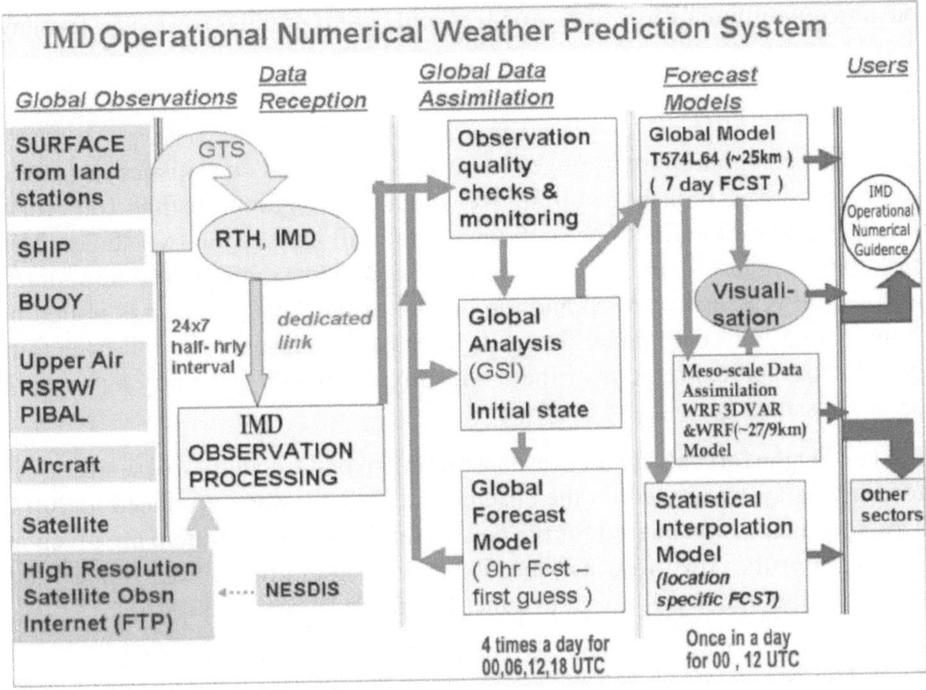

Fig. 2: Operational NWP system

District Level Medium Range Weather Forecast

IMD implemented a Multi-model Ensemble (MME) based district level quantitative forecasts in the operational mode since 1 June 2008, as required for the Integrated Agro-advisory Service of India. Five NWP models considered for this development work are: (i) IMD GFS T574, (ii) ECMWF T799, (iii) JMA T899, (iv) UKMO and (v) NCEP GFS. Model outputs of the constituent models are interpolated at the uniform grid resolution of $0.25^0 \times 0.25^0$lat/long. Further, the weight for each model at each grid is determined objectively by computing the correlation co-efficient between the predicted rainfall and observed rainfall. High resolution gridded rain- gauge data produced operationally at National Centre of IMD Pune are used for development and validation of the forecasts. The ensemble forecasts (day 1 to day 5 forecasts) are generated at the $0.25^\circ \times 0.25^\circ$ resolution. The ensemble forecast fields are then used to generate district level forecasts by taking average value of all grid points falling in a particular district. The products comprise of quantitative forecasts for 7 weather parameters viz., rainfall, maximum temperature, minimum temperatures, wind speed, wind direction, relative humidity and cloudiness. These products are disseminated to Regional Meteorological Centres and Meteorological Centres of IMD located in different states. These offices undertake value addition to these products using local conditions and synoptic interpretation of model out -

put and communicate to 130 Agro-Met Field Units (AMFUs) on every Tuesday and Thursday.

Long Range Forecast

IMD has been issuing long-range forecasts (LRF) based on statistical methods for the southwest monsoon rainfall over India (ISMR) for more than 100 years. The forecast for the South-West monsoon rainfall is issued in two stages. The first stage forecast for the seasonal (June to September) rainfall over the country as a whole is issued in April and the update of the April forecast is issued in June. Along with the update forecast, forecast for seasonal rainfall over four broad geographical regions of India and July rainfall over country as a whole are also issued.

For issuing the forecast for the seasonal rainfall over the country, a new statistical forecasting system based on the ensemble technique was introduced in 2007. The 8 predictors considered for the new ensemble forecast system are given in Table-1. For the April forecast, the first 5 predictors given in this Table 1 are used. For the updated forecast in June, last 6 predictors are used.

Table 1: Details of the 8 predictors used for the new ensemble forecast system

Details of 8 predictors	Predictor	Used for forecasts in
1	NW Europe Land Surface Air Temperature	April
2	Equatorial Pacific Warm Water Volume	April
3	North Atlantic Sea Surface Temperature	April and June
4	Equatorial SE Indian Ocean Sea Surface Temperature	April and June
5	East Asia Mean Sea Level Pressure	April and June
6	Central Pacific (Nino 3.4) Sea Surface Temp.Tendency	June
7	North Atlantic Mean Sea Level Pressure	June
8	North Central Pacific wind at 1.5 Km above sea level	June

Dynamical (Numerical) Forecasting System for Month and Season

The experiments of this study were carried out with the National Centers for Environmental Prediction (NCEP) seasonal forecasting model (SFM), which provides regional details for the Global Spectral Model (GSM). The SFM (Kanamitsu *et al.*, 2002) uses the spectral transform method to solve the primitive equations. The model is a non-hydrostatic global spectral model with a triangular truncation of 62 spherical harmonics (T62), equivalent to a horizontal resolution of about 280 km was chosen. The model of this study used a vertical sigma coordinate system which contains 28 layers. The basic model Physical processes and parameterization schemes were developed from reanalysis-II version of NCEP Medium Range Forecast model.

The ensemble simulation was carried out for 31 year period from 1982 to 2012 to prepare model climatology. Each forecast was initialized at first 10 days of each month and run continuously for next five months. The initial conditions corresponding to 0000Z UTC were obtained from NCEP reanalysis. The forecasted SST from Coupled model, NCEP Climate Forecast System version 2 (CFSv2) were used as SST condition. SST Boundary condition prepared by averaging of CFSv2 SST forecast for first 10 days initial conditions of each month (Total 40 members). For making real time forecast, initial and boundary conditions are taken from NOAA operational Modal archival and distribution system.

Special Weather Forecast for Extreme Events

Special weather forecast for agriculture provides the necessary meteorological input to assist farmers in making decisions. The requirements for these special forecasts will vary from season to season and crop to crop. Special forecast issued are as follows:

- Tropical storms (cyclones, hurricanes, typhoons, etc.) associated with high winds, flooding and storm surges.
- Floods (other than those related to tropical storms) heavy rains due to monsoons, water logging and landslides.
- Severe thunderstorms, hail storms, tornadoes and squalls.
- Drought and heat waves.
- Cold spells, low temperature, frost, snow and ice-storms.
- Dust storms and sand storms.
- Pest and diseases of crop and livestock.

Early Warning Systems

IMD has developed early warning systems for extreme weather events like drought, flood, tropical cyclone, heat and cold wave, fog, hail. For drought monitoring and assessment, IMD has established system for real-time monitoring of rainfall on daily, weekly, monthly and seasonal basis for the districts/ meteorological subdivisions/ state in the country. IMD is preparing indices based aridity anomaly maps on weekly basis that gives information about the moisture stress experienced by growing plant.

This analysis would indicate qualitatively retardation in the plants growth and so poor yields. Indirectly, this may also be helpful for irrigation scheduling, the amount and the time at which the water is badly needed by the plant. IMD is also preparing Standard Precipitation Index maps which describe dry and wet periods in the same way. SPI is an appropriate tool for monitoring agricultural drought but in combination with other information: water balance, crop condition, etc. Computation of SPI is done at a monthly time scale.

भारत मौसम विज्ञान विभाग अनावृष्टि अनुसंधान अनुभाग–पुणे

Indian Meteorological Department Drought Research Unit-Pune

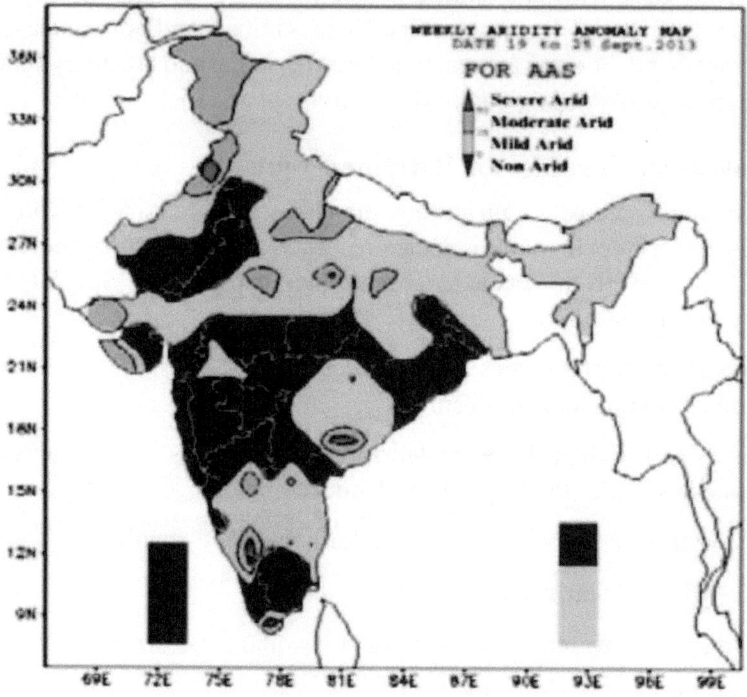

Fig. 3: Weekly aridity anomaly map for AAS

Agromet advisory during cyclonic storm 'NILAM' (October, 2012) for livestock

Due to heavy rain along with high wind-speed, farmers in North Eastern Zone of Tamil Nadu are advised to house animals in well-constructed shed, protect the young animals from chillness and rain, avoid stagnation of water nearby animal shed and clean the bushes to avoid multiplication of files so that vector borne diseases like anaplasmosis, theilerosis, bebesisosis and trypanosomisis can be prevented.

Agromet Advisory Service in India

Agrometeorological Advisory Service (AAS) rendered by India Meteorological Department (IMD), Ministry of Earth Sciences (MoES) in active collaboration with ICAR, State Agriculture Universities, State Department of Agriculture etc. is a step to contribute to weather information based crop/livestock

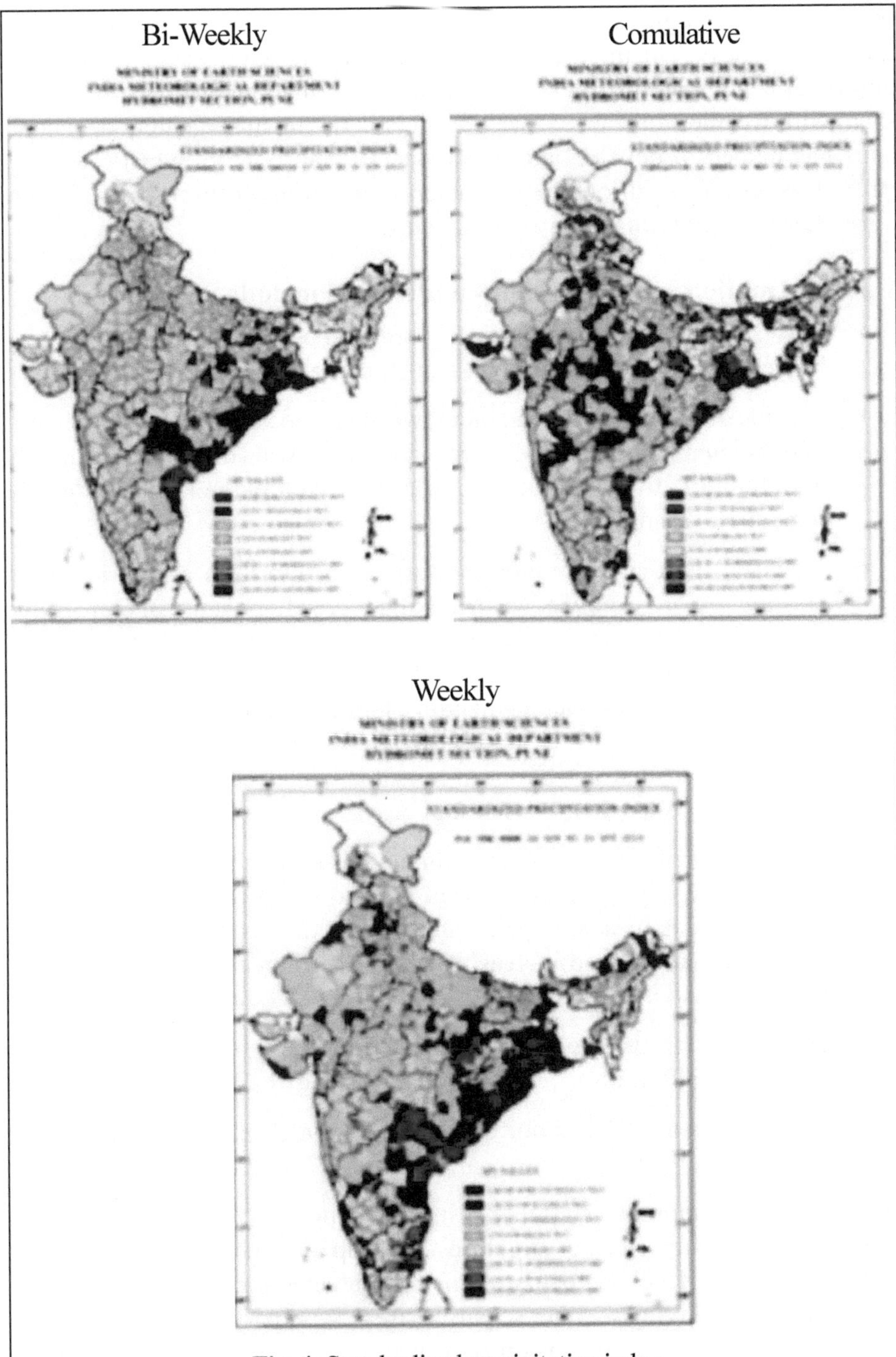

Fig. 4: Standardized precipitation index

management strategies and operations dedicated to enhancing crop production and food security. The main emphasis of the existing AAS system is to collect and organize climate/weather, soil and crop information, and to amalgamate them with weather forecast to assist farmers in taking management decisions. This has helped to develop and apply operational tools to manage weather related uncertainties through agro-meteorological applications for efficient agriculture in rapidly changing environments.

The information support systems under AAS include

- Provision of weather, climate, crop/soil and pest disease data to identify biotic and abiotic stress for on-farm strategic and tactical decisions.
- Provide district specific (Pan India) weather forecast (rainfall, cloudiness, maximum/ minimum temperature, wind speed, wind direction, maximum/ minimum relative humidity) up to 5 days with outlook for rainfall for remaining two days of a week.
- Translate weather and climate information into farm advisories using existing research knowledge on making more efficient use of climate and soil resources through applications of medium range weather forecast. A broad spectrum of advisories include weather sensitive farm operations such as sowing/ transplanting of crops, fertilizer application based on wind condition & intensity of rain, pest and disease control, intercultural operations, quantum and timing of irrigation using meteorological threshold and advisories for timely harvest of crops.
- Introduction of the state-of-art technologies such as remote sensing, GIS and crop simulation model based decision support system for agro-meteorologists to adapt agricultural production systems to weather & climate variability and increasing scarcity of input such as water, seed, fertilizer, pesticide etc.
- Develop effective mechanism to on time dissemination of agromet advisories to farmers,
- Effective training, education and extension on all aspects of agricultural meteorology.

Being a multidisciplinary and multi- institutional project, AAS project is being implemented through five tier structures to set up different components of the service spectrum. It include meteorological (weather observing & forecasting), agricultural (identifying weather sensitive stress & preparing suitable advisory using weather forecast), extension (two way communication with user) and information dissemination (Media, Information Technology, Telecom) agencies.

Agrometeorological Support for Farm Management

Weather based farm advisories as support system has been organized after characterization of agro-climate, including length of crop growing period, moisture availability period, distribution of rainfall and evaporative demand of the regions, weather requirements of cultivars and weather sensitivity of farm input applications. All this knowledge is used as background information. Following are the ingredients of a typical Agromet Advisory Bulletin to reap benefits of benevolent weather and minimize or mitigate the impacts of adverse weather;

i. District specific weather forecast, in quantitative terms, for next 5 days for rainfall, cloud, max/min temperature, wind speed/ direction and relative humidity, including forewarning of hazardous weather event likely to cause stress on standing crop and suggestions to protect the crop from them.

ii. Weather forecast based information on soil moisture status and guidance for application of irrigation, fertilizer and herbicides etc.

iii. The advisories on dates of sowing/planting and suitability of carrying out intercultural operations covering the entire crop spectrum from pre-sowing to post harvest to guide farmer in his day–to-day cultural operations.

iv. Weather forecast based forewarning system for major pests and diseases of principal crops and advises on plant protection measures.

v. Propagation of techniques for manipulation of crop's microclimate e.g. shading, mulching, other surface modification, shelter belt, frost protection etc. to protect crops under stressed conditions.

vi. Advisory on contingency plan under extreme weather situations.

vii. Reducing contribution of agricultural production system to global warming and environment degradation through judicious management of land, water and farm inputs, particularly pesticides, herbicides and fertilizers.

viii. Advisory for livestock on health, shelter and nutrition.

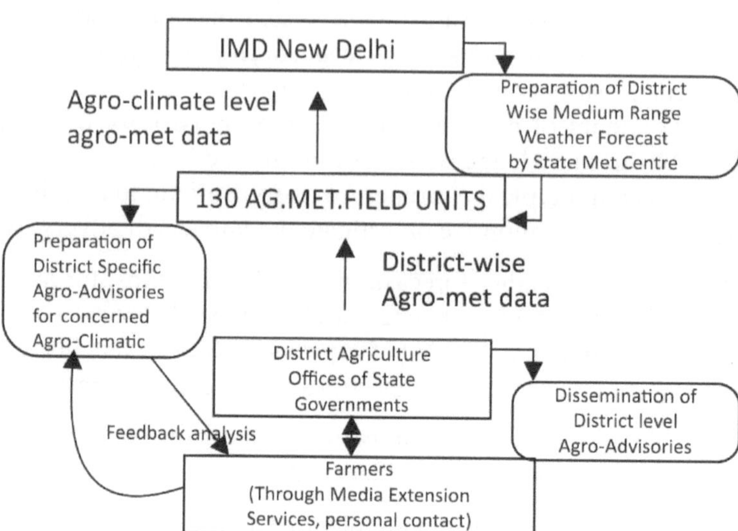

Fig. 5: The flow chart for District Level Agro-Met Advisory Service (DAAS) System

The support on above is rendered through preparing district specific agrometeorological advisory bulletins by Agro-Meteorological Field Units (AMFUs). Inter-disciplinary group of agricultural and extension specialists at AMFUs formulates weather based farm advisory bulletin. These bulletins contain location specific and crop specific farm level advisories tailored to meet the farmers' need. The bulletins are encoded in a format and language which is easy to comprehend by the farmer in his decision making processes. The Agrometeorologists play a vital role in the encoding and decoding of the messages from the meteorologists to the agricultural sector. Crop weather calendars initially developed by IMD are used in preparation of weather based advisory bulletin. Recently All India Coordinated Research Project on Agrometeorology (AICRPAM) units in different states coordinated by CRIDA, Hyderabad have refined these calendars including pest & disease component which is very useful in crop specific advisory preparation.

The Agro-met Advisory Bulletins are issued at district, state and national levels to cater the needs of local level to national level. The State Level bulletin is a composite of district bulletins helping to identify the distressed district of the state as well as plan the supply of appropriate farm inputs such as seeds, irrigation water, fertilizer, pesticides etc. These bulletins are jointly prepared by State Meteorological Centre of IMD and AMFUs and mainly used by State Government functionaries. National Agromet Advisory Bulletins are prepared by National Agromet Advisory Service Centre, IMD, Pune, using inputs from various states. This bulletin helps identify stress on various crops for different regions of the country and suitably incorporate advisories. Ministry of Agriculture is prime user of these bulletins, which help take important decisions in Crop Weather Watch Group (CWWG) meetings at national level. The bulletins are also used by a large number of other agencies including fertilizer, pesticide industries.

Agricultural scientists at Agrometeorological Field Units have started using water balance, crop simulation models as a decision support tool for helping with weather forecast based farm management decision making. Agro- meteorological Field unit can objectively assess the impact of skipping irrigation at a particular phenophase of a crop on its dry matter yield though with some uncertainties. Agrometeorologists can consider many of the factors involved, and answer the question with a reasonable estimate. The crop models are also being used as technique for prediction of different phenophases and final yield.

System of Agrometeorological Advisory Dissemination

The task is to provide information to help farmers make the best possible use of weather and climate information (Aggarwal, 2002). Critical factors for successful dissemination include relevance information to weather and climate sensitive decision making in agriculture, followed by good outreach and uptake at grass root level. To ensure delivery of information to the farming community, a multi-mode dissemination system for agro-meteorological advisories is essential in which besides the conventional modes (radio, television & print media), the emerging modes of communication such as mobile phones and internet are also deployed. To smarten the service through continuous up-gradation of service one needs back flow of information on quality and relevant of information or demand for specific product, hence two way communication has to be the integral part of the dissemination system.

AAS has considered different aspects pertinent to the flow and content of information and accordingly evolved a strategy for dissemination of agro-meteorological information. Although concerted efforts are being made to set up two way communications, but as of now the information flow is largely one-way. Although, Agro-meteorological Field Units (AMFUs) have limited interaction with the farmers, good communication and working relationships have been set up with the agricultural extension, Krishi Vigyan Kendra (Agriculture Science Centres), Kisan (Farmer) Call Centre etc. to promote participatory methods for interactions with farmers. Information is disseminated through multi-modes of delivery including mass and electronic media. It include, All India Radio, Television, Print Media (local news paper in different vernacular languages), internet (Web Pages) as well as group and individual relationships through email, telephone etc. The agrometeorological bulletins always contain dynamic information and hence, repetitive dissemination is being made. This reiterative process also helps to address large temporal and spatial variability having significant influence of weather & climate on agriculture.

The use of electronic media such as e-mail or the Internet depend on the availability and access of these methods to the users, which is picking up in India particularly through initiative of Department of Information technology, who is in the process of setting up a very strong net work of common service centres (CSC). AAS is a scalable system, which can be incrementally developed and extended to cover all the farmers (crops) of India in a cost-effective manner. It enables the farmer to receive both crop and location-specific expert advice in a timely manner. With the advent of computers and Internet, emphasis is often being given to electronic communication systems. However, TV and radio services are still the best ways of communicating advisories among rural people as these are not only fast methods, but also large and illiterate masses can be contacted by these methods. Broadcasting of advisories in vernacular language provides an edge on other means of communication. Under Integrated Agromet Advisory Service (IAAS) scheme at IMD/MoES efforts are being made to strengthen the outreach of the agromet advisory as per the need of the farmers. Under this project advisories are primarily disseminated to the farmers by mass mode, outreach at village level and human face for advisory dissemination. Advisories are being disseminated to farmers through the following multi-channel systems;

i. All India Radio (AIR) and Doordarshan
ii. Private TV and radio channels
iii. Newspaper
iv. Mobile phone / SMS
v. Internet
vi. Virtual Academy / Virtual Universities / NGOs
vii. Kisan Call Centres / ICAR and other related Institutes / Agricultural Universities / Extension network of State
viii. Krishi Vigyan Kendra (KVKs)

For dissemination through SMS on mobile phone, an information platform has been created, which allows the existing Agromet Field Units (AMFUs) to provide the information in a convenient and timely manner. The advisories are crop and location specific and delivered within actionable time to the farmers. Under IVR system the information from AMFUs for each state are collected and then stored, and converted into voice where the farmer would be calling and receiving the desired information. IMD is sending these SMS to farmers with the help of many private and government companies. IMD has so far partnered with Reuter Market Light (RML), Handygo, NOKIA, IFFCO Kisan Sanchar Ltd. (IKSL), National Bank for Agriculture and Rural development (NABARD), Maharashtra State Agriculture Department (Government of Maharashtra) and KISAAN SMS portal. Number of farmers covered under SMS is presently more than 10.0 million.

Crop Yield Forecasting

Reliable and timely forecast of agricultural commodity provides important and useful inputs for proper planning. This enables planners and decision makers to predict how much to import in case of shortfall or optionally, to export in case of surplus. It also enables government to put in place strategic contingency plans for redistribution of food during times of famine. In India, there is also growing need for micro level planning and particularly the demand for crop insurance, which increases the need for field level yield statistics. Forecasting Agricultural Output using Space, Agrometeorology and Land based observations (FASAL) involves developing econometric, remote sensing and agromet based model to generate multiple crop yield forecasts at national, state and district level starting with crop sowing to end of season for 11 major *Kharif* and *Rabi* crops *viz.*, Rice, Jowar, Maize, Bajra, Jute, Ragi, Cotton, Sugarcane, Groundnut, Rapeseed & Mustard and Wheat.

Forecasts envisaged under the FASAL Scheme

In IMD the crop yield forecasting models are developed at different stages of the crops combining statistical, crop growth simulation and other methods/ techniques, which take into account the influence of weather and technological advances on crop yield. The model requires the meteorological data (rainfall, temperature, humidity, bright hours of sunshine with wind speed, wet spell etc.) as well as technological data as the inputs to estimate the yield. Using agromet models and database developed, IMD is providing the yield forecast at district/ state/ national scale to DAC since *Rabi* 2010-2011.

Statistical Model

IMD has developed crop yield forecasting models based on multiple correlation and regression technique. It is a linear combination of predictors (both meteorological parameters and technological parameters), which takes into account the influence of weather and technological advances on yield. Based on the above methodology, pre-harvest crop yield forecast models have been developed for *kharif* rice & *rabi* wheat for all the meteorological subdivisions where these crops are grown predominantly.

Crop Growth Simulation Model

IMD is also developing Crop growth simulation models to estimate crop yield as a function of complex interaction of different physiological processes with the environment, biotic and abiotic factors. These models estimate biomass production using daily crop growth simulator. These models will be developed for major crops of the country to provide yield forecast.

Although, a good progress has been made in this regard but much is required to be done. There is need to develop methodologies for Remote Sensing and Conventional Data Merging. Concerted efforts are needed for ground based data collection, satellite data collection, GIS software applications, operational applications of meteorological satellite data, weather radars and the monitoring of cropping season by meteorological and remote sensing data to equip AAS units for yield forecasting and to generate better advisories. The system is working to improve these aspects of the service (K.K.Singh *et al.,* 2012).

IMD is implementing Agromet component of the scheme in coordination with 23 State Agromet Centres (SAMCs), 46 principal Agro-Met Field Units (AMFUs) and IASRI in the country to develop agromet models and issue in-season crop yield forecast based on statistical and crop simulation models. This information is shared with Mahalanobis National Crop Forecast Centre (MNCFC), New Delhi.

Future Needs and Strategies

Though Agro met Advisory service is being provided efficiently at district level in the country, still there is need to strengthen in terms of observation, seamless weather forecast, manpower, real time information flow, research and development (R&D), dissemination etc. There is need to develop methodologies for remote sensing and conventional data merging. Concerted efforts are needed for ground based data collection, satellite data collection, GIS software applications, operational applications of meteorological satellite data, weather radars and the monitoring of cropping season by meteorological and remote sensing data to equip AAS units to generate better advisories. Though district level medium range weather forecast is being prepared for agromet services, there is an urgent need to develop and issue high resolution accurate weather forecasts to help tailor crop plan for the farmers at village level. Thus there is a need to scale up the service from district to block level with dissemination up to village level across the country to meet the end user's requirement. Such idea leads to take up the new concept of establishing the Garmin Krishi Mausam Sewa (GKMS) scheme under Integrated Agromet Advisory Services in the country during the XIIth Five Year Plan. Some of the important activities being planned under GKMS are as follows.

Seamless Weather Forecast

a. Block level forecast

Though district level medium range weather forecast is being prepared for agromet services, there is an urgent need to prepare and issue high resolution (9 km × 9 km or finer using WRF model) accurate weather forecast at sub-district level. Thus it is proposed to generate block level forecast during XIIth Five Year Plan. On pilot mode block level weather forecast for the parameters like rainfall, maximum and minimum temperature, cloud amount, maximum and minimum relative humidity, wind speed and wind direction for 342 blocks in the country considering one district per state is being developed. This procedure would be followed until the bias removal techniques and high resolution (10-12 km) GCM becomes operational. Neural networks would be tried for MME forecasts and Kalman filter would be tried for maximum and minimum temperature as the bias removal techniques. As such, this forecast has been started for 37 selected districts in the country. By the end of XIIth FYP, block level forecast along with advisories will be scaled to cover the maximum districts in the country.

b. Extended range forecast

b.1. Under National Monsoon Mission set up by Ministry of Earth Sciences, Indian Institute of Tropical Meteorology (IITM), Pune in collaboration with IMD and National Centre for Medium Range Weather Forecasting (NCMRWF) has started experimental forecast for monsoon rainfall during next four pentads (up to 20 days) issued every week. IMD has started National Agromet Advisory bulletins based on Extended Range Weather Forecast during southwest monsoon 2013 to fulfill the needs of farmers and other users. Bulletin was prepared and issued for the next fortnight with update on every Friday. Successful implementation of this initiative would help the farmer to get more lead time to manage his weather based farm management.

b.2. In another effort entitled "Development and Application of Extended Range Forecast System (ERFS) for Climate Risk Management in Agriculture" supported by Ministry of Agriculture, Government of India, the key objectives to develop the experimental extended range test forecast from month to season scale for Indian region and its application for the development of climate risk management tools to address these risks with the involvement of agricultural scientists, farmers and service support agencies at field levels. This project integrates risk management and climate science research, involving leading institutions of India and abroad. The project adopts a very strong demonstration approach, focusing on select districts in nine states that face significant livelihood impacts due to variability in the southwest monsoon: Andhra Pradesh, Gujarat, Maharashtra, Madhya Pradesh, Orissa, Himachal Pradesh, Rajasthan, Tamil Nadu, and Uttarakhand.

These forecast products include both type of forecasts *viz*., deterministic as well as probabilistic forecasts. Methods used to generate the deterministic rainfall and mean temperature forecast include Singular Value Decomposition (SVD) based Regression, Supervised Principal Component Regression, Canonical Correlation Analyses. In these regression techniques, predictors (independent variables) are different general circulation models' rainfall and temperature products while the predicted (dependent variables) is IMD rainfall and temperature. All these above stated forecasts were combined using some statistical techniques and has been converted in probabilistic forecast to see the chance of occurrence of particular event. Probabilistic forecast is made in three categories *viz*., Below Normal, Near Normal and Above Normal. These categories are based on the previous year's observed data set (IMD's gridded data). The experimental forecast products thus generated since monsoon 2009 were disseminated to all the selected demonstration sites through IMD for its verification and application purposes. The performance of experimental test forecast products were evaluated at each selected pilot demonstration sites with the observed data of temperature and rainfall during the month. Performance of real time experimental forecast for summer monsoon seasonal rainfall (JJAS) is shown in the Fig. 4.

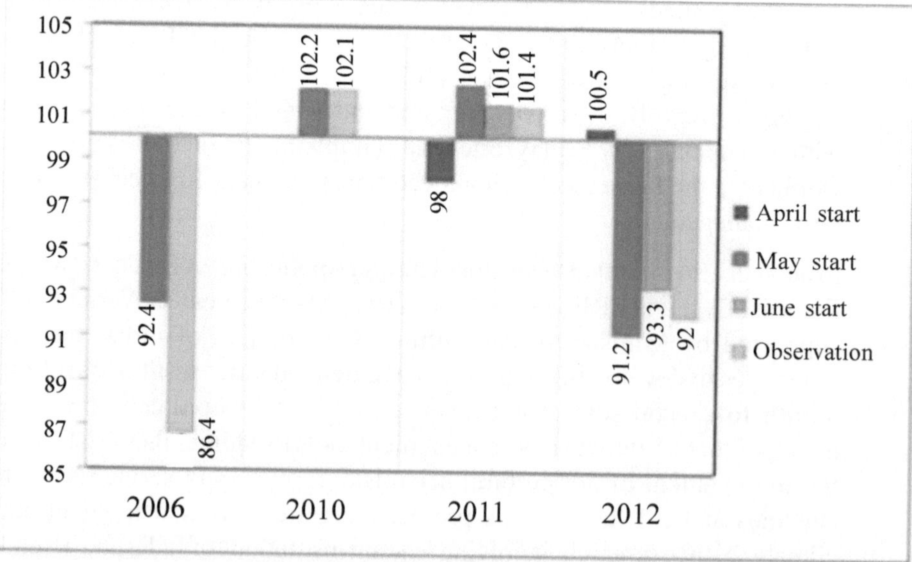

Fig. 6: Performance of real time experimental forecast for summer monsoon seasonal eAgromet System

The idea of informed decision making at farm/ farming system level gains more reality if integrated with reliable seamless weather forecast during the crop season. Crop simulation models and DSS are being increasingly used in decision making for risk minimization in agriculture. These models need daily weather data for crop season. The major challenge in linking seasonal forecast to crop model is mismatch between spatio- temporal scale of forecast and crop model. On the one hand, these forecasts are available at bigger spatial domain for a month or season and on the other, crop simulation models can be run at point scale or smaller grid space. Forecast needs to be downscaled spatially and temporally. Crop model based generated viable management options against a given forecast require frequent validation through farmers participatory meetings. In order to make use of extended range forecast for decision making on strategic agriculture planning and risk management and integrate with ongoing advisory service system, a range of climatic risk management tools were developed.

1. Extended/ seasonal climate forecast.
2. Climatic risk matrices: Climate risk analysis for crop under various farming situations showing the impact of soil (type and depth), time of planting and weather requirement and extremes at various phenopahese.
3. Stochastic Weather generator: Downscaling can be achieved by linking a seasonal forecast with a weather generator. Nonhomogeneous hidden Markov model (NHMM) or HMM are used as a promising approach to constructing multi-station weather generators.
4. Crop simulation model and Decision support system: Crop models embedded in Decision Support System for Agro-technology Transfer (DSSAT), Info-Crop, APSIM etc. are process- oriented, management-level model of crop growth and development and can be linked to daily sequence of desegregated weather data to generate crop growth performance against a number of weather realizations.
5. Farmers participatory meeting: A very critical process to validate/refine the above findings as viable options for farm management.

Pilot study demonstrated use of monthly forecast on rainfall and temperature by establishing an end-to-end information generation and application. This was implemented with participation of agricultural universities/research centres, service support agencies at field levels and farmers for better crop planning and management in the following areas:

- Land preparation
- Nursery raising
- Irrigation scheduling
- Urea application to rainfed crops in normal rain forecast
- Frequent intercultural operation to conserve soil moisture
- Contingency plan under delayed monsoon onset rains
- Temperate apple crops- Chilling hours to help skip hormonal spray for bud breaking/ flower setting
- Nomadic farming hills- sheep, goat
- Livestock- malnutrition and disease infection under deficit rainfall

Future plan includes the integration of ERFS experimental system in IMD for its operationalization, generation of seamless weather forecast for agriculture risk management; expand to all parts of the country and R&D for continuous improvement in weather forecast products at fortnightly to monthly scale.

As there is a need to develop a computer- based decision support system and to automate the process of advisory bulletin preparation and dissemination, IMD in collaboration with Indian Institute of Information Technology, Hyderabad is generating a software for preparing advisory bulletin under the scheme "IT-based eAgromet software for automated AAS bulletin preparation" to prepare district/block level AAS bulletins. It simplifies the preparation of district level AAS bulletin as it acts as an organizing tool for weather forecasts and agromet advisories with the help of repository of past AAS bulletins. A server has been launched by IIIT (Indian Institute of Information Technology) for the said purpose. Number of AMFU scientists has already been trained and remaining AMFU scientists will be trained in utilization of the software and it will be implemented in GKMS.

Economic Impact Assessment

Impact assessment framework made by National Centre for Agricultural Economics and Policy Research (NCAP) was adapted to assess economic impact by taking field experiments conducted from 2003-2007 at 15 locations spread in country covering filed crops, vegetables and horticulture (Rathore and Maini, 2008). Forecast based agro-advisories were issued by NCMRWF to AAS users. An objective comparison between AAS and non AAS farmers at village level shows that impact depends on the risk taking ability of farmers. In quantitative terms, study reports that the AAS farmers were able to reduce the cost of cultivation by 2-5% except in case of fruits where the cost of cultivation has increased by 5-10%. This shows that the right selection of fertilizers and seed due to organization of awareness programme in the village and spraying of appropriate pesticide due to advisory saved input cost. It is also observed that the yield increased by almost 10-25% in most of the crops with maximum increase in the fruit crops.

National Council of Applied Economic Research (NCAER), an independent agency, during 2009-2010 estimated that the economic benefit from the use of weather information as the product of the percentage of farmers receiving information, scenario-wise, times the percentage of farmers benefiting from the information times average profit, crop-wise, attributable to weather information times the total national production of crops. Conversion factors, crop-wise, were used to convert farmers' financial profits to economic profits (NCAER, 2010). At present only 10 to 15 percent of the farmers are benefitting from the SMS services. The economic profit estimates vary between Rs. 50,000 crore (where currently 24 percent farmers receive weather information by all means) to Rs. 211,000 crore (where all farmers receive weather information).

In order to judge worthiness and role of the agromet service in improving the agricultural production system and farm management, its economic impact will be assessed through conducting extensive feedback surveys as guided by an independent agency. Studies will be undertaken at different representative sites covering principal crops. The study will comprise of *kharif* (monsoon) and *rabi* (winter) crop seasons including livestock. The objective of the studies will be to observe whether weather based advisories have a positive impact on the overall yield and also help in reducing the cost of cultivation.

References

Aggarwal, P.K., 2002. Climate Change and Agriculture: Information Needs and Research Priorities. International Conference on Capacity Building for Climate Change 21 October 2002. www.ficci.com/media-room/speeches-presenta- tions/2002/oct/oct-climate-aggarwal.ppt.

Kanamitsu, M., A., Kumar, H., Ming, H., Jung, J., Kyung, W.I., Wang, F., Yang, S., Hong, , P., Peng W., Chen, S. Moorthi, and M. Ji, 2002, NCEP Dynamical forecast System 2000., Bull. Amer. Meteor.Soc.,83: 1019-1037.

National Council of Applied Economic Research (NCAER). 2010. Impact assessment and economic benefits of weather and marine services. Supported by Ministry of Earth Sciences, New Delhi.

Singh K.K., A.K.Baxla, R.K.Mall, P.K.Singh, R.Balasubramanian and Shikha Garg. 2012. Wheat yield predication using CERES-Wheat model for decision support in agro-advisory and Journal of Vayu Mandal. Vol.No 35 &36 (1-4): page No. 97-109; 2009 &2010.

Rathore, L.S. and P. Maini. 2008. Report on Economic impact assessment of agrometeorological advisory service of NCMRWF. Ministry of Earth Sciences, New Delhi.

15

Wastewater: A Wealth in the Face of Water Scarcity Due to Climate Change

P. Dey

The UN designation of 2013 as the International Year of Water Cooperation aptly demonstrates growing gap between water supply and demand. These symptoms, which are already visible in a few regions around India, are soon to assume a national proportion and may become permanent feature of the water sector in the country, unless suitable policies are adopted quickly to manage water demand and supply at different levels. Water demand is growing fast due to a rapid population growth and economic activity, but water supply is not growing at the same rate because of serious ûnancial and physical limits for supply augmentation. Water resources developed at present amount to about 680 km^3, which constitutes 61% of the utilizable potential of 1122 km^3. However, it is difficult to add supply beyond this level due to heavy costs, environmental concerns and interstate water conûicts. In contrast, the total demand for water is projected to reach 784–843 km3 by 2025 and 973–1180 km^3 by 2050 (Ministry of Water Resources, 2000). Recent research studies predict that, if the demand–supply gap continues to increase, nine basins that have over four-ûfths of the total water use in India will face physical water scarcity by 2050 (see Amarasinghe *et al.*, 2007). For a heavily populated, monsoon-dependent and rural-based country, such as India, water scarcity of this magnitude will not only lead to serious water conûicts among sectors and regions, but also have a devastating effect on the food and livelihood fronts.

With the current emphasis on environment, health and water pollution issues, wastewaters generated need safe disposal. Due to scarcity of good quality water and competition from other remunerating sectors, availability of good quality water to agriculture is reducing at an alarming rate (Dey, 2011). Under such conditions, water from any source including wastewater is to be used effectively and judiciously in agriculture not only to avoid the pollution of valuable resources but also to exploit their irrigation and nutrient potential. India with its 4.2% share of global water resources is supporting 16.7 per cent of global population. Nearly 85% of India's fresh water resources are being utilized in agriculture and the balance 15% in domestic and industrial sectors.

Despite of abundant global availability, only a miniscule fraction of the water available on the earth surface is used by the mankind; other being too saline to use is locked at places and in forms that man is unable to make its use under the current technologies available to him (Figure 1). Although, agriculture is the major user of water claiming more than 70 per cent of water withdrawn from rivers, yet it is unable to compete economically for scarce water resource as industry and domestic sectors can pay more for water and can earn more per unit of water used. But food security being closely linked with water availability to agriculture, it will continue to be the major beneficiary of water resource development. At the same time, the sector would be increasingly called upon to use waters of low quality. Although both the quantity and the quality aspects of water are important but in this paper we review national scene with respect to water supply as related to its quality and its use in crop production.

In agriculture or for that matter in any sector, there is no such thing as "pure" water because even the rainwater will have many impurities as it falls through the air to the earth or flows on the surface of the earth or moves underground to meet the groundwater. The rainwater usually takes on the characteristics of the materials it has encountered on its way till it meets its final place of storage.

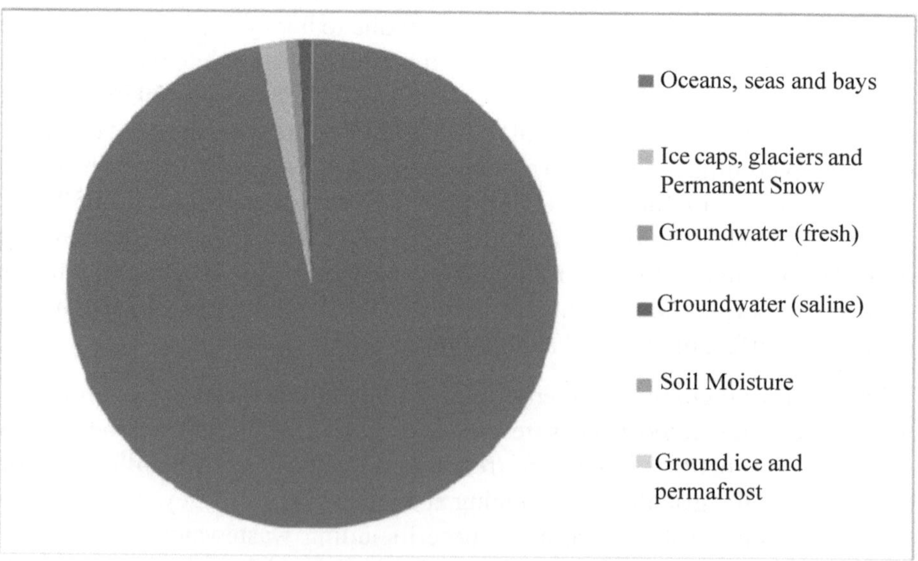

Fig. 1: Global distribution of water (after Gleick, 1996)

Generally speaking, surface water doesn't have enough impurities that would affect the utilization of these water for most practical purposes in domestic, industrial or agriculture. But on the other hand, groundwater due to excessive dissolution of salts could turn saline/sodic and may become unfit for use in domestic, industrial or even agriculture unless managed with care. Such groundwater is generally referred as naturally occurring wastewaters and a characteristic of the arid and semi-arid regions where groundwater is usually saline/sodic in nature. Saline and sodic water contains salts that impair plant growth but rarely contains pathogens. However, short and long-term use of such water can lead to soil salinization/sodication. Quality of water is usually defined with respect to specific use (Table 1). These are only preliminary guidelines but more specific and detailed guidelines are also available for such uses or some other uses such as land disposal and disposal in the water bodies. Central Soil Salinity Research Institute has developed detailed guidelines for use of saline/sodic including toxic groundwater for irrigation. Any water to be useful to the society has to have the parameters characterizing it, in a desirable proportion depending upon its use.

Table 1: Use based classification of surface waters in India

Designated Best use	Class of water	Criteria
Drinking Water Source without conventional treatment but after disinfection	A	• Total Coliforms Organism MPN/100ml shall be 50 or less • pH between 6.5 and 8.5 • Dissolved Oxygen 6mg/*L* or more • Biochemical Oxygen Demand (5 days 20°C) 2mg/*L* or less
Outdoor bathing(Organized)	B	• Total Coliforms Organism MPN/100ml shall be 500 or less • pH between 6.5 and 8.5 • Dissolved Oxygen 5mg/*L* or more • Biochemical Oxygen Demand (5 days 20°C) 3mg/*L* or less
Drinking water sourceafter conventionaltreatment and disinfection	C	• Total Coliforms Organism MPN/100ml shall be 5000 or less • pH between 6 to 9 • Dissolved Oxygen 4mg/l or more • Biochemical Oxygen Demand (5 days 20°C) 3mg/*L* or less
Propagation of wild lifeand fisheries	D	• pH between 6.5 to 8.5 • Dissolved Oxygen 4mg/*L* or more • Free Ammonia (as N) 1.2 mg/*L* or less
Irrigation, Industrial Cooling, Controlled Waste disposal	E	• pH between 6.0 to 8.5 • Electrical Conductivity at 25°C dS/m (maximum) 2.25 • Sodium absorption Ratio (maximum) 26 • Boron (maximum) 2mg/*L*

Sources of wastewater

In general wastewater is grey domestic (without human excreta), black domestic with human excreta, or industrial wastewater. Domestic wastewater consists of discharges from households, institutions and commercial buildings while industrial wastewater is the effluents discharged by variable manufacturing units and food processing plants.

Besides the naturally occurring saline/sodic ground waters, other kinds of wastewaters from domestic and industrial sectors have received prominence mainly because of shrinking water resources across the globe as well as their nuisance value in polluting the surface and groundwater resources. In the real term, the term "wastewater" is a broad, descriptive term that designates liquids and waterborne solids from domestic, industrial or commercial uses that have been adversely affected in quality by anthropogenic influence and have limited applications for use unless appropriately treated/managed.

Although not separable, two categories of wastewaters i.e. domestic wastewaters and industrial wastewaters have emerged mainly because of major differences in their quality. Domestic wastewaters originate principally from domestic, household activities but will usually include waters discharged from commercial and business buildings and institutions. Nearly 80 % water supplied to domestic purposes drains back into municipal sewer system. Surface and storm waters may also be present. Domestic wastewaters are usually of a predictable quality and quantity. These domestic wastewaters are sometimes sub-grouped as dark and grey depending upon the source in the household. Human wastes consisting mostly of feces and urine, which becomes part of the wastewater through toilet flushing is called dark water. On the other hand, household wastes derived from home laundry operations, bathing, kitchen wastes, from washing and cooking foods and dishwashing etc. is termed as grey water. The grey water will contain synthetic detergents but the bacterial contamination would be to a much lesser degree. Whereas grey water would pose minimal problems in its reuse particularly in agriculture such as irrigating lawns and kitchen gardens, the disposal assumes significance because dark water is usually discharged with grey water.

Wastewater quantity

Total wastewater generated from all major industrial sourced in India is 83,048 MLD that includes 66,700 MLD of cooling water generated from thermal power plants and out of remaining 16,348 MLD another 7,275 MLD is generated as boiler blow down water and overflow from ash ponds. Second bigger contributor is small scale engineering industries. Of this, electroplating units are most polluting units. Other significant contributors to industrial effluents are paper, textile, steel and sugar industries. Latest estimates suggest that about 22900 MLD of domestic wastewater is generated in different cities of our countries. Information on wastewater characteristics and its flow rates are important in designing and operation of collection and disposal facilities as well as in knowing their irrigation potential for agricultural purposes. Use of wastewater for irrigation will avoid pollution of water bodies and environmental hazards like public health and foul smell.

Wastewaters generated in huge quantities from municipalities and agro-based industries in Haryana were assessed for their quality, irrigation potential, nutrient supply and possible health hazards for formulating efficient, eco friendly technologies and measures required for their sustained use in agriculture. The farmers in peri-urban areas largely depended upon these waters for their livelihood and meeting most of the crop nitrogen and phosphorus requirements. Estimates are that sewage waters can annually irrigate about 1.5 Mha (million hectares) of land area and has a potential to contribute about one million tonnes of nutrients and 130 million man-days of employment. Irrigation with various sewage or sewage mixed with industrial effluents resulted in saving of 25 to 50 per cent of N and P fertilizer and higher crop productivity over the normal waters. However, long-term sustainability of irrigation with such waters depends on several factors such as site-specific soil, climate, crop, application techniques and socio-political environment. Indiscriminate and unscientific application of wastewater led to higher pathogen load, deterioration in crop quality and contamination of soil and groundwater..

Opportunities to reduce wastewater in industries

- Adoption fibre recovery system in paper mill helps reducing wastewater generation to just one-fourth (from 200 m3 to 50 m3 per ton of paper produced).
- To achieve the goal of "Zero-discharge" to get the label of clean processing/industry - some additional incentives could help to spread this concept.
- Sugar industry - Recycling techniques help reduce effluent (300 liters to 50 liters per ton of cane crushed).
- Distilleries - Continuous fermentation reduce wastewater generation to just half (7 liters) per liter of alcohol produced compared to alcohol produced in batch fermentation process.

Composition of wastewater

Though actual composition of wastewater may differ from community to community but all municipal wastewater contain; organic matter, nutrients (N, P, K and micronutrients), dissolved inorganic minerals, toxic chemicals and pathogens. However, the final composition of raw wastewater depends on the source of water supply, types and numbers of industrial units discharging effluents, and level of treatment given. Routine measurements of municipal wastewater pertain to water pollution parameters like BOD, suspended solids and COD. But agriculturally important chemical characteristics as elemental composition and compounds that affect soil properties and crop growth are sporadically monitored. Chemical characteristics of effluents of some industries are given in table 2.

The major contributors of pollution in terms of BOD are distilleries followed by paper mills. Distillery effluents are very concentrated and thus difficult to treat. Paper and board mills also generate heavy organic pollution load. Other significant contributors of organic load are sugar mills and tanneries. Industries generating chemical pollution can be divided in two categories *i.e.* those generating high TDS as wastes of pharmaceuticals, rayon plants, chemicals, caustic soda, soap, detergents and smelters *etc.* while second type include those units which generate toxic wastes like pesticides, smelters, inorganic chemicals, organic chemicals, steel plants and tanneries. Distilleries, textile units, pharmaceuticals and rayon plants contribute to TDS; whereas thermal power plants followed by paper mills and tanneries generate suspended solids loads.

Fertilizer plants generate toxic wastes as cyanide and arsenic. Steel plants and oil refineries contribute to phenols while engineering units, refineries and vanaspati (hydrogenated vegetable oil) industry release oils and greases in the environment. Tanneries add Cr and fertilizer units also add fluoride to the system. Similarly, caustic soda units release Hg in environment.

Wastewater disposal and use for irrigation

Developed countries with surplus funds regard wastewater treatment as vital to protect human health, environment and prevent the pollution of surface water bodies such as lakes and rivers. But for most developing countries, this solution presents insurmountable problems and is prohibitively expensive. Under these conditions, applying wastewater to agricultural lands is a more lucrative/ economical alternative as it provides a reliable source of water for agriculture. Water does not lose its utility after one/two use(s) and has the potential of recycling for few activities that don't require good quality water. Thus, use of wastewater in agriculture is a low cost method to dispose of municipal wastewaters compared to its land disposal or disposal in lakes/rivers/sea. In

Table 2: Chemical characteristics of some industrial wastewater

Industry	pH	EC (dS/m)	SS ('000)	BOD ('000)	COD ('000)	N (ppm)	P (ppm)	K (ppm)
Dairy	8.0	1-5.14	0.69-1.32	0.9-2.2	-	43-180	14-59	13-39
Distillery	4.5	46.0	1.2- 4.0	45-75	27-110	1000-1900	280-310	6600- 10000
Dye	11.8	8-11	0.68-1.0	1.16-1.8	-	30-35	7-9	8-13
Electro-plating	6-8.9	2.4	0.06-0.38	-	-	-	-	-
Fertiliser	9.5	-	2.23	0.259	-	108	3.4	15
Food	5.2	0.81	-	-	-	12-20	2-4	9-12
Paper	10.1	1.5	0.62-1.25	1.1-1.6	0.80	168	-	15-43
Refinery	7.1	1.8	1.94	0.048	0.32	140	-	1.8
Sugar	6.0	0.95	1.5-1.8	0.65-0.82	0.06-0.09	11-15	7-8	30-40
Tannery	9.5	22	3.2	1.0-3.0	-	-	-	-
Textile	11.0	-	1.0-1.5	0.22-2.0	0.75-8.0	-		

the Indian context many of its rivers have already turned into drains during season other than monsoon. The positive and negative effects of wastewater irrigation are summarized in Table 3. If the negative issues could be appropriately addressed then many of the reservations for wastewater use could be countered.

Millions of farmers around the globe irrigate with wastewaters either because they have no other source or it is cheap source of their livelihood. No precise estimates of wastewater irrigated area are available for India yet the widespread use could be gauged from the fact that along one river, the Musi in Andhra Pradesh, alone an approximately 40,500 ha area is irrigated with wastewater. According to another report, the rural areas downstream of Vadodara in Gujarat, India, present an interesting case where wastewater supports annual agricultural production worth Rs. 266 million (Bhamoria, 2005). Critics however, aver that irrigation with untreated wastewater can represent a major threat to public health (of both humans and livestock), food safety, and environmental quality. The microbial quality of wastewater usually measured by the concentration of the two primary sources of water-borne infection – faecal coliforms and nematode eggs might not suffice in the long run as a range of viruses and protozoa pose additional health risks. In sum, wastewater is a resource of growing global importance and its use in agriculture must be carefully managed in order to preserve the substantial benefits while minimizing the serious risks.

Table 3: Positive and negative effects of wastewater irrigation

Positive	Negative
• It conserves water resource and provides more reliable water supply to the cultivator • It is low-cost method for disposal of municipal/industrial wastewater • It helps to reduce pollution of rivers, canals and other surface water bodies • It helps to conserve nutrients and reduces the need for fertilizer application • Crop yields are increased	• It is a health risk to the agricultural field workers and their families, crop handlers, consumers (of crops, meat and milk) and those living near affected fields. • Likely contamination of groundwater • Soil build-up of chemical pollutants (heavy metals) • Excessive growth of algae and vegetation in drains/canals carrying wastewater

Pollution vis-à-vis public health

Studies at Haryana revealed that TDS in sewage ranged between 0.6 – 3.3 g/ *L* and these are on higher side than permissible levels, this could be because of collection wastewater through open drains containing considerable quantities of clays and silt getting mixed with sewage. BOD and COD of wastewater ranged from 176 to 345 ppm and 233 to 457 ppm, respectively. Ambala sewage had maximum BOD and COD. Most of nitrogen in wastewater is found in

NH_4-N (39 ppm). These parameters indicate that these should not be allowed to be disposed in water bodies as these may deteriorate aquatic environment. Considering the agroclimatic conditions of Haryana, the most of wastewater generated are suitable for irrigation purposes.

Health hazards associated with disposal of raw or partially treated sewage are monitored in terms of *E. coli, F. coli*, total bacterial counts, *Salmonella* and *Shigella*. The populations of *E. coli* and *F. coli* are considerably higher 4 x 106 per 100 ml of wastewater. Bacterial counts and fungi were also high than permissible levels. Although other pathogenic bacteria were not detected but presence of high levels of coliform bacteria indicates towards health hazards from use of these wastewater.

Nutrient potential

As a matter of fact, additions of macro- and micro-nutrients and organic matter for conditioning of soils are inseparable from sewage irrigation and thus their use can diminish their requirements for fertilizers, sewage from different districts in general are rated high in terms of plant nutrients contents. Contents of major plant nutrients, i.e., N, P and K, averaged 45.9, 6.9 and 62.4 ppm, respectively. Samples analyzed for micronutrients status indicate 0.17, 1.01 and 0.024 ppm, respectively of Zn, Fe and Cu. Thus potential for the supply of major nutrients like N, P, K with irrigation is 34.4, 5.2 and 46.7 kg/ha in addition to 130, 760 and 20 g of Zn, Fe and Cu, respectively (Table 4).

Supplied nutrients are expected to be utilized more efficiently as these are added in splits and the total nutrients added during crop growth period will be sufficient for successful crop production. Also it is a fact these effluents contain appreciable amounts of organic matter that improves soil conditions. From the concentrations of respective elements and quantum of sewage, it is estimated that these effluents have potential to contribute about 56.4 Mg/day and 20,872 Mg/year cumulatively of all nutrients. In Haryana 662679, 17178 and 3950 Mg/year of N, P and K were used during 2001-2002, whereas the potential of sewage is worked out to be 8140, 121 and 11046 Mg/year. Thus approximately 2.4% of total micronutrients used can be supplied through sewage irrigation.

Utilization of sewage water through drip irrigation

Water shortage associated with intensive depletion of underground aquifers has prompted the search for alternative water sources. It has led to secondary treated domestic wastewater being considered for irrigation of field crop and raw eaten vegetable crops (Oron *et al.*, 1991). Secondary domestic wastewater is now being used on a relatively large scale, mostly in developed countries, for field crops and landscape irrigation, groundwater recharge, and storage in

Table 4: Nutrient potential of sewage in Haryana

Nutrient	Contents (ppm)	Nutrient added (kg/irrigation)	Contribution (Mg/year)	Price (Rs./Mg)	Value (Million Rs.)
N	45.9	34.4	8140	10434	84.93
P	6.9	5.2	1221	15625	19.08
K	62.4	46.7	11046	7330	80.97
Zn	0.17	0.13	3.1	48000	1.67
Fe	1.01	0.76	178.8		
Cu	0.024	0.02	4.3		

recreational centers. In a few cases, tertiary or advanced treatment of the wastewater is required (Kirkpatrick and Asano, 1986). The concept of water saving might seem to contradict the idea of maintaining maximum yield from irrigated crops. The conflict might be more significant in arid zones with limited natural, high quality, permanent water source. A possible remedy to this conflict is to use non-conventional water, such as domestic treated wastewater, applied by drip irrigation. Using a subsurface drip irrigation system can further increase the efficiency of water application (Phene *et al.*,1985). The other advantages of use of drip irrigation system with sewage water are that no aerosols are formed, water logging due to runoff and deep percolation is negligible and the only contact with the water occurs when the product to be consumed touches the soil; the product of the plants growing above the soil being practically devoid of pathogens when the drip system is buried in the soil or covered by the plastic sheets (Capra and Scicolone, 2004).

Current scenario of sewage water utilization in India

Currently, about 30% of untreated sewage water is being utilized to grow vegetable crops around urban center using surface method of irrigation. This practice, besides a health risk to the farmers and the consumers of the product, it is causing enormous ground water contamination since excessive deep percolation losses cannot be avoided. Moreover the productivity of land and water is quite less, which could be increased substantially by adapting drip irrigation. Around 60 % of sewage water is directly disposed off in surface water bodies and low lying areas causing groundwater and surface water contamination and inefficient use of our water resources. Visualizing the alarming level of environmental pollution, around 10 % of sewage water is being treated in conventional sewage treatment plants generating mostly primary treated sewage water. These waters are also utilized for irrigation purpose with surface method of irrigation.

Bioremediation of wastewater

Use of wastewater in agriculture has been increased in recent years. However, wastewater, particularly from industries contains high concentration of heavy metals which enter into human beings and animals through food chain. Therefore, before its use in agriculture, it is desirable to remove these heavy metals from wastewater through low cost technology like efficient microbial culture. Biomass of microbes acts as adsorbent to remove heavy metals from wastewater. The ability to remove heavy metals from wastewater varies greatly among microbes. This needs to be exploited for removal of heavy metals from wastewater through efficient microbes. Laboratory experiments conducted at CSSRI, Karnal for bioremediation of heavy metals through microorganisms showed encouraging results. Removal of lead and cadmium from liquid medium by fungi cultures (*Aspergillus awamorii, Trichoderma viride* and *P. chryosporium*) was found to be substantial. The above results indicate potential of some of the fungi for removal of heavy metals like Pb and Cd which can be used for bio-treatment of wastewater at low cost and in eco-friendly way.

Epilogue

Disposal of the wastewater is a serious problem especially in developing countries, which is causing groundwater and surface water contamination and creating environmental pollution. The conventional method of disposal of sewage water (sewage treatment plants) are cost intensive and beyond the reach of many municipalities. The use of sewage water through subsurface drip irrigation may help to solve the disposal problem and finding a solution, which may be economically viable. Though its adaptation for efficient utilization of good quality and saline water is steadily increasing in India still there are few challenges. Investment needs being high, the technology could be popularized through one window system of financial assistance including subsidy. Its large-scale expansion will reduce many of its shortcomings, which will also encourage the use of sewage water through drip irrigation. Wastewaters become more dangerous and even unsuitable when contaminated with significant loads of industrial effluent. Some of the low cost technologies which reduced the crop contact with wastewater like planting on raised beds, washing the produce with clean water and discarding the most contaminated part resulted in significant reduction of pathogen load. Viable options include disposal of wastewater in tree plantations and phyto-remediation through flowers and aromatic plants where produce is non-edible but economical.

References

Amarasinghe, U.A., *et al.*, 2007. India's water future to 2025–2050: Business-as-usual scenario and deviations. Colombo, Sri Lanka: International Water Management Institute, Research Report No. 123.

Capra, A. and B. Scicolone, 2004. Emitters and filter tests for wastewater reuse by drip irrigation. Agricultural Water Management, 68:135-149.

Dey, P. (2011). Irrigation with poor quality water vis-à-vis soil health management. The IUP Journal of Soil and Water Sciences, 4(2): 64-75.

Dey, P. (2011). Safe Use of Wastewater in Agriculture-A Water-Sparing Initiative. Water Supply and Sanitation Collaborative Council, Geneva Switzerland, Link: http://www.wsscc.org/resources/resource-publications/safe-use-wastewater-agriculture-water-sparing-initiative?rck=18d5f053aef50fa23a104c9873a0c212

Dey, P. (2013). Wastewater Use in Agriculture: Issues and Strategies. Invited paper presented in 100th Indian Science Congress, Agriculture & Forestry Section, Kolkata, India held on January 3-7.

Dey, P. (2015). Wastewater Management for Livelihood and Food Security: Challenges and Opportunities. Lead paper presented in the International Conference on Natural Resource Management for Food Security and Rural Livelihoods, New Delhi, India held on February 10-13, 2015.

Gleick, P. H., 1996: Water Resources. In Encyclopedia of Climate and Weather, ed. by S. H. Schneider, Oxford University Press, New York, 2: 817-823.

Kirkham, W.R.,and Asano,T.1986. Evaluation of tertiary treatment system for wastewater reclamation and reuse. Water Sci. Tech., 18 (10), 83-95.

Ministry of Water Resources, 2000. Annual Report 1999–2000. New Delhi: Government of India.

Oron, Gideon, Joel Demalach, Zafrir Hoffman and Rodica Cabotaru, 1991. Subsurface microirrigation with effluents. Journal of Irrigation and Drainage Engineering, 117(1): 25-37.

Phene, C.J., Hutmacher, R.B.., Davis, K.R.,and McCormic, R.L.1985. Subsurface drip irrigation offers success. Calif.-Ariz. Farm press, 7/8 (40): 24-31.

Saleth, R. M. (2011) Water scarcity and climatic change in India: the need for water demand and supply management. Hydrol. Sci. J. 56(4): 671–686.

Colour Plates

Chapter 1: Impact of Climate Change on Indian Agriculture

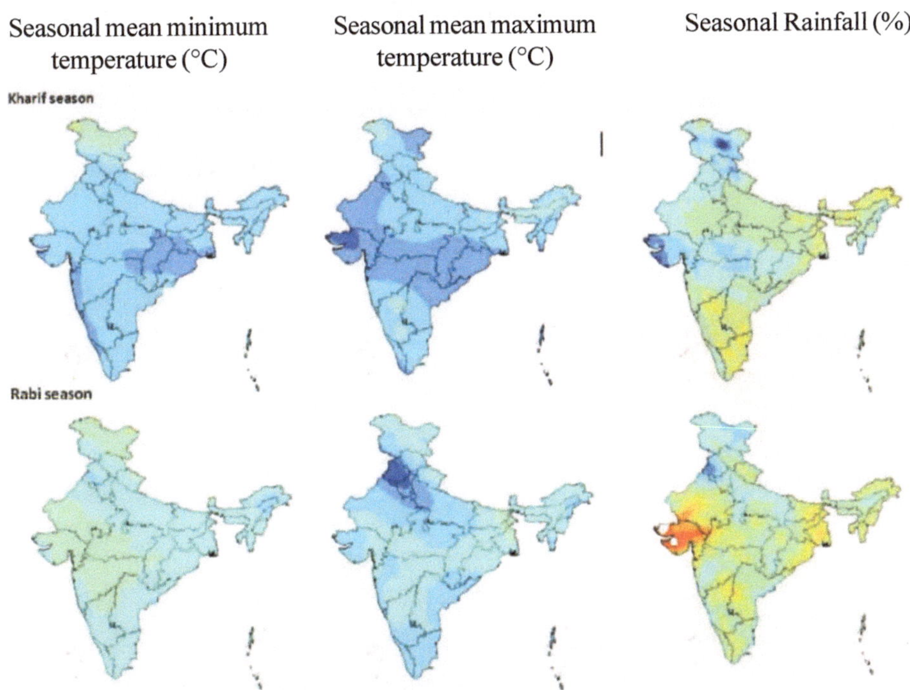

Fig. 1: Projected change in seasonal maximum and minimum temperatures and rainfall in India in A1B 2030 scenario

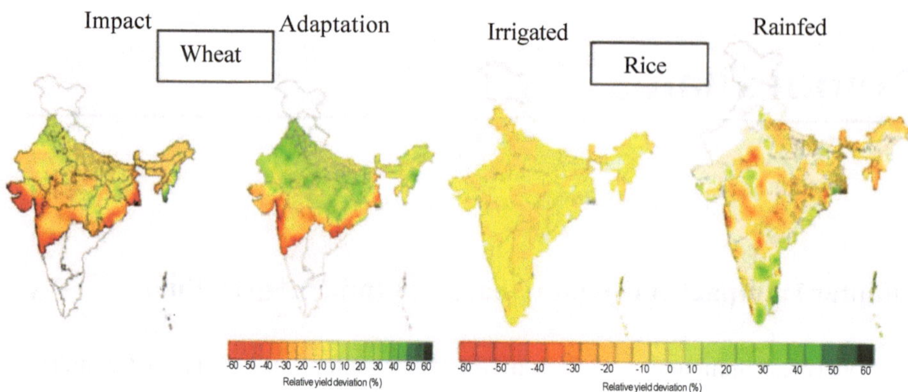

Fig. 2: Impacts on Wheat and Rice-2020 scenario

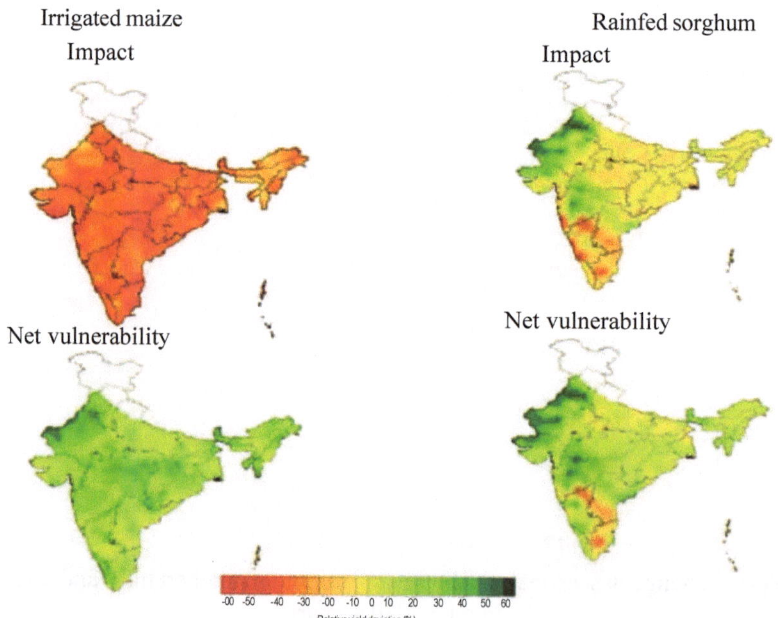

Fig. 3: Climate change impacts and net vulnerability of Maize and Sorghum

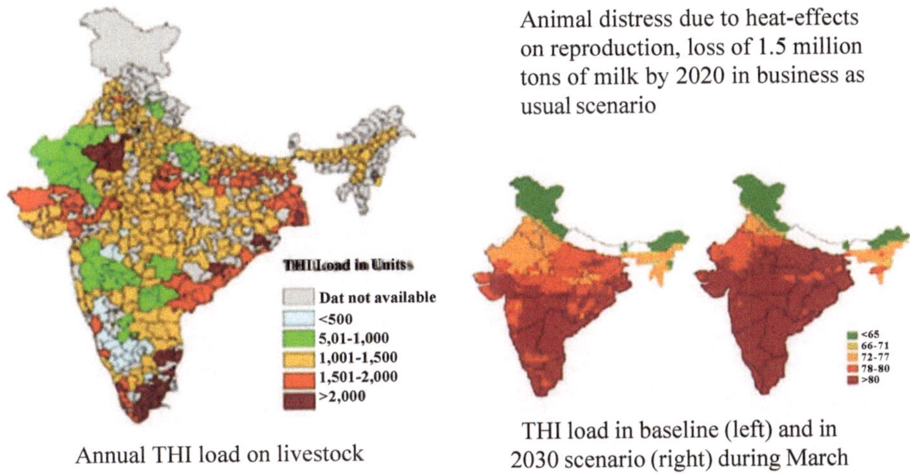

Fig. 7: Annual Temperature Humidity Index load on cattle in India

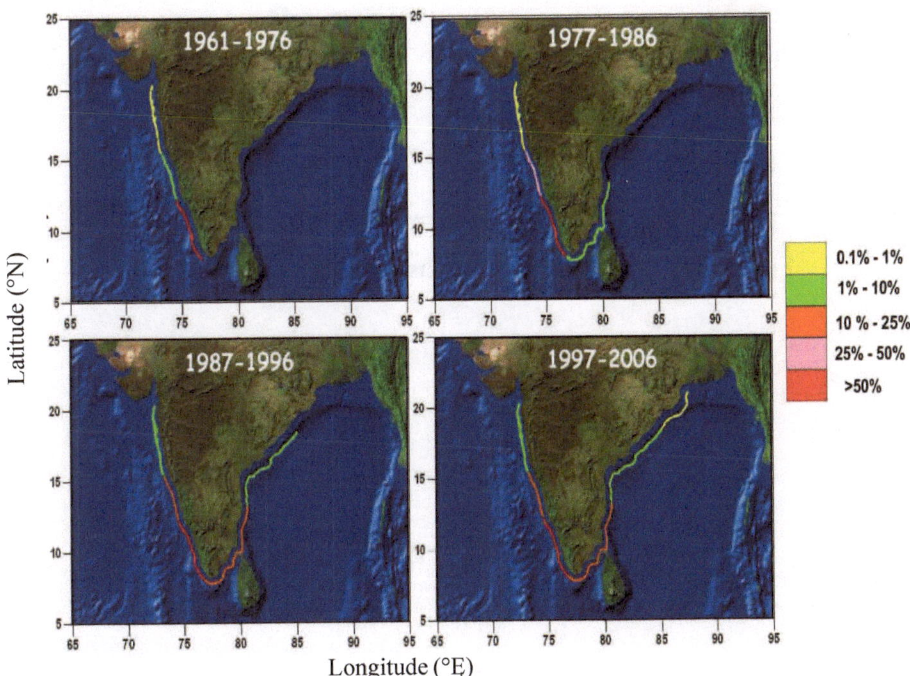

Fig. 8: Extension of abundance of Oil sardine along Indian coast due to rise in sea surface temperature

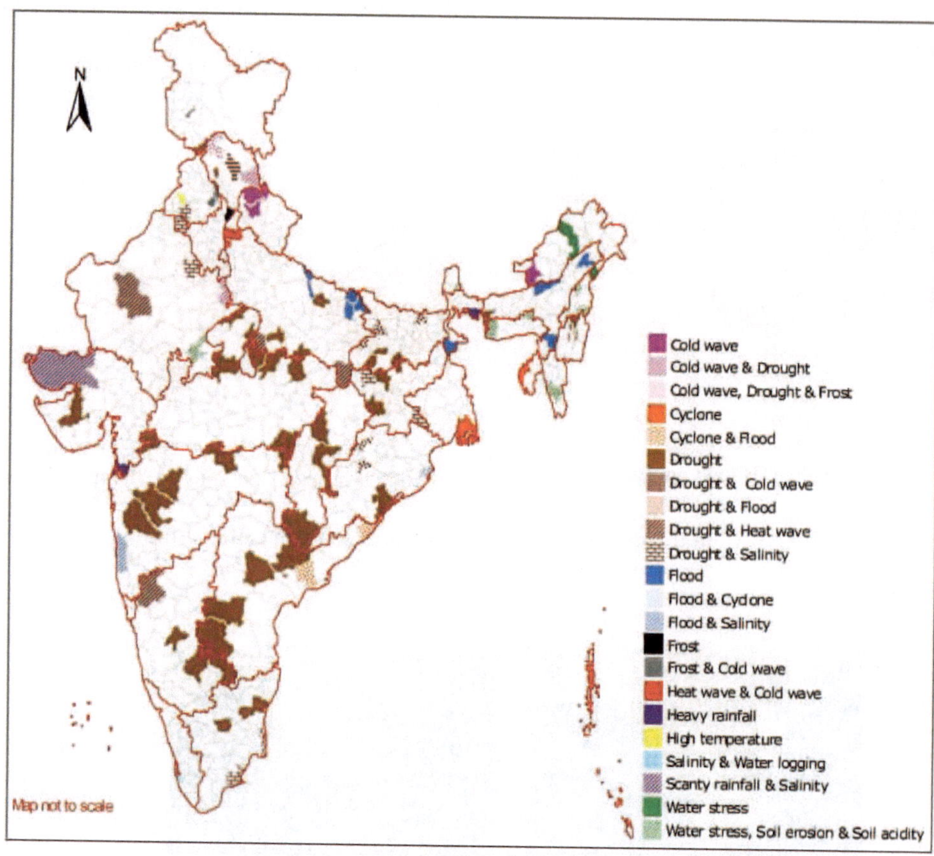

Fig. 11: The location of 130 districts being covered in NICRA

Chapter 2: Climate Change, El Niño, Monsoon and Impact on Crops with Special Reference to M.P.

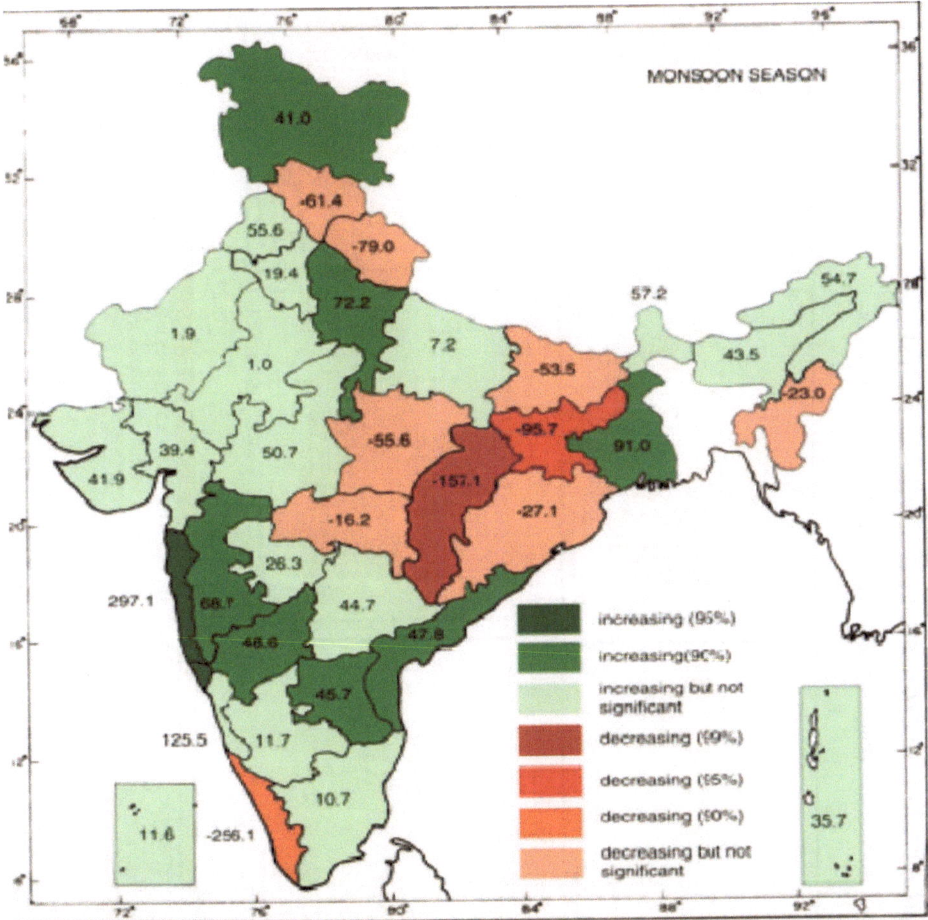

Fig. 11: Increase/decrease in rainfall in mm in 100 year for each of 36 sub-divisions for the south-west monsoon season

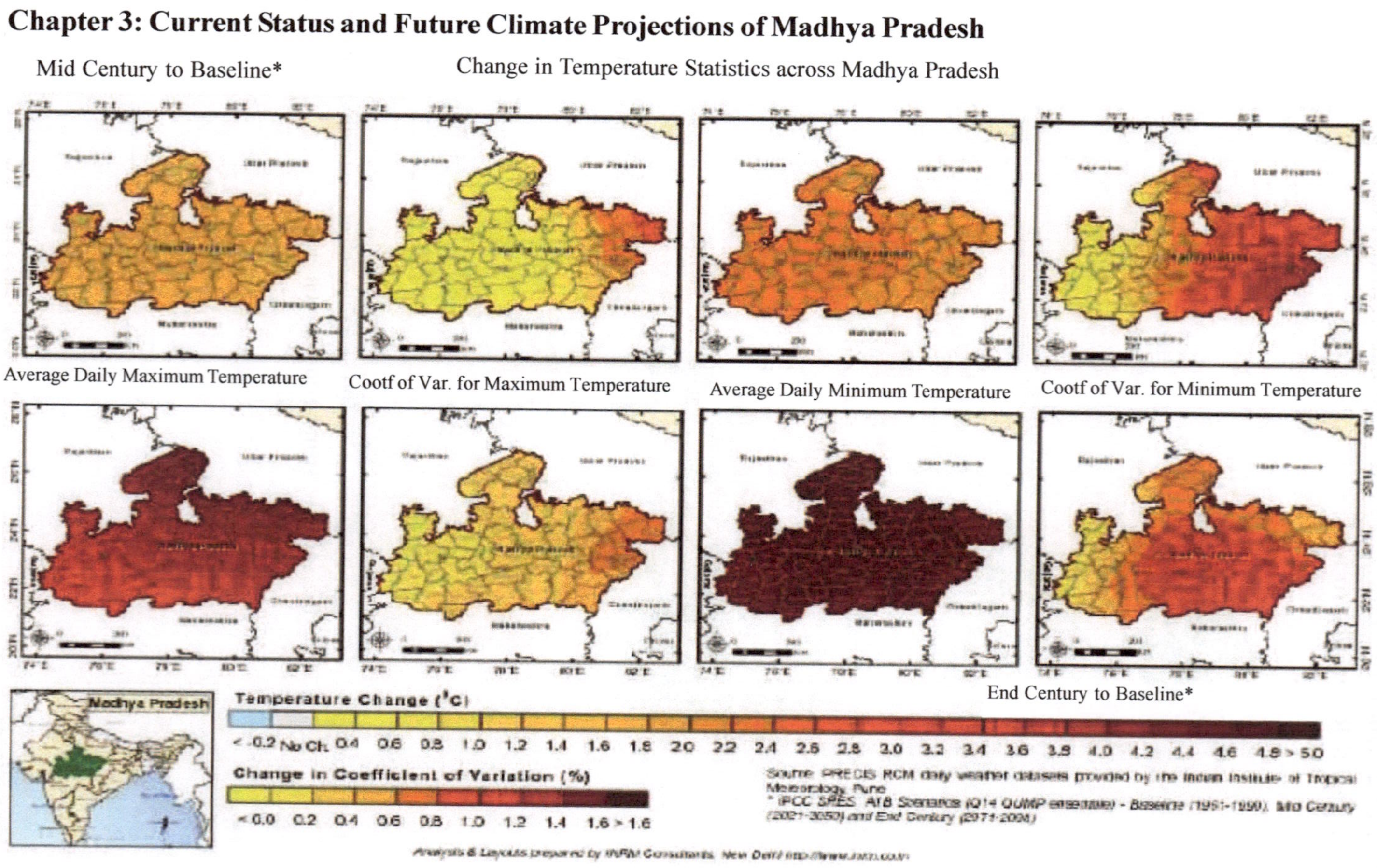

Chapter 3: Current Status and Future Climate Projections of Madhya Pradesh

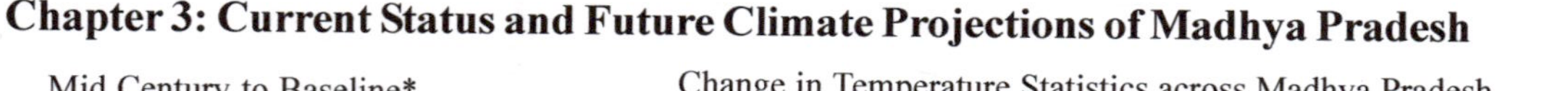

Fig. 1: Average temperature across of Madhya Pradesh

Average Observed* Rainfall during Winter, Pre-monson, Monsoon & Post-monsoon Seasons across Mandhya Pradesh State

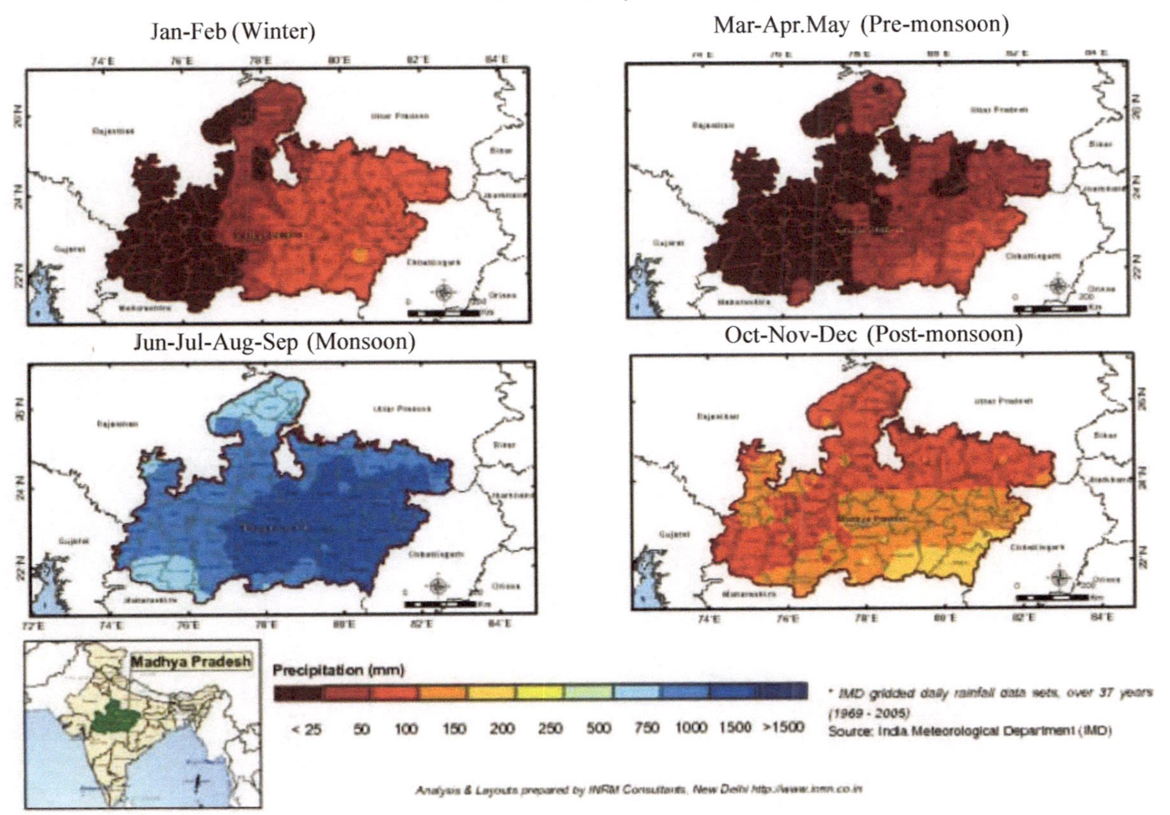

Fig. 2: Average rainfall across Madhya Pradesh

Chapter 6: Climate Change Adaptation and Mitigation Strategies in Rainfed Agriculture

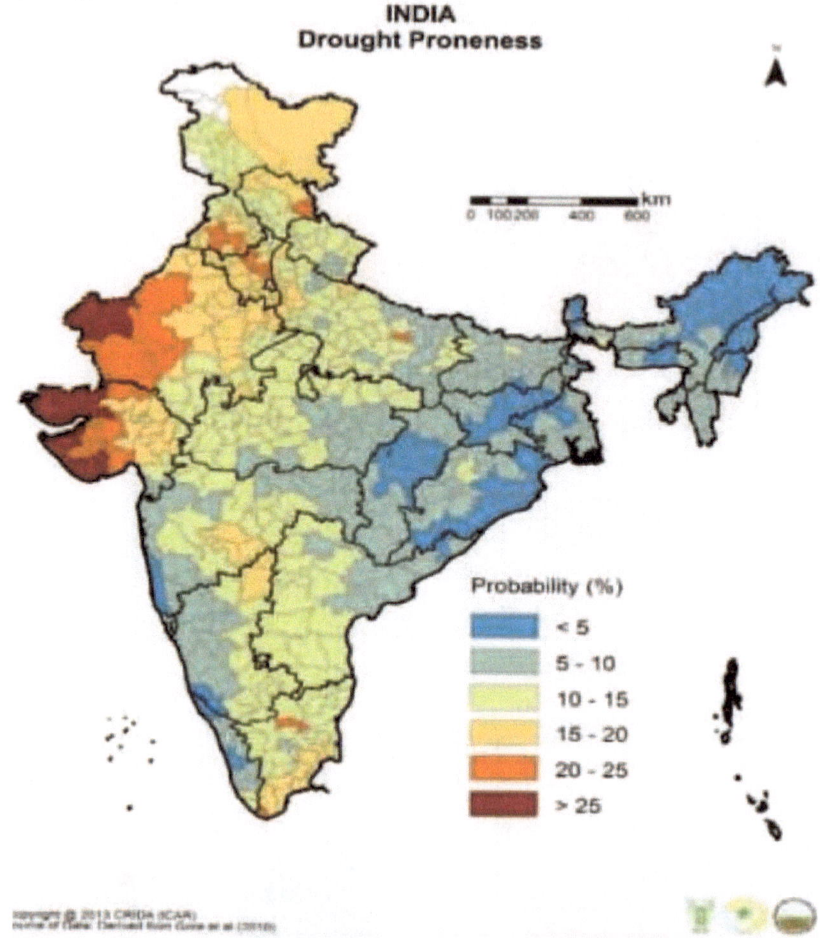

INDIA
Drought Proneness

km
0 100 200 400 600

Probability (%)
< 5
5 - 10
10 - 15
15 - 20
20 - 25
> 25

Fig. 1: Probability of drought occurrence in different parts of India

Maize-horsegram cropping sequence on Alfisols. Rainy season maize (left) and post-rainy season horsegram on stored moisture (right)

Chapter 14: Agromet Advisory Services and Its Importance in Climate Change Scenarios

भारत मौसम विज्ञान विभाग अनावृष्टि अनुसंधान अनुभाग–पुणे

Indian Meteorological Department Drought Research Unit-Pune

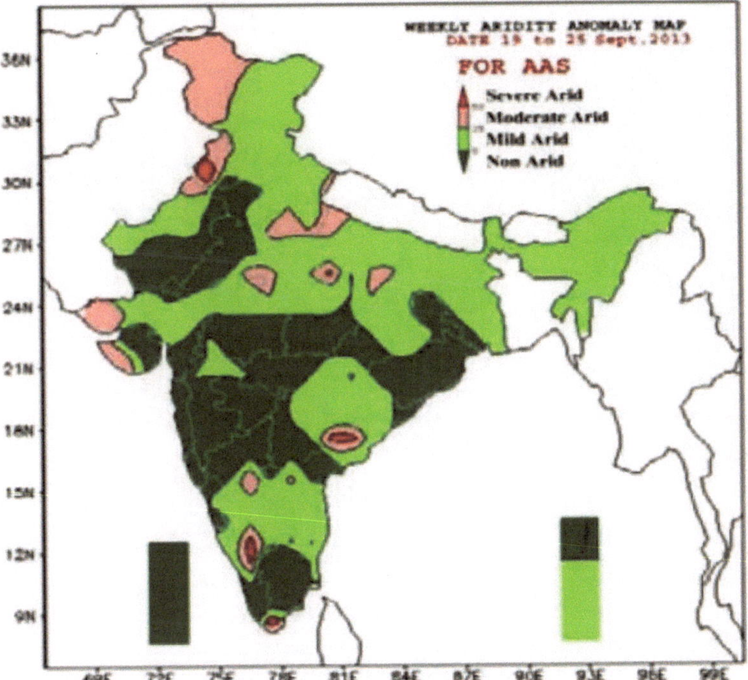

Fig. 3: Weekly aridity anomaly map for AAS

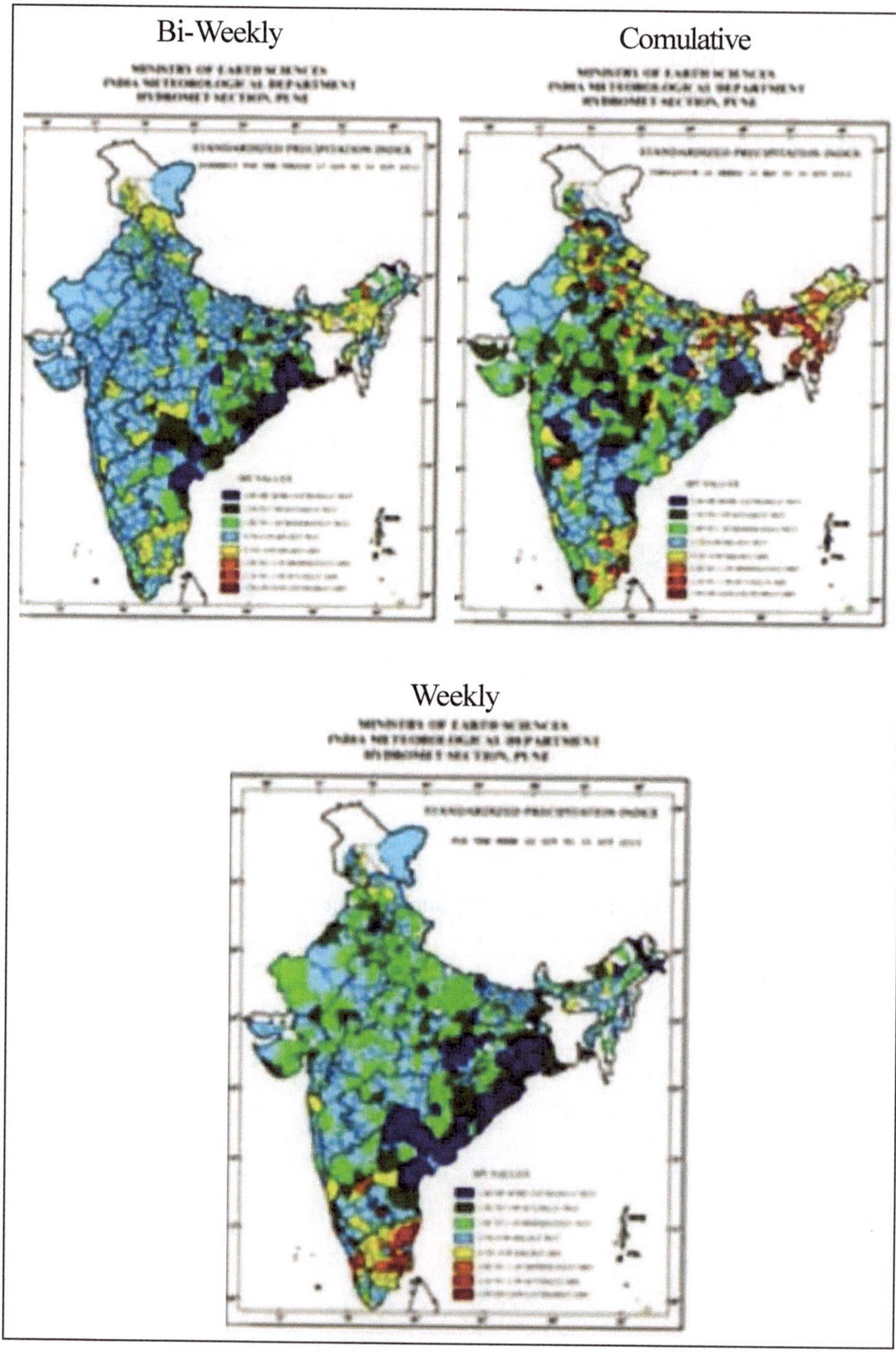

Bi-Weekly

Comulative

Weekly

Fig. 4: Standardized precipitation index

सी.वी. रमन

कुछ प्रमुख जीवनियाँ

ईश्वरचंद्र विद्यासागर
नेपोलियन बोनापार्ट
चंद्रशेखर आजाद
लाला हरदयाल
राजा राममोहन राय
आनंदमूर्ति
बेंजामिन फ्रैंकलिन की आत्मकथा
जगदीशचंद्र बसु
लोकमाता अहिल्याबाई
लोकमान्य बाल गंगाधर तिलक
लियोनार्दो द विंची
शिखर भारतीय महिलाएँ
मैडम भीखाजी कामा
प्रथम अंतरिक्ष यात्री यूरी गागरिन
लियो टॉलस्टॉय
आचार्य विनोबा भावे
स्वामी रामदेव
आर्यभट
अन्ना हजारे
नेल्सन मंडेला
महर्षि अरविंद घोष
कस्तूरबा गांधी
कर्नल जिम कॉर्बेट
मौलाना अबुल कलाम आज़ाद
होमी जहाँगीर भाभा
नेताजी सुभाषचंद्र बोस
भगिनी निवेदिता
भारत कोकिला सरोजिनी नायडू
स्टीफन हॉकिंग
गोपाल कृष्ण गोखले
रवींद्रनाथ टैगोर
स्वामी दयानंद सरस्वती
बिपिनचंद्र पाल
निकोलस कॉपरनिकस

सी.वी. रमन

तेजेन कुमार बसु

विद्या विहार, नई दिल्ली

प्रकाशक : विद्या विहार,
19, संत विहार (पहली मंजिल) गली नं. 2, अंसारी रोड, नई दिल्ली–110002
 / संस्करण : 2024 / मूल्य : तीन सौ पचास रुपए
मुद्रक : प्रिंट मीडिया, नई दिल्ली ISBN 978-93-80186-40-5

C.V. RAMAN *(biography)* by Shri Tejen Kumar Basu ₹ 350.00
Published by **VIDYA VIHAR**
19, Sant Vihar (First Floor), Street No.2, Ansari Road, New Delhi-110002

वैज्ञानिक पत्रिका का प्रकाशन शुरू किया, जिससे शोध कार्यों की जानकारी सबको मिल सके। भारतीय विज्ञान संस्थान, बैंगलौर के वे प्रथम भारतीय निदेशक थे। उन्होंने 'रमन अनुसंधान केंद्र' का निर्माण किया, जिसके वे प्रथम निदेशक बने एवं आजीवन बने रहे। उनके कई शिष्यों ने विज्ञान के क्षेत्र में महत्त्वपूर्ण कार्य किया।

रमन ने भारत में वैज्ञानिक अध्ययन एवं अनुसंधान हेतु सदैव कार्य किया। इस पुस्तक में लेखक ने सर सी.वी. रमन के जीवन एवं कृतित्व के बारे में अत्यंत ही सरल भाषा में वर्णन किया है, जो पाठकों को प्रेरणा देगा, ऐसी मेरी आशा है। साधारणत: हिंदी भाषा में उपलब्ध पुस्तकों में वैज्ञानिकों के शोध कार्यों के बारे में विस्तृत जानकारी कम ही मिलती है, परंतु इस पुस्तक में रमन के कार्यों के उल्लेख के साथ उनका विवरण भी दिया गया है, जो सभी वर्गों के लोगों में विज्ञान के प्रति उत्सुकता जगाने में सफल होगी।

—प्रद्युम्न काव

निदेशक

प्लाज्मा अनुसंधान संस्थान,

(गुजरात)

प्राक्कथन

सी. वी. रमन विज्ञान की दुनिया में एक ऐसा नाम है जिनका न केवल भारत में बल्कि समूचे विश्व में एक विशिष्ट स्थान है। वे ही एकमात्र भारतीय वैज्ञानिक हैं जिन्हें देश में किए गए आविष्कार के लिए विश्व के सर्वोच्च सम्मान नोबेल पुरस्कार से विभूषित किया गया था। यह पुरस्कार उन्हें प्रकाश प्रकीर्णन पर की उनकी मौलिक खोज के लिए सन् 1930 में प्रदान किया गया था। 'रमन प्रभाव' पर आधारित शोध आज भी जारी है तथा विज्ञान के हर क्षेत्र में इसका उपयोग हो रहा है।

रमन का जीवन संघर्षपूर्ण था। उन्होंने 19 वर्ष की अल्पायु में आजीविका के लिए कलकत्ता में अर्थ विभाग में सहायक संचालक पद पर कार्य करना शुरू किया; परंतु उनकी रुचि सदैव विज्ञान में थी। उन्हें विज्ञान के क्षेत्र में काम करने का अवसर जल्द ही मिल गया, जब वे अचानक कलकत्ता के एक वैज्ञानिक संस्थान इंडियन एसोसिएशन फॉर दि कल्टिवेशन ऑफ साइंस के संपर्क में आए। नौकरी के साथ-साथ वे संस्थान की प्रयोगशाला में नियमित तौर पर शोध कार्य करने लगे। अपने मौलिक कार्य की वजह से शीघ्र ही उनकी ख्याति देश-विदेश में फैल गई, जिसके फलस्वरूप कलकत्ता विश्वविद्यालय ने उन्हें सीधे प्राध्यापक पद पर नियुक्त कर लिया। उस समय वे केवल 29 वर्ष के थे। उन्होंने

प्रस्तावना

चंद्रशेखर वेंकट रमन या सर सी.वी. रमन विश्व के महानतम वैज्ञानिकों में से एक हैं। उनका नाम याद आते ही सबसे पहले नोबेल पुरस्कार का ध्यान आता है, क्योंकि वे ऐसे अकेले भारतीय वैज्ञानिक हैं जिन्हें देश में किए गए आविष्कार पर विश्व का सर्वश्रेष्ठ पुरस्कार दिया गया था। उन्हें यह पुरस्कार पदार्थों से प्रकाश के प्रकीर्णन पर मूल शोध करने तथा प्रकीर्णित प्रकाश में आपाती प्रकाश के अलावा अन्य रंगों की उपस्थिति का पता लगाने के लिए मिला था, जिसे 'रमन प्रभाव' के नाम से जाना जाता है। रमन को भारत सरकार ने देश के सर्वोत्तम नागरिक सम्मान 'भारतरत्न' से विभूषित किया था। रमन को आज से अस्सी वर्ष पहले नोबेल पुरस्कार मिला था, जब अनुसंधान क्षेत्र में इतनी सुविधाएँ नहीं थीं। अपनी लगन और निष्ठा के बल पर उन्होंने 'रमन प्रभाव' की खोज की, जिसके नए उपयोग आज भी इजाद हो रहे हैं। तकनीकी विकास के साथ रमन वर्णक्रममापी उपकरणों की संवेदनशीलता व आकार में भारी परिवर्तन हुआ है। विज्ञान का शायद ही कोई ऐसा क्षेत्र बचा हो जहाँ 'रमन प्रभाव' का इस्तेमाल नहीं हो रहा है। यहाँ तक कि ब्रह्मांड के किसी अन्य ग्रहों में जीवन की संभावना का पता लगाने के लिए भी अंतरिक्ष यानों में 'रमन स्पेक्ट्रोमीटर' का इस्तेमाल किया जा रहा है। 'रमन प्रभाव' पर

आधारित नए यंत्रों का विकास आज भी जारी है।

रमन ने ध्वनिक, पराध्वनिक, प्रकाशिकी, चुंबकत्व, स्फटिक भौतिकी आदि कई क्षेत्रों में मौलिक अनुसंधान किए। भारतीय वाद्ययंत्रों पर रमन ने गहन अध्ययन किया। इन यंत्रों की बनावट मधुर आवाज उत्पन्न करने में कैसे सहायक होती है, इसका विस्तृत विवरण उन्होंने दिया। रमन एक प्रख्यात वैज्ञानिक ही नहीं बल्कि स्वप्नदृष्टा भी थे। विज्ञान से वे सचमुच प्यार करते थे। रमन के दफ्तर का दरवाजा हमेशा खुला रहता था, जिससे कोई भी आकर उनसे मिल सके। उन्होंने विज्ञान के माध्यम से देश की प्रगति का सपना देखा था।

रमन का जीवन संघर्ष और सफलता का एक ज्वलंत उदाहरण है। अपने जीवनकाल में उन्हें काफी अपमान सहने पड़े थे। परंतु इन सबके बावजूद उन्होंने कभी विज्ञान का साथ नहीं छोड़ा। जीवन की साँझबेला में भी वे शोधकार्य से जुड़े रहे। उन्होंने 'रमन अनुसंधान केंद्र' का निर्माण किया था तथा 'विज्ञान अकादमी' की स्थापना की थी। अकादमी ने सन् 1935 से शोधपत्रों के प्रकाशन के लिए 'प्रोसिडिंग्स ऑफ इंडियन एकेडमी ऑफ साइंस।' नामक पत्रिका का प्रकाशन शुरू किया था। वे स्पष्ट वक्ता थे तथा अपनी बात कहने में बिलकुल भी नहीं झिझकते थे, चाहे वह दूसरों को कितनी भी कड़वी क्यों न लगे।

आम जनता में विज्ञान के प्रति जागरूकता पैदा करने के लिए वे अकसर सभाओं का आयोजन करते थे, जिसमें वह विज्ञान के क्षेत्र में हुए शोध के बारे में विस्तार से बताते थे। व्याख्यान के दौरान विषय को सरलता से समझाने के लिए वह कुछ प्रक्रियाओं का प्रदर्शन भी दिखाते थे। लोगों की रुचि विज्ञान में बनी रहे, इसके लिए रमन ने सन् 1959 में गांधी जयंती के दिन व्याख्यान श्रृंखला की शुरुआत की थी, जो आज भी जारी है। वे बड़े ही अच्छे वक्ता थे।

आज के युवा वर्ग में धैर्य की कमी है तथा वे जल्द ही सबकुछ हासिल करना चाहते हैं। उनके लिए रमन की जीवनी एक प्रेरणा स्रोत है, जिन्होंने

कठिन परिस्थितियों में भी हौसला बुलंद रखते हुए अपने इरादे को पाया।

इस पुस्तक में वर्णित सामाग्री एवं तथ्यों का संकलन करने में जिन पुस्तकों एवं पत्र-पत्रिकाओं से सहायता ली गई है, उनके प्रति मैं अपना आभार व्यक्त करता हूँ। परमाणु ऊर्जा विभाग, भारत सरकार की ओर से मुझे राजा रामन्ना अध्येता वृत्ति मिली, जिसकी वजह से मैं इस पुस्तक को लिखने का साहस जुटा पाया, इसके लिए मैं उसका विशेष तौर पर आभारी हूँ। अंत में मैं अपनी पत्नी मिताली एवं पुत्री मिष्टुन को इस कार्य में भरपूर सहयोग देने के लिए धन्यवाद देता हूँ।

—तेजेन कुमार बसु

अनुक्रमणिका

1
संक्षिप्त जीवनी

रमन का जन्म 7 नवंबर सन् 1888 को दक्षिण भारत के कावेरी नदी के किनारे बसे तंजावुर जिले के तिरुचिरापल्ली शहर के पास थिरुवानैक्कावाल गाँव में एक अय्यर ब्राह्मण परिवार में हुआ था। उनकी माता का नाम पार्वती अम्मल तथा पिता का नाम चंद्रशेखरन था। पार्वती अम्मल व चंद्रशेखरन के पाँच बेटे तथा तीन बेटियाँ थीं, जिनमें रमन द्वितीय पुत्र थे। यद्यपि उनके पूर्वजों का पारंपरिक व्यवसाय खेतीबारी व जमींदारी था, परंतु चंद्रशेखरन इसे न अपनाकर प्रारंभिक शिक्षा पूरी करके एक स्थानीय विद्यालय में अध्यापक की नौकरी करने लगे। बाद में वे सन् 1891 में भौतिक विज्ञान विषय में स्नातक की परीक्षा उत्तीर्ण करके कॉलेज में नौकरी करने लगे। रमन के दादा एवं नाना दोनों ही महान् पंडित व विद्वान् थे। रमन के बड़े भाई सुब्रमण्यम के बेटे एस. चंद्रशेखर विश्वविख्यात खगोलशास्त्री थे, जिन्हें सन् 1983 में भौतिकी क्षेत्र में नोबेल पुरस्कार मिला था।

प्रारंभिक जीवन

रमन जब चार वर्ष के थे, उस समय उनके पिता प्राध्यापक की नौकरी लेकर सपरिवार विशाखापत्तनम चले गए। उनके पिता की रुचि विज्ञान व गणित के अलावा खेल-कूद व संगीत में भी थी। वे वायलिन

वादन और संगीत का अच्छा ज्ञान रखते थे। उन्हें पुस्तक पढ़ने का बेहद शौक था तथा उनके पास पुस्तकों का अच्छा संग्रह था। अतः बचपन में रमन को इन सब पुस्तकों को पढ़ने का सुअवसर प्राप्त हुआ, विशेषकर प्रसिद्ध जर्मन वैज्ञानिक हेल्महोल्ज (1821-1891) और ब्रिटिश वैज्ञानिक लॉर्ड रैले (1842-1919) की ध्वनि से संबंधित पुस्तकों से उनका जीवन काफी प्रभावित हुआ। शुरू से ही रमन विज्ञान की ओर आकर्षित थे। स्वभाव से ही उनमें काम करने का जोश तथा अपने आस-पास की चीजों के बारे में जानने की प्रबल इच्छा थी। प्रकृति ने भी उन्हें एकाग्रता की शक्ति तथा कुशाग्र बुद्धि दी थी।

युवावस्था में अपने पिता की तरह रमन बलवान नहीं थे। उनका शरीर दुबला-पतला था तथा उनका स्वास्थ्य भी बहुत अच्छा नहीं रहता था। वे शारीरिक तौर पर कमजोर जरूर थे परंतु उनकी दिमागी काबिलियत बहुत अधिक थी। वे पढ़ाई में काफी होशियार थे तथा शुरू से ही इसके आसार नजर आने लगे थे जब उन्हें पुरस्कार एवं छात्रवृत्तियाँ मिलने लगी थीं। स्कूल के समय से ही उनका रुझान भौतिक विज्ञान की ओर होने लगा था। उस समय उन्होंने खुद एक डायनामो बनाया था।

रमन ने माध्यमिक परीक्षा यानी दसवीं कक्षा सिर्फ ग्यारह वर्ष की अल्पायु में उत्तीर्ण की, जिसमें वे अव्वल आए। उच्च माध्यमिक शिक्षा के लिए उन्होंने कॉलेज में दाखिला लिया, जहाँ फिर से उनका बेहतरीन नतीजा रहा। अब स्नातक की पढ़ाई करने के लिए छात्रवृत्ति प्राप्त कर सन् 1903 में मद्रास रवाना हुए, जहाँ वे प्रेसिडेंसी कॉलेज में भरती हुए। उस समय वहाँ के सभी प्राध्यापक विदेशी थे। रमन यहाँ आकर भौतिकी के साथ-साथ अंग्रेजी साहित्य में भी रुचि लेने लगे। स्नातक की परीक्षा में वे प्रथम आए तथा भौतिकी एवं अंग्रेजी विषय में स्वर्ण पदक प्राप्त किया।

स्नातक की परीक्षा उत्तीर्ण करने के बाद रमन के प्राध्यापकों ने उन्हें उच्च शिक्षा के लिए इंगलैंड जाने की सलाह दी। परंतु मद्रास के

सिविल अस्पताल के चिकित्सक ने इसकी इजाजत नहीं दी, क्योंकि छोटे तथा कमजोर रमन के लिए इंग्लैंड की कठोर जलवायु को सह पाना मुश्किल होता, जिसका अर्थ जीवन को जोखिम में डालना था। अतः उन्होंने भौतिकी में स्नातकोत्तर (एम.ए.) की पढ़ाई करने के लिए फिर से प्रेसिडेंसी कॉलेज में दाखिला ले लिया।

उस समय भौतिकी के प्राध्यापक आर. लेवलीन जोंस थे। उन्होंने रमन की योग्यता के कारण उन्हें काफी छूट दे रखी थी, जिससे उन्हें कक्षा में नियमित उपस्थित रहने की आवश्यकता नहीं थी। रमन ने इसका भरपूर फायदा उठाया तथा ज्यादातर समय कॉलेज के प्रयोगशाला में विभिन्न प्रयोग करने में बिताने लगे। यहीं उन्होंने एक नया प्रयोग किया। उन्हें इस बात का ज्ञान तो था कि जब प्रकाश की किरणें सीधी रखी धातु की प्लेट में बनी एक पतली दरार से होकर गुजरती हैं तब आवर्तित प्रकाश एक पैटर्ण का निर्माण करती है। रमन के दिमाग में इस बात का खयाल आया कि यदि प्लेट को सीधी न रखकर तिरछी रखें तो पैटर्ण में क्या परिवर्तन होगा? इसका अध्ययन करने के लिए उन्होंने एक सामान्य उपकरण बनाकर प्रयोग किए। इस शोध पर उन्होंने एक लेख तैयार किया, जिसे लंदन की सुप्रसिद्ध पत्रिका फिलॉसाफिकल मैगजीन में प्रकाशनार्थ भेज दिया। यह लेख सन् 1906 के नवंबर अंक में प्रकाशित हुआ, जिसका शीर्षक था—'अनसिमेट्रिकल डिफ्रेक्शन बैंड्स ड्यू टू ए रेक्टेंगुलर ऐपरचर'। उस समय रमन की आयु मात्र अठारह वर्ष ही थी तथा वे एम.ए. की पढ़ाई कर रहे थे। इसके कुछ महीने बाद उनका दूसरा लेख 'द कर्वेचर मेथड ऑफ डेटरमाइनिंग दि सर्फेस टेन्शन ऑफ लिक्विड्स' भी फिलॉसाफिकल मैगजीन में प्रकाशित हुआ, जो तरल पदार्थ के सतही खिंचाव के मापन से संबंधित था। खुद के नाम से लिखे इन दोनों शोधपत्रों में रमन ने किसी का भी आभार व्यक्त नहीं किया था, क्योंकि इन्हें करने में उन्होंने किसी की भी मदद नहीं ली थी। रमन के इस लेख ने लॉर्ड रैले का ध्यान आकर्षित किया तथा दोनों में

कुछ पत्राचार भी हुए। पत्र में रैले ने रमन को 'प्रोफेसर' नाम से संबोधित किया था। उन्हें क्या पता था कि इस लेख के रचयिता एक विद्यार्थी हो सकता है। ताज्जुब की बात तो यह है कि रमन को लंदन से प्रकाशित होनेवाली इस पत्रिका के बारे में मालूम कैसे हुआ, क्योंकि उस समय न तो प्रेसिडेंसी कॉलेज और न ही मद्रास विश्वविद्यालय में यह पत्रिका उपलब्ध थी। निश्चित ही रमन विज्ञान के क्षेत्र में तात्कालीन गतिविधियों से वाकिफ रहने के लिए अतिरिक्त प्रयत्न करते थे।

रमन ने जनवरी 1907 में एम.ए. की परीक्षा सर्वप्रथम स्थान प्राप्त कर और गौरवपूर्ण ढंग से उत्तीर्ण की। उन्होंने सभी पुरस्कार जीते। यद्यपि उनकी इच्छा अनुसंधान करने की थी, परंतु उस समय भारतीयों के लिए इस क्षेत्र में प्रवेश मिलना आसान नहीं था। स्वास्थ्य अच्छा न रहने की वजह से लंदन जाकर उच्च्व शिक्षा पाना भी संभव नहीं था। सरकारी नौकरी करना ही उनके लिए एकमात्र विकल्प रह गया था, जो उन्हें इज्जत एवं सुरक्षित जीविका दे सकती थी। लेकिन रमन की विलक्षण प्रतिभा किसी उच्च्च पद के योग्य थी, जिसके लिए प्रतियोगी परीक्षा पास करना जरूरी था। सबसे उच्च्च राजकीय पद आई.सी.एस. के लिए लंदन में जाकर पढ़ाई करके वहीं परीक्षा देनी पड़ती थी जो रमन के लिए संभव नहीं था। परंतु वित्त विभाग के लिए परीक्षा (एफ.सी.एस.) का आयोजन भारत में ही होता था, जिसमें बैठने से पहले सरकार द्वारा गठित समिति के सदस्यों के सामने साक्षात्कार देना आवश्यक था। इसके लिए रमन को साहित्य, इतिहास और राजनीति शास्त्र जैसे नए विषयों का अध्ययन करना पड़ा। रमन में आत्मविश्वास कूट-कूटकर भरा था। उन्होंने परीक्षा में बैठने के पहले ही कह दिया था कि वे ही सर्वप्रथम आएँगे, और ऐसा ही हुआ, रमन ने अपने स्तर को बनाए रखते हुए इस परीक्षा में भी सर्वाधिक अंक प्राप्त किए।

प्रथम नौकरी

उन्नीस वर्ष की उम्र में सन् 1907 में रमन ने वित्त विभाग की परीक्षा पास करके अपनी नियुक्ति की प्रतिक्षा कर रहे थे। इसी बीच 6 मई 1907 को उनका विवाह मदुरै की तेरह वर्षीया सुशील कन्या लोकसुंदरी के साथ हुई। उस समय शादी माता-पिता तय किया करते थे तथा वर-वधु की जन्मकुंडली का मेल खाना भी जरूरी होता था। परंतु रमन ने अपनी पसंद की लड़की से शादी की। यद्यपि लोकसुंदरी ब्राह्मण परिवार से थीं, परंतु उनका गोत्र रमन के गोत्र से भिन्न था। अतः परिवारवालों को यह रिश्ता मंजूर नहीं था। परंतु रमन के पिता स्वतंत्र विचारों के थे, जिन्हें इस शादी से कोई एतराज नहीं था, यद्यपि उनकी माता को सख्त आपत्ति थी। रमन ने अपनी शादी में दहेज लेने से भी इनकार किया था। लोकसुंदरी सही अर्थों में रमन की अर्धांगिनी थी, जिन्होंने हर वक्त उनका साथ दिया। रमन के आखिरी दम तक वे उनकी सेवा करती रहीं।

चित्र : 1.1 कलकत्ता में रमन व उनकी धर्मपत्नी लोकसुंदरी

चित्र : 1.2 कलकत्ता में सहायक अकाउंटेंट जनरल रमन

शादी के कुछ ही समय बाद रमन की नियुक्ति कलकत्ता में हो गई, जहाँ वे वित्त विभाग के सहायक अकाउंटेंट जनरल पद पर जून 1907 से काम करने लगे। यहाँ उनका वेतन 400 रुपए प्रतिमाह था, जिसमें 150 रुपए शादी भत्ते के रुप में मिलता था। उस जमाने में यह रकम काफी अधिक थी। लेडी रमन अकसर मजाक में कहा करती थीं कि रमन ने जल्दी शादी इसलिए की ताकि वे इस भत्ते को पा सकें।

इस प्रकार रमन पश्चिम की ओर जाने के बजाय पूरब की ओर गए। ब्रिटिश साम्राज्य की राजधानी लंदन जाने के बजाय वे ब्रिटिश इंडिया की राजधानी कलकत्ता गए। विज्ञान क्षेत्र में प्रवेश करने के वजाए रमन सरकारी मुलाजिम बन गए। यानी अँधेरे में उनके लुप्त हो जाने के सभी आसार नजर आ रहे थे, परंतु रमन ने ऐसा नहीं होने दिया। उन्होंने दिखा दिया कि यदि लगन सच्ची हो तो कोई कार्य असंभव नहीं होता।

'भारतीय विज्ञान प्रगति संस्थान' से नाता

कलकत्ता में रमन ट्रॉम से दफ्तर आया-जाया करते थे। ट्रॉम बो बाजार स्ट्रीट से होकर गुजरती थी। कलकत्ता पहुँचने के एक सप्ताह के अंदर ही दफ्तर जाते समय एक दिन रमन की नजर बाहर लगे एक साइनबोर्ड पर पड़ी, जिसमें लिखा था—इंडियन एसोशिएशन फॉर दि कल्टिवेशन ऑफ साइंस, 210, बो बाजार स्ट्रीट। बस फिर क्या था! उनके मन में जिज्ञासा उत्पन्न हुई और उसी दिन घर वापस लौटते समय

रमन वहाँ उतर गए और एसोसिएशन के दरवाजे पर जाकर दस्तक दी। दरवाजा एक नौजवान ने खोला, जिसका नाम था आशुतोष दा। रमन ने अंदर प्रवेश कर चारों ओर दृष्टि दौड़ाई। उन्हें वह जगह विरान सी लगी, जैसे लंबे समय से वहाँ कोई आया ही न हो। उन्होंने आशुबाबू से उस संस्था में होने वाले विज्ञान संबंधी गति-विधियों के बारे में जानना चाहा तो वे बिना कुछ बोले उन्हें सचिव श्री अमृतलाल सरकार से मिलवाने ले गए। रमन ने उनसे पुनः वही पूछा परंतु उत्तर देने के बजाय अमृतलाल ने पलटकर रमन से उनकी दिलचस्पी का कारण जानना चाहा। रमन ने अपना परिचय देते हुए कहा कि वे सरकारी नौकरी करते हैं तथा खाली समय का इस्तेमाल विज्ञान के क्षेत्र में करना चाहते हैं, क्योंकि उन्हें शोध कार्य में काफी दिलचस्पी है। इस बात के प्रमाणस्वरूप उन्होंने प्रेसिडेंसी कॉलेज में प्रकाशित अपने लेखों का उल्लेख किया। अमृतलाल को मानो साँप सूँघ गया हो। एक पल के लिए उन्हें विश्वास ही नहीं हुआ कि यह उच्चपदस्थ व्यक्ति इस जगह काम करना चाहता है और वह भी बिना किसी वेतन या मानदेय के। उनके चाचा स्वर्गीय महेंद्रलाल सरकार की दूरदृष्टि का परिणाम था कि सन् 1876 में यह संस्थान (भारतीय विज्ञान प्रगति संस्थान) ब्रिटिश एसोसिएशन तथा रॉयल इंस्टीट्यूट ऑफ लंदन की तर्ज पर स्थापित किया गया, जिसके विस्तार के लिए सन् 1891 में विजयानगरम के महाराजा ने एक बड़ी रकम दान में दी थी। भारत की यह पहली विज्ञान संस्थान था, जिसका निर्माण मूल अनुसंधान करने के लिए किया गया था। यद्यपि महेंद्रलाल एक चिकित्सक थे, परंतु उनका मानना था देश के विकास के लिए विज्ञान में मौलिक अनुसंधान करना जरूरी है, जो अपनों द्वारा संचालित खुद के संस्थान में होना चाहिए। अपने जीवनकाल में महेंद्रलाल कोशिश करते रहे कि कोई यहाँ आकर उपलब्ध सुविधाओं का इस्तेमाल करके विज्ञान के क्षेत्र में प्रयोगात्मक शोध कार्य की शुरुआत करें, परंतु वे इस काम में नाकाम

रहे। सन् 1904 में अपने अधूरे स्वप्नों के साथ उदास महेंद्रलाल स्वर्ग सिधार गए एवं तब से यह संस्थान और भी उपेक्षा का शिकार होता चला गया।

महेंद्रलाल की मृत्यु के तीन वर्ष बाद अनजाने ही रमन उस दरवाजे पर आ खड़े हुए थे जिसे उन्हीं की प्रतीक्षा थी। रमन की मंशा सुनकर अमृतलाल भाव-विभोर हो गए और रमन को बाँहों में भरकर रुँधे कंठ से कहा—'अब तक हम तुम्हारे जैसे व्यक्ति की प्रतीक्षा कर रहे थे। आज से यह संस्थान तुम्हारे हवाले है'। आशूबाबु की मदद से रमन तुरंत काम में जुट गए। उन्होंने पूरे हॉल की सफाई करवाई और फिर प्रयोगशाला के यंत्रों को साफ कर उन्हें ढंग से लगवाया। इसके बाद वे घर लौटे, जो संस्थान के निकट स्कॉट लेन में ही था। रमन जब तक कलकत्ता में रहे, इस संस्थान से जुड़े रहे एवं आशूबाबू उनके विश्वस्त सहायक के तौर पर काम करते रहे।

कलकत्ता में अब रमन काफी व्यस्त जीवन बिता रहे थे परंतु वे खुश थे। उनकी पत्नी के अनुसार रमन की दिनचर्या इस प्रकार थी—सुबह साढ़े पाँच बजे संस्थान जाना, पौने दस बजे घर लौटकर जल्दी से नहाकर दो मिनट में खाना खाकर टैक्सी से कार्यालय जाना, जिससे वहाँ पहुँचने में देर न हो। उसके बाद कार्यालय में काम करके शाम को पाँच बजे सीधे संस्थान जाना, और फिर वहाँ से रात दस बजे घर लौटना। रविवार को भी वे संस्थान जाते थे। निश्चय ही एक नई दुलहन के लिए यह कोई रोमांचक जीवन नहीं था। परंतु लोकसुंदरी ने कभी शिकायत नहीं की। वह उद्यमी थी, अतः समय का सदुपयोग कर अंग्रेजी बोलना व लिखना सीख लिया। यह कोई छोटी बात नहीं थी कि क्योंकि उन दिनों लड़कियाँ स्कूली शिक्षा से बंचित रहती थीं और लोकसुंदरी के साथ भी ऐसा हुआ था।

अपने कठोर परिश्रम से रमन ने संस्थान में कई महत्वपूर्ण शोध कार्य किए, जिनका प्रकाशन प्रसिद्ध वैज्ञानिक पत्रिकाओं जैसे 'नेचर',

'द फिलोसॉफिकल मैगजीन' (लंदन से प्रकाशित) तथा 'फिजिकल रिव्यू' (अमेरिका से प्रकाशित) में होने लगा। इससे उनकी कीर्ति यूरोप व अमेरिका में फैलने लगी, जबकि अबतक उन्होंने देश के बाहर कदम भी नहीं रखा था। कलकत्ता में भी लोग रमन को जानने लगे थे, क्योंकि वे अकसर आम जनता के लिए विज्ञान संबंधित लोकप्रिय व्याख्यान दिया करते थे। रमन एक बहुत ही अच्छे वक्ता थे तथा व्याख्यान के दौरान प्रयोगों का प्रदर्शन करके विषय को सरल व सरस बना देते थे। उस समय लंदन में लोकप्रिय वार्ता के दौरान प्रयोग करके दिखाने का प्रचलन आम था जिसे रमन ने अपनाया था।

सन् 1909 में रमन का तबादला रंगून हो गया जहाँ उन्हें करेंसी अधिकारी के पद पर भेजा गया। परंतु रमन ने विज्ञान से नाता नहीं छोड़ा। रंगून में गए कुछ ही महीने हुए थे कि पिता के गंभीर रूप से बीमार पड़ने की खबर मिली। रमन छुट्टी लेकर उनसे मिलने आए। इस दौरान रमन ने समय निकालकर मद्रास के प्रेसिडेंसी कॉलेज में जाकर कुछ प्रयोग भी किए। कुछ दिनों बाद पिता का देहांत हो गया। उनके क्रिया-कर्म करने के बाद रमन काम में लौट गए, परंतु फिर से उनका स्थानांतरण नागपुर हो कर गया। नागपुर में उन्हें ज्यादा समय नहीं बिताना पड़ा, क्योंकि सन् 1911 में अकाउंटेंट जनरल के पद पर वे कलकत्ता वापस आ गए।

कलकत्ता लौटते ही रमन ने दुबारा विज्ञान संस्थान की प्रयोगशाला में काम करना आरंभ कर दिया। इस बार उन्होंने मकान संस्थान के एकदम निकट लिया था, जिससे वहाँ ज्यादा समय गुजार सकें। अपने मकान और संस्थान के बीच उन्होंने एक दरवाजा भी बनवा लिया था, जिससे वे प्रयोगशाला में इच्छानुसार आ जा सकें। रमन के सहायक के तौर पर आशू बाबू हर समय मौजूद रहते थे। रमन ने अपने कई शोधपत्र उनके साथ लिखा था।

रमन का झुकाव विज्ञान की ओर था, परंतु वे अपने नौकरी के प्रति कर्तव्यशील भी थे। उन्होंने इन दोनों के बीच अद्‌भुत सामंजस्य बनाए रखा। वाइसराय सभासद के वित्त विभाग के सदस्य ने उनके बारे में लिखा था—'इस विभाग में वेंकटरमन सबसे उपयोगी कार्यकर्ता हैं, वास्तव में वे हमारे सबसे अच्छे अधिकारियों में से एक हैं।' विभाग को उनसे काफी उम्मीदें थी, परंतु रमन के जीवन में भारी बदलाव आने वाला था। इस दौरान यानी सन् 1907 से 1917 के दशक में जब रमन सरकारी पदाधिकारी थे, अकेले 27 शोधपत्र लिखे तथा उन्हें कर्जन अनुसंधान पुरस्कार (1912) एवं वुडबर्न शोधपदक (1913) प्राप्त हुआ। उनका ज्यादातर काम ध्वनि विज्ञान व प्रकाश विज्ञान के क्षेत्र में था। उन्होंने वाद्ययंत्रों जैसे वायलिन, वीणा, तबला, मृदंगम आदि सुरीले आवाज पैदा करनेवाले यंत्रों का गहन अध्ययन किया, जिससे वे इस विषय के विशेषज्ञ माने जाने लगे।

विज्ञान जगत् में औपचारिक रूप से आगमन

सन् 1916 में प्रसिद्ध वकील सर आशुतोष मुखर्जी कलकत्ता विश्वविद्यालय के उपकुलपति नियुक्त हुए। वे विज्ञान शिक्षा के क्षेत्र में कुछ खास करना चाहते थे। सरकारी मदद के बिना ही उन्होंने कई लाख रुपए की धनराशि इकट्‌ठा करके 'विज्ञान विश्वविद्यालय' की स्थापना की। जिन लोगों ने अधिक आर्थिक सहायता दी थी, उनके नाम से आचार्य पदों का निर्माण किया गया था, जैसे सर तारकनाथ पालित चेयर एवं सर रासबिहारी घोष चेयर। आशुतोष इन पदों में केवल उच्च काबिलियतवालों को ही रखना चाहते थे और वे रमन के काम से वाकिफ थे। उनकी दिली इच्छा थी कि भौतिकी विषय में पालित चेयर का पद रमन जैसे प्रतिष्ठित भौतिकविद् सँभालें परंतु उनके मन में शंका थी कि रमन उच्च पदस्थ एवं अधिक वेतनवाली सरकारी नौकरी छोड़कर उनका प्रस्ताव स्वीकार करेंगे या नहीं। यहाँ वेतन सिर्फ 600 रुपए

प्रतिमाह था, जबकि रमन को उस समय 1100 रुपए मिल रहा था। फिर भी उपकुलपति ने औपचारिक तौर पर रमन को अपना प्रस्ताव भेजा। रमन ने तुरंत अपनी सहमति दे दी, परंतु वित्त विभाग उनके इस निर्णय से खुश नहीं था। वे एक कर्मठ पदाधिकारी खोना नहीं चाहते थे। उन्होंने रमन को आगाह किया कि वाइसराय के सभासद में उनका चयन वित्त विभाग के सदस्य के तौर पर होने की प्रबल संभावना है (जो वर्तमान वित्त मंत्री के पद जैसा है), परंतु रमन ने इन सबकी परवाह नहीं की। ऐसी उम्मीद थी कि रमन कुछ समय के लिए छुट्टी लेकर आचार्य पद पर कार्य करके उसे परखने के बाद ही नौकरी से इस्तीफा देंगे, परंतु सर हारकोर्ट बटलर, जो उस समय शिक्षा विभाग के सदस्य थे, यह राहत देने से इनकार कर दिया। विज्ञान के प्रति सच्चे प्रेम की वजह से रमन ने नौकरी छोड़कर आचार्य पद स्वीकार कर लिया। परंतु एक अड़चन और था—इस पद के लिए विदेश में शिक्षित होना आवश्यक था; अतः रमन को लंदन जाने को कहा गया। रमन ने ऐसा करने से साफ इनकार कर दिया, क्योंकि वे भारतीय की प्रतिभा को मनवाने के लिए विदेशी प्रमाणपत्र की आवश्यकता को नकारते थे। सर आशुतोष की मध्यस्थता के बाद उन्हें इस शर्त से छूट दी गई। इस प्रकार सन् 1917 में रमन का विज्ञान जगत् में औपचारिक आगमन हुआ। सरकारी नौकरी से सीधे प्राध्यापक पद पर नियुक्ति केवल उनकी योग्यता पर मिली थी और वह भी सिर्फ 29 वर्ष की अल्पायु में।

पालित प्रोफेसर का दायित्व अध्यापन कार्य करने का नहीं था; उनका काम अपने विषय में शोध करना, शोधकर्ताओं का मार्गदर्शन करना और विज्ञान की प्रयोगशालाओं का निरीक्षण करना था। फिर भी रमन एम.ए. व एम.एससी. के विद्यार्थियों को नियमित रूप से पढ़ाने लगे। वे कहा करते थे कि किसी भी विषय को ठीक से समझने का सबसे आसान उपाय उस विषय पर वार्ता देना होता है। रमन को व्याख्यान देना इतना अच्छा लगता था कि वे लगातार दो-तीन घंटे बिना

रुके बोलते रहते थे तथा विद्यार्थी मंत्रमुग्ध होकर सुनते थे। रमन ने विज्ञान संस्थान से भी अपना नाता बनाए रखा, विश्वविद्यालय में विद्यार्थी एवं शोधकर्ताओं से निपटने के बाद वे संस्थान में प्रयोगात्मक काम करने जाया करते थे। यह आश्चर्य की बात है कि रमन खुद डॉक्टरेट की उपाधि प्राप्त नहीं थे, परंतु उनके पर्यवेक्षण में कई शोधकर्ता इस उपाधि को प्राप्त करने के लिए काम कर रहे थे। आखिरकार सन् 1921 में कलकत्ता विश्वविद्यालय ने उन्हें सम्मानिक डॉक्टरेट की उपाधि प्रदान की।

सन् 1919 में अमृतलाल के देहांत के बाद रमन को संस्थान का सचिव नियुक्त किया गया, जिससे उनकी जिम्मेदारी व व्यस्तता और भी बढ़ गई। पंद्रह वर्षों तक वे कलकत्ता विश्वविद्यालय में पालित प्रोफेसर के पद पर रहे एवं इस दौरान प्रकाश विज्ञान तथा प्रकाश के प्रकीर्णन पर मौलिक कार्य किया, जिससे उनकी ख्याति विश्वभर में फैल गई। रमन पहली बार सन् 1921 में विदेश की यात्रा की, जब वे लंदन में हो रहे ब्रिटिश साम्राज्य विश्वविद्यालयों के सम्मेलन में कलकत्ता व बनारस विश्वविद्यालय के प्राध्यापकों के प्रतिनिधि के रूप में गए। वहाँ उनकी भेंट कई विख्यात वैज्ञानिकों जैसे जे.जे. थॉमसन, ब्रैग व रदरफोर्ड से हुई। लंदन में पहली बार रमन जब रॉयल इंस्टीट्यूट में रदरफोर्ड की वार्त्ता सुनने गए, वे चुपचाप पीछे की पंक्ति में जाकर बैठ गए। इसके पहले उनकी आपसी मुलाकात नहीं हुई थी। हमेशा मद्रासी पगड़ी पहननेवाले रमन पर जब रदरफोर्ड की नजर पड़ी, उन्होंने तुरंत पगड़ी की वजह से रमन को पहचान लिया और उन्हें आगे की पंक्ति में आकर बैठने का इशारा किया, जहाँ जानी-मानी हस्तियाँ बैठी थीं। लंदन में भी उन्होंने कुछ प्रयोगात्मक अध्ययन किए। सितंबर 1921 में लंदन से समुद्री जहाज एस.एस. नारकुंडा द्वारा वापस लौटते वक्त रमन का ध्यान भूमध्यसागर की गहरी नील जल ने आकर्षित किया और वे 15 दिनों के यात्रा के दौरान जहाज में ही कई प्रयोग किए तथा उस पर

टिप्पणी लिखकर 'नेचर पत्रिका' में प्रकाशन के लिए भेज दिया। उनका मानना था कि समुद्री जल इसलिए नीले रंग का दिखता है क्योंकि प्रकाश की किरणें जल के अणुओं से प्रकीर्णन करती हैं, ठीक वैसे ही जैसे वायुमंडल में व्याप्त गैस के अणुओं के द्वारा प्रकाश के प्रकीर्णन होने से आसमान का रंग नीला दिखता है। आसमान के नीले रंग की यह व्याख्या लॉर्ड रैले ने दी थी एवं उनका मानना था कि समुद्री जल का नीलापन केवल आकाश का प्रतिबिंब मात्र है। रमन की पैनी नजर ने देखा कि आकाश के बादलों से घिरे रहने के बावजूद समुद्र का पानी नीला ही दिखता है। बस, फिर क्या था, कलकत्ता लौटते ही उन्होंने प्रायोगिक प्रमाण जुटाने के लिए ढेर सारे प्रयोग शुरू कर दिए, परंतु सबूत उन्हें सात साल बाद मिला।

सन् 1924 में ब्रिटिश विज्ञान प्रगति संस्थान के आमंत्रण पर रमन कनाडा गए जहाँ टोरेंटो में प्रकाश विकीर्णन संगोष्ठी की अध्यक्षता की। कनाडा से वे अमेरिका गए जहाँ फ्रैंकलीन संस्थान के शताब्दी समारोह में भारत का प्रतिनिधित्व किया। उसके बाद वे रॉबर्ट मिल्लीकन के आमंत्रण पर चार महीने कैलिफोर्निया प्रोद्योगिकी संस्थान (कैलटेक) में अतिथि प्राध्यापक रहे, जो काफी प्रतिष्ठित पद माना जाता है। रमन से पहले इस पद पर सोम्मरफेल्ड, लॉरेंज तथा आईनस्टाइन जैसे महान वैज्ञानिकों को आमंत्रित किया गया था। अमेरिका-प्रवास के दौरान रमन वहाँ के कई वैज्ञानिक संस्थानों में गए जहाँ कॉम्प्टन, लैंगमूर, कुलीज. जैसे कई विशिष्ट वैज्ञानिकों से उनकी मुलाकात हुई। अमेरिका से वापस लौटते समय वे यूरोप के कई जगहों पर रूके एवं वहाँ के वैज्ञानिकों से मिले। मार्च 1925 में वे भारत लौट आए। इसके बाद अगस्त में रमन लेनिनग्राद विज्ञान अकादमी की शताब्दी समारोह में भाग लेने सोवियत संघ गए। वहाँ से वे बर्लिन, पेरिस, जिनेवा एवं रोम होते हुए लौटे। ऐसा सोचना गलत होगा कि रमन अब केवल विदेश यात्रा ही कर रहे थे। अगले तीन-चार वर्षों तक वे विदेश नहीं गए एवं इसी दौरान 'रमन

प्रभाव' की खोज हुई। परंतु विदेश दौरे के बाद रमन के काम करने के तरीके पर जरूर परिवर्तन आया। जहाँ पहले वे काम करते वक्त समय का बिलकुल खयाल न रखकर खाना-पीना तथा आराम करना भी नजरअंदाज. करते थे, वहीं अब उनकी दिनचर्या में नियमितता आई। लेडी रमन को इससे जरूर राहत महसूस हुई होगी।

'रमन प्रभाव' का आविष्कार

रमन के साथ अब कई शोधकर्ताओं के होने से उनकी देख-रेख में कई क्षेत्रों में काम होने लगा। परंतु रमन खुद प्रकाश प्रकीर्णन के काम को ज्यादा गुरुत्व देने लगे। सन् 1923 में उनके छात्र रामनाथन ने एक प्रयोग करते समय पानी द्वारा विवर्तित प्रकाश में कुछ अलग रंग के किरणों को देखा था, जिसे वे क्षीण प्रतिदिप्त समझ रहे थे। उसके दो साल बाद रमन के दूसरे शिष्य कृष्णन ने भी ऐसा ही कुछ देखा। सन् 1927 के आखिर में एक और छात्र वेंकटेश्वरन अपने अन्वेषण में फिर से विवर्तित प्रकाश में आवर्तित प्रकाश के रंग से अलग रंग की कुछ किरणें दिखाई दीं। रमन अब विश्वास करने लगे थे कि यह रहस्यमयी प्रकाश प्रतिदिप्त की वजह नहीं हो सकता। तो यह क्या था? क्या यह 'कॉम्प्टन प्रभाव' जैसा ही कुछ था जो द्रव्य प्रकाश से उत्पन्न हो रहा था? रमन अब अपने सबसे अच्छे शिष्य कृष्णन, जो उस समय चुंबकीय प्रभाव पर शोध कर रहे थे, के साथ इस विषय में गहरी तहकीकात करने जुट गए। उन्होंने एक सरल उपकरण का निर्माण किया जिसमें विवर्तित प्रकाश को आँख से देखना पड़ता था। चूँकि यह प्रकाश अत्यंत धुँधला था, अत: उसे अच्छी तरह देखने के लिए चारों तरफ से बंद एक बक्से का निर्माण किया गया, जिससे बक्से के अंदर बाहरी प्रकाश न जा सके। बक्से के भीतर प्रेक्षक खड़े होकर एक छिद्र से विवर्तित किरणों को देख सकता था। कई महीनों के अथक परिश्रम के बाद 28 फरवरी सन् 1928 में रमन ने निश्चित तौर पर यह प्रमाणित किया कि क्षीण

प्रकाश की किरणें एक प्रकार का प्रकीर्णन हैं, जिसे रूपांतरित विकिरण कहा गया। इसके दूसरे ही दिन रमन की इस ऐतिहासिक खोज के बारे में सारी दुनिया को पता चल गया। 'रमन प्रभाव' का आविष्कार हो चुका था। इस विधि से किसी भी पदार्थ की आणविक संरचना के बारे में जानकारी प्राप्त होती है। इस खोज के बारे में विस्तृत जानकारी अन्य अध्याय में दी गई है।

ब्रिटिश सरकार ने पहले ही रमन के कार्यों की मान्यतास्वरूप उन्हें सन् 1924 में एफ.आर.एस. तथा सन् 1929 में 'सर' के खिताब से नवाजा था। इसके बाद सन् 1930 में उन्हें भौतिक विज्ञान के क्षेत्र में नोबेल पुरस्कार देकर सम्मानित किया गया। वे पहले एशियाई थे जिन्हें विज्ञान हेतु यह पुरस्कार मिला था। उन्हें देश-विदेशों से अनेक पुरस्कार एवं उपाधियाँ मिलीं। कलकत्ता आने के 23 वर्ष बाद उन्होंने एक इतिहास रच डाला था। जिस समय उन्हें यह पुरस्कार मिला, उनकी उम्र केवल 42 वर्ष थी। नोबेल पुरस्कार के इतिहास में आविष्कार के इतने कम समय के अंदर यह पुरस्कार बहुत कम लोगों को मिला है। आमतौर पर किसी भी खोज के बाद उसकी उपयोगिता के बारे में पता लगते लगते कम-से-कम दस वर्ष तो बीत ही जाते हैं, तब जाकर नोबेल पुरस्कार दिया जाता है। रमन ने श्रॉडिंगर, डिराक व हाइजेनबर्ग के मुकाबले यह पुरस्कार जीता था जिन्हें उसी वर्ष नामांकित किया गया था। इन विभूतियों को बाद में नोबेल पुरस्कार मिला। रमन की खोज बीसवीं सदी के आरंभ में हुए श्रेष्ठ प्रायोगिक खोजों में गिनी जाती है जिससे क्वांटम सैद्धांतिकी की पुष्टि होती है। अस्सी वर्ष बीत गए हैं परंतु रमन अब भी प्रथम भारतीय हैं जिन्हें अपने देश में किए गए कार्य के लिए विश्व का सर्वश्रेष्ठ पुरस्कार मिला।

लगभग दो दशकों (1918 से 1938) तक यूरोप के अलावा भारत में कलकत्ता ही एक ऐसा शहर था जहाँ भौतिक विज्ञान के क्षेत्र में मौलिक अनुसंधान का कार्य जोरों से चल रहा था और इन सबके केंद्रबिंदु

थे सर रमन। यही नहीं, उस समय कलकत्ता में हो रहे शोध कार्य अमेरिका के प्रसिद्ध विश्वविद्यालयों, जैसे एम.आई.टी., हार्वर्ड, प्रिंसटन में हो रहे शोध कार्यों से भी श्रेष्ठ था। रमन ने 1926 में 'इंडियन जर्नल ऑफ फिजिक्स' का प्रकाशन शुरू किया, जिससे शोध-कार्यों को जल्द-से-जल्द प्रकाशित किया जा सके। इससे पहले सन् 1913 में इंडियन साइंस कांग्रेस की शुरुआत हुई थी, जिसका उद्देश्य था वर्ष में एक बार विज्ञान के क्षेत्र से जुड़े सभी शोधकर्ताओं को एक जगह इकट्ठा करना, जिससे वे आपस में विचार- विनिमय कर सकें। रमन कई वर्षों तक इस संस्था के सचिव रहे। यह दुर्भाग्य की बात है कि उस दो दशक के स्तर को कायम नहीं रखा जा सका। विज्ञान में उत्कृष्टता बनाए रखने के लिए मुख्यत: तीन बातों को ध्यान में रखने की आवश्यकता होती है— शोधकार्य में आत्मविश्वास, विज्ञान की शिक्षा देने के लिए महान् गुरु तथा अच्छे विश्वविद्यालयों का निर्माण, जिसमें नई पीढ़ी के लिए शिक्षक तैयार किए जा सकें। रमन ने इस बात का ध्यान रखा कि उनके शिष्य प्रशिक्षण पाकर देशहित के लिए दूसरे जगहों में जाकर काम करे, न कि उसी संस्थान में स्थायी रूप से रह जाएँ। यही कारण है कि रमन के कई शिष्यों ने विभिन्न क्षेत्रों मे प्रसिद्धता हासिल की।

कलकत्ता से विदाई

जिस कलकत्ता ने रमन की प्रतिभा को पहचाना एवं उसे पनपने का मौका दिया, अब उनकी असाधारण सफलता को मानो बर्दाश्त नहीं कर पा रहा था। सफलता हमेशा ईर्ष्या को जन्म देती है एवं रमन के साथ भी कुछ ऐसा ही हुआ। रमन भी स्पष्टवादी व खुद्दार व्यक्ति थे, जिससे उनके हितैशियों की संख्या में कमी आ रही थी। रमन पर यह गलत इलजाम लगाया गया कि वे केवल दक्षिण भारतीयों को प्राधन्यता देते हैं। इससे उनके अहंकार को गहरी ठेस पहुँची। यह सच है कि कुछ छात्रों के लिए, जो सुदूर मद्रास से कलकत्ता सिर्फ उनके साथ काम

करने के उद्देश्य से आए थे, आरंभ में रमन ने उनके आवास आदि का बंदोबस्त किया था, परंतु वे बंगाली विरोधी कतई नहीं थे। कुछ लोग तो यह भी कहने लगे थे कि यह आविष्कार रमन नहीं बल्कि उनके छात्र कृष्णन ने किया था। घटनाचक्र ने कुछ ऐसा मोड़ लिया कि रमन के खिलाफ साजिश करके उन्हें 'भारतीय विज्ञान प्रगति संस्थान' के सचिव पद से हटाया गया। रमन भी ऐसी हालत में कलकत्ता छोड़ना चाहते थे। इसी दौरान 'भारतीय विज्ञान संस्थान', बैंगलौर के निदेशक पद के लिए योग्य व्यक्ति की तलाश हो रही थी। इसके लिए समिति लंदन से किसी को लाना चाहती थी जैसाकि उस समय रिवाज था। जब लॉर्ड रदरफोर्ड से इस बारे में सलाह माँगी गई तो उन्होंने रमन का नाम सुझाते हुए कहा कि 'जब प्रार्थी आपके देश में मौजूद है तो बाहर क्यों ढूँढ़ रहे हैं ? रमन जैसा योग्य व्यक्ति आपको नहीं मिलेगा।' अत: रमन का चयन हो गया एवं स्वीकृति के लिए उनके पास प्रस्ताव भेजा गया। रमन इस शर्त पर बैंगलौर जाने के लिए तैयार हुए कि वहाँ जाकर वे भौतिक विज्ञान के क्षेत्र में एक नए विभाग का गठन करेंगे। उनकी इस माँग को भी मंजूरी मिल गई।

बैंगलौर में आगमन

अप्रैल 1933 में रमन ने बैंगलौर में 'भारतीय विज्ञान संस्थान' के निदेशक का पदभार ग्रहण किया। वे इस संस्थान के प्रथम भारतीय निदेशक थे। उनसे पहले सभी निदेशक विदेशी थे। इस संस्थान की स्थापना सर जमशेदजी टाटा ने सन् 1911 में की थी। आरंभ में यहाँ केवल तीन क्षेत्रों में काम शुरू हुआ था—सामान्य व अनुप्रयुक्त रासायनिकी, कार्बनिक रासायनिकी तथा विद्युत् प्रौद्योगिकी। रमन के आने से पहले जीव रासायनिकी विभाग में भी काम आरंभ हो गया था। संस्थान में ज्यादातर काम व्यावहारिक क्षेत्र में होता था तथा मौलिक अनुसंधान की तरफ कम ध्यान दिया जाता था। रमन ने आते ही नए भौतिक विज्ञान विभाग का

गठन किया, जिसकी अनुमति उन्हें पहले ही मिल चुकी थी, तथा कुछ विभागों में प्रशासनिक दृष्टिकोण से फेरबदल किया। उन्होंने कार्यशाला को उपयोगी बनाने के लिए उसमें भी कुछ परिवर्तन किए, जिससे अनुसंधान के लिए आवश्यक उपकरणों का निर्माण वहीं किया जा सके। रमन का मानना था कि उत्कृष्ट काम करने के लिए संस्थान में श्रेष्ठ व योग्य लोगों को लाने की जरूरत है, अतः उन्होंने जर्मनी के कुछ प्रसिद्ध वैज्ञानिकों से संपर्क किया, जो उस समय हिटलर के डर से देश छोड़कर भाग रहे थे। यद्यपि रमन के विचार में संस्थान की उन्नति के लिए यह सब करना अत्यंत आवश्यक था परंतु संस्थान के पुराने लोगों को यह सब अच्छा नहीं लगा। कुछ वरिष्ठ प्राध्यापक, जो खुद निदेशक बनने की तमन्ना रखते थे, रमन के कार्य करने के ढंग की आलोचना करने में जुट गए तथा उनके विरुद्ध षड्यंत्र रचने लगे। ऐसा कहा गया कि उनमें प्रशासनिक दक्षता नहीं है। उनपर केवल भौतिकी विषय को बढ़ावा देने का आरोप लगाया गया। रमन भी कहाँ हार माननेवाले थे। वे खुद को औरों से श्रेष्ठ मानते थे तथा उन्हें इस बात का गर्व था। अतः आपसी समझौता होने के बजाए तनाव बढ़ता गया। बात इतनी बढ़ गई कि रमन की कार्यप्रणाली की समीक्षा करने के लिए जनवरी सन् 1936 में एक समिति गठित की गई जिसने रमन के खिलाफ रिपोर्ट दी। रमन एकदम अकेले पड़ गए थे। उनके पास निदेशक पद से इस्तीफा देने के सिवा और कोई उपाय नहीं बचा था। अंततः सन् 1938 में वे निदेशक पद से हट गए, लेकिन प्राध्यापक पद पर बने रहे। यद्यपि रमन में मनोबल की कमी नहीं थी, लेकिन कलकत्ता तथा बैंगलौर के हादसों ने उन्हें झकझोरकर रख दिया। परंतु उनकी वैज्ञानिक गतिविधियों में कोई कमी नहीं आई। उनका शोधकार्य जारी रहा। उन्होंने 'भारतीय विज्ञान अकादमी' की स्थापना की तथा विज्ञान पत्रिका प्रोसिडिंगस ऑफ दि एकेडेमी का प्रकाशन शुरू किया। यहाँ इस बात का जिक्र करना उचित होगा कि रमन की यह सोच थी कि देश में विज्ञान की जड़ मजबूत करने व इसकी प्रगति के लिए

प्रसिद्ध विदेशी वैज्ञानिकों को भारत में आमंत्रित करना जरूरी है, परंतु उनके इस प्रयास को सफल नहीं होने दिया गया; जबकि इसी विचारधारा को अपनाकर अमेरिका आज विश्व में महाशक्ति के रूप में उभरा है।

उस समय विश्व के कई देशों में नाभिकीय भौतिकी पर जोरों से काम चल रहा था। रमन भी बैंगलौर में इस विषय पर अनुसंधान कार्य शुरू करना चाहते थे, परंतु उसके लिए आवश्यक धनराशि उपलब्ध न होने की वजह से वे ऐसा नहीं कर सके। नवंबर 1948 में साठ वर्ष के रमन 'भारतीय विज्ञान संस्थान' से सेवा-निवृत्त हुए, परंतु उनका शोधकार्य जारी रहा। भारत सरकार द्वारा देश के उपराष्ट्रपति पद ग्रहण करने के प्रस्ताव को उन्होंने ठुकरा दिया था। उन्हें किसी भी पद का लालच नहीं था, इसलिए सेवानिवृत्त होने के बाद वे किसी भी संस्था से जुड़े नहीं रहे।

सेवानिवृत्त होने के पहले ही रमन बैंगलौर में एक 'विज्ञान अनुसंधान केंद्र' के निर्माण की योजना बना चुके थे एवं उसके लिए भवन का निर्माण भी हो चुका था। इसे बनाने के लिए उन्होंने स्वयं धन इकट्ठा किया था। उन्होंने सरकार से कोई सहायता नहीं ली थी। अवकाश प्राप्त करने के तुरंत बाद उन्होंने उस केंद्र में अनुसंधान कार्य शुरू कर दिया, जिसके परिसर में ही उनका निवास स्थान था। इस केंद्र का नाम 'रमन अनुसंधान संस्थान' रखा गया, जिसके वे प्रथम निदेशक बने तथा आजीवन इस पद पर बने रहे। उनका ध्यान अब भी विज्ञान के प्रति रुचि जगाने तथा वैज्ञानिक प्रतिभाओं की खोज में लगा था। भारत सरकार ने उन्हे पहला राष्ट्रीय प्राध्यापक पद से गौरवांवित किया। सन् 1954 में भारत सरकार ने देश के सुप्रसिद्ध नागरिकों के सम्मानार्थ 'राष्ट्रीय गौरव पुरस्कार' देने का निर्णय लिया। रमन उन व्यक्तियों में से थे जिन्हें सर्वप्रथम भारत के सर्वोच्च सम्मान 'भारतरत्न' के लिए चुना गया। उस समय रमन विदेश में अपने किसी शिष्य के शोधकार्य की समीक्षा करने गए थे। भारत के प्रथम राष्ट्रपति डॉ. राजेंद्र प्रसाद ने रमन को इस बारे में सूचित

करते हुए उन्हें देश में आकर पारितोषिक ग्रहण करने का आमंत्रण दिया। रमन अपने काम के प्रति इतने लगनशील थे कि उन्होंने राष्ट्रपति को धन्यवाद देते हुए समारोह में उपस्थित न रह पाने की असमर्थता जाहिर की, क्योंकि उसी समय उनका एक शिष्य शोध-प्रबंध पूरा करने के काम में जुटा था, जिसके निरीक्षण के लिए उनका वहाँ रहना जरूरी था। भारतरत्न उन्हें उनकी अनुपस्थिति में दिया गया। ऐसी थी काम के प्रति उनकी प्रतिबद्धता!

समय गुजरता गया और रमन अकादमी के कार्य करने के साथ साथ नित्य नए प्रयोग भी करते रहे। अब वे प्राकृतिक वस्तुओं—जैसे फूल, पत्थर, पक्षी, तितली, पतंगों आदि के मनमोहक रंगों के विश्लेषण में रुचि लेने लगे। विज्ञान के प्रति उनका प्रेम अटूट था और वे खाली बैठना नहीं चाहते थे। जीवन में उतार-चढ़ाव तो आते रहते हैं और रमन भी इससे अछूते नहीं रहे। उनके करीबी सभी लोग एक-एक करके उनका साथ छोड़ दिया। समय के साथ भौतिक विज्ञान में भी नए क्षेत्रों में शोध हो रहे थे, अतः नई पीढ़ी के युवा इन विषयों की ओर अधिक आकर्षित हो रहे थे। देश स्वतंत्र होने के बाद विज्ञान व प्रौद्योगिकी के विकास के लिए जिस रास्ते को अपनाया जा रहा था, विशेषकर आयात का मार्ग, उससे रमन बेहद खफा थे एवं इसके लिए वे अकसर सरकार की कड़ी आलोचना किया करते थे। उनका कहना था कि विज्ञान की तरक्की के लिए धन की आवश्यकता अवश्य होती है, परंतु विज्ञान में बेहतर शोध करने के लिए इससे भी अधिक जरूरत है उद्‌दमी एवं योग्य व्यक्तियों की। वे सीधी बात कहने से नहीं डरते थे, जो कुछ लोगों को पसंद नहीं आया। इस वजह से लोगों से उनकी दूरी बढ़ती गई। उदासीपन ने उन्हें जकड़ लिया। अब वे अकेलापन महसूस करने लगे। वे भी किसी से मिलना नहीं चाहते थे। अतः अपने संस्थान के सामने उन्होंने एक बोर्ड टँगवा दिया जिसमें लिखा था 'आगंतुकों का अंदर

आना मना है।' यही नहीं, संस्थान की चाहरदीवारी की ऊँचाई भी बढ़ा दी। परंतु कुछ समय बाद सबकुछ भुलाकर रमन फिर से जीवन में उत्साह लेने लगे। 'रमन अनुसंधान संस्थान' में उन्होंने सुंदर बगीचे बनवाए थे, जिसमें विभिन्न प्रकार के गुलाब व अन्य फूलों का समावेश था। इन फूलों से अर्क निकालकर उनपर शोध करना, पंखुड़ियों से प्रकाश के प्रकीर्णन पर प्रयोग करना आदि उनका शौक बन गया। इसके अतिरिक्त वे नियमित रूप से छोटे स्कूली बच्चों को अपने संस्थान में बुलाते तथा उनके साथ घुल-मिल कर बातें करते। उन्हें विज्ञान संबंधी कहानियाँ सुनाते, उनके अंदर प्रकृति से प्रेम करने की लालसा पैदा करते, उनकी जिज्ञासाओं का समाधान करते, उन्हें संस्थान के संग्रहालय में ले जाकर वहाँ रखे विभिन्न किस्म के मणि व खनिज पदार्थों के बारे में बताते। इस तरह उन्हें बढ़ती उम्र में उन्हें जीने का सहारा मिल गया था। बच्चों की किलकारियों से संस्थान गूँजने लगा था। ये बच्चे रमन के पीछे एक जगह से दूसरी जगह भागते नजर आते।

अंतिम घड़ी

रमन ने अपनी अंतिम लोकप्रिय वार्त्ता 2 अक्तूबर सन् 1970 को 'गांधी स्मरण व्याख्यान श्रृंखला' के अंतर्गत दी थी। उसी महीने के अंत में अपनी प्रयोगशाला में काम करते वक्त अचानक वे बेहोश हो गए। परीक्षण करने पर पता चला कि उनके हृदय के वॉल्व खराब हो चुके थे। डॉक्टरों के अनुसार वे चंद घंटों के मेहमान थे, परंतु वे जीवित रहे। स्वास्थ्य में थोड़ा सुधार होने पर वे अस्पताल में और नहीं रहना चाहते थे, अतः वे संस्थान लौट आए। जब उन्हें पता चला कि भविष्य में सामान्य तौर पर जीवन व्यतीत करना उनके लिए असंभव-सा है और बाकी जिंदगी बिस्तर पर लेटकर बितानी पड़ेगी तो उन्होंने दवाइयाँ लेना बंद कर दिया। उन्होंने कहा कि ऐसी जिंदगी जीने से तो मरना बेहतर है। 19 नवंबर को वे अपने एक पुराने शिष्य, जो वहाँ मौजूद थे, उन्हें

पास बुलाकर कहा कि अकादमी से प्रकाशित हो रही पत्रिका को बंद नहीं होने देना, क्योंकि इसके माध्यम से लोगों को देश में हो रही विज्ञान की गतिविधियों का पता चलता है। उसके बाद संस्थान के भविष्य के बारे में अपनी राय देते हुए उन्होंने कहा—'सन् 1948 में इस संस्थान का निर्माण मैंने मूल शोध करने के उद्देश्य से किया था तथा मैं चाहता था कि देश में एक उच्चकोटि की विज्ञान संस्था हो। उसी को ध्यान में रखते हुए इसकी प्रशासन व्यवस्था मैने अपने ढंग से की है। यहाँ का वातावरण दूसरे संस्थानों से भिन्न है, जो मूल शोध करने की प्रेरणा देता है। विज्ञान का अनुशीलन मेरे लिए सुरुचिपूर्ण एवं आनंददायक अनुभव रहा है। यह संस्थान मेरा आश्रयस्थल रहा है मैं जहाँ अपनी रुचि के मुताबिक अनुसंधान किए हैं। जाहिर है, मेरे जाने के बाद इसकी कार्यपद्धति में बदलाव आएगा। मैं चाहता हूँ कि यह संस्थान ज्ञान अर्जन करने का एक विशाल केंद्र बनकर उभरे, जिससे देश व विदेश के वैज्ञानिक इसकी ओर आकर्षित हों। इसके सुंदर बाग-बगीचे, बड़ा पुस्तकालय, विशाल संग्रहालय आदि उच्च शिक्षा प्राप्त करने के केंद्र बनने में मददगार होंगे। विज्ञान तभी पनप सकता है जब इस क्षेत्र में काम करने की आंतरिक इच्छा हो। बाहरी दबाव से इसका विकास संभव नहीं है। विज्ञान में मौलिक अनुसंधान किसी के निर्देश व जोर-जबरदस्ती से नहीं हो सकता। यही कारण है कि मैंने तय किया था कि जहाँ तक संभव हो, सरकार से किसी प्रकार की सहायता न लेने का। मैंने अपनी सारी संपत्ति संस्थान के नाम कर दी है। परंतु शायद यह पर्याप्त नहीं होगा जिससे संस्थान को ज्ञान उपार्जन करने के केंद्र के रूप में विकसित किया जा सके। अतः सरकारी सहायता न लेने की शर्त पर कायम न रहकर अब संस्थान को उन सभी से मदद लेनी होगी, जो बिना किसी शर्त के सहायता करने को तैयार हैं।' इसके दो दिन बाद, 21 नवंबर, 1970 के तड़के 82 वर्ष की उम्र में शांतचित्त रमन ने सदा के लिए अपनी आँखें बंद कर लीं। उनके इच्छानुसार संस्थान में ही उन्हें दफनाया

गया, जिसके चारों ओर फैले परिवेश में वे हमेशा खुश रहते थे तथा इससे उनका गहरा लगाव भी था। कोई धार्मिक अनुष्ठान भी नहीं किया गया। उस जगह एक पौधा लगाया गया था, जो आज पेड़ बनकर फलने-फूलने लगा है। वहाँ कोई स्मारक चिन्ह नहीं है।

रमन के शोधकार्य क्षेत्र व सम्मान—एक झलक

भारत में रमन प्रभाव के आविष्कार का दिन 28 फरवरी 'राष्ट्रीय विज्ञान दिवस' के रूप में मनाया जाता है। अपने लंबे जीवन में रमन ने विभिन्न विषयों पर कार्य किया। अपने 66 वर्ष के वैज्ञानिक जीवनकाल में उन्होंने 450 से भी अधिक मूल शोधपत्र व निबंध लिखे। इसके अलावा उनके शिष्यों द्वारा प्रकाशित शोधपत्रों की संख्या इससे तीन गुणा है। आज भी 'रमन प्रभाव' पर आधारित शोधपत्र छपते हैं। यदि उनके कार्यकाल को छह दशकों में बाँटा जाए तो हम पाते हैं कि उनकी रुचि में लगातार परिवर्तन होता गया, जैसाकि नीचे वर्णित है—

सन् 1910 से 1920—ध्वनिकी

सन् 1920 से 1930—दृष्टि विद्या एवं प्रकाश का बिखरना

सन् 1930 से 1940—पराश्रव्य विवर्तन, तरल पदार्थ में ब्रीलोईन प्रकीर्णन की उपयोगिता तथा स्फटिकों से रमन प्रकीर्णन।

सन् 1940 से 1950—हीरा तथा स्फटिक लैटीसों में कंपन

सन् 1950 से 1960—खनिज पदार्थों की प्रकाशिकी

सन् 1960 से1970—रंगो का बोध

रमन के इन क्षेत्रों में शोधों का विस्तृत विवरण अगले अध्यायों में दिया गया है।

अपने कार्यकाल में रमन को अनेक सम्मान मिले, जिनकी सूची नीचे दी गई है—

1912 कर्जन अनुसंधान पुरस्कार

1913 वुडबर्न अनुसंधान पदक

1924 रॉयल सोसाइटी लंदन की सदस्यता

1928 मैट्टेउच्ची पदक, रोम

1929 सर का खिताब

1930 ह्यूग्स पदक, रॉयल सोसाइटी, लंदन

1930 नोबेल पुरस्कार

1935 राज सभा भूषण, मैसूर के महाराजा द्वारा सम्मान

1941 फ्रैंकलीन पदक, फ्रैंकलीन संस्थान, फिलाडेल्फिया

1948 राष्ट्रीय प्राध्यापक नियुक्त

1954 भारतरत्न से विभूषित

1957 लेनीन शांति पुरस्कार, सोवियत महासंघ

निम्नलिखित विश्वविद्यालयों ने उन्हें मानद डॉक्टर की उपाधि से सम्मानित किया —

इलाहाबाद, बनारस, मुंबई, कलकत्ता, ढाका, देहली, कानपुर, लखनउ , मद्रास, मैसूर, पटना, उसमानिया, हैदराबाद, श्रीवेंकटेश्वर, तिरुपति, फ्रायबुर्ग, ग्लासगो, पेरिस।

निम्न संस्थानों ने उन्हें सम्मानित सदस्यता प्रदान की थी—

दॉयचे अकादमी, म्युनिख
हंगरीयन विज्ञान अकादमी
भारतीय विज्ञान कांग्रेस प्रतिष्ठान तथा भारत के
अनेका विज्ञान संस्थान
रॉयल आइरीश अकादमी
रॉयल फिलोसोफिकल सोसाइटी, ग्लासगो
जूरीख फिजिकल सोसाइटी
रोमानिया समाजवादी गणतंत्र अकादमी
कैटगुट अकाऊस्टिक सोसाइटी

निम्न संस्थाओं के रत्न-सदस्य मनोनीत हुए थे—

सी.वी. रमन

ऑप्टिकल सोसाइटी, अमेरिका
मिनरलॉजिकल सोसाइटी, अमेरिका
विज्ञान अकादमी, सोवियत संघ
विज्ञान अकादमी, पेरिस

सन् 1929 में भारतीय विज्ञान कांग्रेस के महाअध्यक्ष तथा 1934 से 1970 तक भारतीय विज्ञान अकादमी के अध्यक्ष पद पर रहे।

रमन के समकालीन ऐसे कई वैज्ञानिक थे, जिन्होंने देश में किए गए अपने अनुसंधान से प्रसिद्धी प्राप्त की थी। इनमें कुछ प्रमुख नाम है—सर जगदीश चंद्र बोस (1858-1937), आचार्य सर प्रफुल्ल चंद्र रे (1861-1944), श्रीनिवास रामानुजनम (1887-1920), मेघनाद साहा (1893-1956), प्रशांत चंद्र महालानोवीस (1893-1972), सत्येंद्र नाथ बोस (1894-1974), शांतिस्वरूप भटनागर (1894-1955), के.आर. रामनाथन (1893-1985), के.एस. कृष्णन (1898-1961), होमी जहाँगीर भाभा (1909-1966), विक्रम साराभाई (1919-1973)। यह सोचने की बात है कि आजादी के बाद अच्छी सुविधाओं के उपलब्ध होने के बावजूद विज्ञान में शायद ही कोई ऐसा मौलिक अनुसंधान हुआ है जो उल्लेखनीय हो।

□

2

कलकत्ता-प्रवास

कलकत्ता में रमन सरकारी नौकरी करने सन् 1907 में पहुँचे एवं वहाँ सन् 1933 तक रहे। इस अवधि को दो भागों में बाँटा जा सकता है—1907 से 1917 तक, जब रमन सरकारी नौकरी करने के साथ खाली समय में शोधकार्य करते थे; एवं 1917 से 1933 तक, जब वे कलकत्ता विश्वविद्यालय में प्राध्यापक थे। रमन के जीवन का यह स्वर्णिमकाल था। यहीं रहकर उन्होंने वह ऐतिहासिक प्रयोग किए, जिससे उन्हें 1930 में विज्ञान के क्षेत्र में विश्व का सर्वश्रेष्ठ सम्मान नोबेल पुरस्कार मिला। यद्यपि रमन उच्चपद पर कार्यरत थे तथा उनका वेतन 400 रुपए प्रतिमाह था, जो उस जमाने में काफी अधिक था; परंतु इससे उनके अंदर का वैज्ञानिक संतुष्ट नहीं था। शुरू से ही वे विज्ञान के प्रति आकर्षित थे, अत: उसी क्षेत्र में शोध करना चाहते थे। जल्द ही उन्हें ऐसा अवसर भी मिल गया, जिसे उन्होंने हाथ से न जाने दिया। कलकत्ता पहुँचने के एक सप्ताह के अंदर ही उन्हें विज्ञान संस्थान 'इंडियन एसोशिएशन फॉर दि कल्टिवेशन ऑफ साइंस' का पता चला, जिसके साथ उनका गहरा नाता बन गया तथा उनके कैरियर को एक नई दिशा मिली। वे इस संस्थान के प्रयोगशाला में भौतिक विज्ञान के क्षेत्र में कई मौलिक अनुसंधान किए, जिससे देश व विदेशों में विख्यात हो गए।

उनके साथ साथ संस्थान का नाम भी सर्वत्र फैलने लगा। इससे नई प्रतिभाएँ संस्थान की ओर आकर्षित होने लगी। जब तक रमन कलकत्ता में रहे, उन्होंने संस्थान का साथ नहीं छोड़ा। यह एक विडंबना है कि जिस व्यक्ति (विज्ञान संस्थान के संस्थापक डॉ. महेंद्र लाल सरकार) की वजह से रमन इतना कुछ हासिल कर सके, उनसे वे नहीं मिल पाए, क्योंकि रमन के कलकत्ता पहुँचने के तीन वर्ष पूर्व उनका निधन हो गया था। रमन के जीवन से गहरा संबंध रखनेवाले एक और व्यक्ति थे—सर आशुतोष मुखर्जी जो रमन के बहुत बड़े हितैशी थे।

'इंडियन एसोसिएशन फॉर दि कल्टिवेशन ऑफ साइंस' में रमन

महेंद्र लाल सरकार का जन्म सन् 1833 में हुआ था। प्रारंभिक शिक्षा समाप्त करके कलकत्ता आयुर्विज्ञान महाविद्यालय से चिकित्साशास्त्र में एम.डी. की परीक्षा में उन्होंने प्रथम स्थान प्राप्त किया था। वे सच्चे देशभक्त थे तथा उनकी दिलचस्पी केवल चिकित्सा संबंधी बातों तक सीमित न रहकर विभिन्न क्षेत्रों में थी। उस समय की शिक्षा व्यवस्था से वे संतुष्ट नहीं थे, क्योंकि विज्ञान विषय को उतनी प्राधन्यता नहीं दी जाती थी, जितनी की कला को। इसके अतिरिक्त भारतीयों के लिए विज्ञान में उच्च शिक्षा पाने व शोध करने के अवसर भी नगणय थे। जो थोड़े-बहुत सरकारी वैज्ञानिक संस्थान थे, उनमें सिर्फ विदेश से शिक्षा प्राप्त करके आए छात्रों को ही प्रवेश मिलता था। महेंद्र लालजी का मानना था कि विज्ञान के क्षेत्र में पीछे रह जाने की वजह से हमारा देश भी पिछड़ा है। इससे उभरने के लिए तथा देश के विकास के लिए विज्ञान में मौलिक अनुसंधान करना जरूरी है, जो हमारे खुद के द्वारा संचालित संस्थानों में किए जाएँ। महेंद्रलाल सरकार के प्रयास से सन् 1876 में ब्रिटिश एसोसिएशन तथा रॉयल इंस्टीट्यूट ऑफ लंदन (सन् 1799 में स्थापित) की तर्ज पर 'भारतीय विज्ञान प्रगति संस्थान' की

चित्र : 2.1 भारतीय विज्ञान प्रगति संस्थान

(A) वो बाजार का पुराना भवन

(B) जाधवपुर का नया भवन

स्थापना हुई। इसके लिए बंगाल सरकार ने 210, बो बाजार स्ट्रीट में अवस्थित एक इमारत संस्थान को दे दी। इस संस्थान में अध्यापन का कार्य शुरू किया गया। इसके विस्तार के लिए सन् 1891 में विजयानगरम के महाराजा ने एक बड़ी रकम दान में दीया जिससे कई प्रयोगशालाओं

का निर्माण किया गया। भारत की यह पहला विज्ञान संस्थान था, जिसका निर्माण मूल अनुसंधान करने के लिए किया गया था। अपने जीवनकाल में महेंद्रलाल कोशिश करते रहे कि कोई यहाँ पर उपलब्ध सुविधाओं का इस्तेमाल कर विज्ञान के क्षेत्र में प्रयोगात्मक शोधकार्य शुरू करे, परंतु वे इस काम में नाकाम रहे। सन् 1904 में अपने अधूरे स्वप्नों के साथ उदास महेंद्रलाल स्वर्ग सिधार गए एवं तब से यह संस्थान और भी उपेक्षा का शिकार होता चला गया।

महेंद्रलाल की मृत्यु के तीन वर्ष बाद अनजाने ही रमन एक दिन भारतीय विज्ञान प्रगति संस्थान के दरवाजे पर आकर दस्तक दी। वहाँ के प्रभारी आशुतोष दा ने दरवाजा खोलकर रमन के आने का

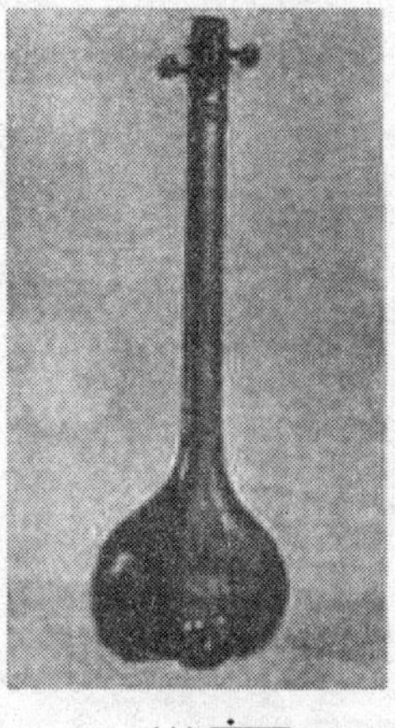

(A) तंबूरा (B) वीणा

चित्र : 2.2 भारतीय वाद्ययंत्र

(C) तबला बाजार

(D) मृदंगम

कारण पूछा। रमन ने अपना परिचय देते हुए विज्ञान के क्षेत्र में शोध करने की इच्छा जताई। रमन की मंशा सुनकर उस समय के सचिव अमृतलाल सरकार, जो महेंद्रलाल के पुत्र थे, भाव-विभोर हो गए और रमन को बाँहों में भरकर रुँधे कंठ से कहा—'अब तक हम तुम्हारे जैसे व्यक्ति की प्रतीक्षा कर रहे थे। आज से यह संस्थान तुम्हारे हवाले'। आशूबाबू की मदद से रमन तुरंत काम में जुट गए। उन्होंने पूरे हॉल की सफाई करवाई और फिर प्रयोगशाला के यंत्रों को साफ कर उन्हें ढंग से लगवाया। दूसरे दिन से वे शोधकार्य में जुट गए। रमन जब तक कलकत्ता में रहे, इस संस्थान से जूड़े रहे एवं आशूबाबू उनके विश्वस्त सहायक के तौर पर काम करते रहे।

अपने कठोर परिश्रम से रमन ने संस्थान में कई महत्त्वपूर्ण शोधकार्य किए, जिनका प्रकाशन प्रसिद्ध वैज्ञानिक पत्रिकाओं जैसे 'नेचर', 'द फिलोसॉफिकल मैगजीन' (लंदन से प्रकाशित) तथा फिजिकल रिव्यू (अमेरिका से प्रकाशित) में होने लगा। इससे उनकी कीर्ति यूरोप व अमेरिका में फैलने लगी, जबकि वे तब तक देश के बाहर कदम भी नहीं रखा था। कलकत्ता में भी लोग रमन को जानने लगे थे, क्योंकि वे अकसर आम जनता के लिए विज्ञान संबंधित विषय पर लोकप्रिय व्याख्यान दिया करते थे। रमन एक बहुत ही अच्छे वक्ता थे तथा व्याख्यान के दौरान प्रयोगों का प्रदर्शन करके विषय को सरल व सरस बना देते थे। रमन का झुकाव विज्ञान की ओर था, परंतु वे अपने नौकरी के प्रति कर्तव्यशील भी थे। उन्होंने इन दोनों के बीच अद्‌भुत सामंजस्य बनाए रखा। इस दौरान यानी सन् 1907 से 1917 के दशक में जब रमन सरकारी पदाधिकारी थे, अकेले 27 शोधपत्र लिखे तथा उन्हें कर्जन अनुसंधान पुरस्कार (1912) एवं वुडबर्न शोध पदक (1913) प्राप्त हुआ। उनका ज्यादातर काम कंपन या तरंग तथा ध्वनि विज्ञान के क्षेत्र में था। उन्होंने वाद्ययंत्रों, जैसे एकतारा, वायलिन, वीणा, तबला, मृदंग

आदि सुरीले आवाज पैदा करनेवाले यंत्रों का गहन अध्ययन किया। सन् 1919 में अमृतलाल के निधन के बाद रमन को भारतीय विज्ञान प्रगति संस्थान का सचिव नियुक्त किया गया, जिस पद पर वे 1933 तक बने रहे। बाद में मेघनाद साहा इस संस्थान के सचिव बने, जिनके पहल करने पर संस्थान का स्थानांतरण बो बाजार से जाधवपुर में बने नए भवन में हुआ।

स्पंदन व ध्वनिकी

सन् 1907 से 1917 की अवधि में रमन ने स्पंदन तथा ध्वनि से संबंधित शोध किए। उनका प्रिय विषय था संगीत वाद्ययंत्रों द्वारा ध्वनि के उत्पन्न होने का अध्ययन। इस विषय में उनकी रूचि हेलम्होल्ज की प्रसिद्ध पुस्तक 'द् सेंसेशंस ऑफ टोन' पढ़ने से हुई थी। इसके अलावा बचपन में अपने पिता के वायलिन बजाना सुनकर भी उनके मन में इस वाद्ययंत्र के बारे में जानने की उत्सुकता पैदा हुई थी। स्पंदन और ध्वनि विज्ञान का आपस में गहरा संबंध है क्योंकि किसी भी वस्तु से आवाज उत्पन्न तभी होती है, जब वह वस्तु अपने स्थिर स्थान से आगे-पीछे होती रहती है, यानी उस वस्तु में स्पंदन होता है। संगीतमय आवाज के स्पंदन में एक आवर्तिता होती है, जिसे सुनकर मन आनंदमय हो उठता है। ध्वनि का अनुभव हमें कानों से सुनकर होता है। ध्वनिकी विज्ञान की वह शाखा है जिसमें आवाज के बारे में अध्ययन किया जाता है। स्पंदन की उत्पत्ति से लेकर उसकी गति, संरचना, ध्वनि तरंग की गुणवत्ता आदि इसके अंतर्गत आते हैं। रमन ने इन सभी क्षेत्रों में काम किया। स्पंदन या कंपन को पैदा करने एवं जारी रखने के लिए कंपित वस्तु में एक बाह्य शक्ति या बल लगाना जरूरी होता है। अब प्रश्न उठता है कि क्या स्पंदन के सभी रूपों को जारी रखा जा सकता है या केवल कुछ को ही। रमन ने इसे समझने के लिए भारतीय वाद्ययंत्रों का गहन अध्ययन किया विशेषकर तार वाले यंत्रों का जैसे एकतारा, वायलिन, तानपूरा,

वीणा, सितार आदि। इन सभी यंत्रों में तारों को तनी अवस्था में रखने के लिए इनके दोनों सिरों को स्थिर रखते हैं। तनी हुई तार में ही कंपन पैदा किया जा सकता है। तार की लंबाई, मोटाई आदि पर कंपनआवृत्ति का मान निर्भर करता है। इसका अध्ययन करने के लिए उन्होंने कार्बन आर्क लैंप, दर्पण, स्वरित्र द्विभुज (ट्यूनिंग फॉर्क) तथा फोटो प्लेटों का इस्तेमाल करके तार में उत्पन्न कंपन का चित्र लिया। यह एक मौलिक उपकरण था, जिससे रमन ने कई नए तथ्यों का पता लगाया। वाद्ययंत्रों में अनुनादी बक्से की भूमिका की वैज्ञानिक व्याख्या दी।

रमन ने वायलिन पर विस्तृत अध्ययन किया। इसके लिए उन्होंने एक यांत्रिक वायलिन की संरचना की, जिसका निर्माण प्रयोगशाला में पड़े कबाड़ की वस्तुओं से की। सिर्फ कुछ ही सामान साइकिल की दुकान से खरीदी। इस यंत्र में धनुष स्थिर था, जबकि वायलिन गतिशील। असल में वायलिन बजाते वक्त वादक वायलिन को स्थिर रखता है तथा धनुष को आगे-पीछे चलाता रहता है। वायलिन से सुरीली आवाज पैदा करने के लिए वादक को कई चीजों का ध्यान रखना पड़ता है, जैसे तार की किस जगह धनुष को रखें, उसे कितना दबाएँ, धनुष की गति एवं दूरी आदि। इन सब बातों पर सही नियंत्रण रखने से ही वायोलिन से मधुर स्वर निकलता है। रमन ने अपने यांत्रिक वायलिन से इन प्राचलों के स्वर की ध्वनि गुणता के प्रभाव पर शोध किया। रमन ने पहली बार ऐसा यंत्र बनाया था। अब भी वायलिन पर शोध हो रहा है, परंतु यंत्रवत् वायलिन में कुछ तबदिली की गई है।

हथेली की ठोक या डंके की चोट से बजाए जानेवाले भारतीय वाद्ययंत्रों जैसे तबला, मृदंगम आदि के मधुर आवाज से भी रमन काफी प्रभावित थे। हमारे पूर्वजों ने अपने अंतरज्ञान से ही इन यंत्रों को बनाया था। वहीं पश्चिमी ड्रमों से पैदा होने वाली आवाज बेसूरी होती है। किसी तनी हुई झिल्ली को ठोकने से विभिन्न विधाओं (मोड) के स्वर पैदा

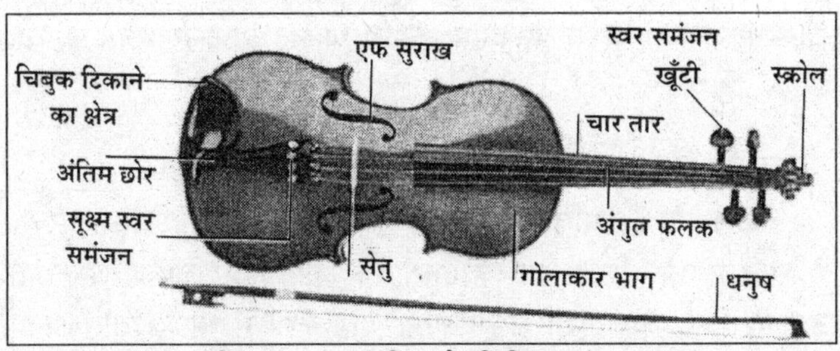

चित्र : 2.3 वायलिन के विभिन्न अंग

होते हैं, जो समस्वर (हारमोनिक) के नहीं होते। भारतीय ड्रमों में झिल्ली की बनावट विशेष ढंग की होती है जिसमें झिल्ली की मोटाई मध्य भाग में अधिक तथा बाहरी भागों में कम होती है। रमन ने विश्लेषण करके दिखाया कि इस प्रकार की झिल्ली में कंपन की उच्च विधाएँ समस्वर के हो जाते हैं, जैसाकि तार वाले वाद्ययंत्रों में होता है। यही कारण है कि हमारे झिल्ली वाले वाद्ययंत्र भी संगीतमय आवाज पैदा करते हैं, जिसे सुनना अच्छा लगता है।

रमन ने अन्य कई वाद्ययंत्रों की ध्वनिकी पर शोध किया, जैसे पियानो, जलतरंग आदि। वे इस विषय के विशेषज्ञ बन गए। जर्मनी से प्रकाशित भौतिकी विश्वकोश 'हैंडबुक डेर फीजिक' में इस विषय पर उनका एक समीक्षा लेख छपा था, जिसे लिखने के लिए उनसे विशेष अनुरोध किया गया था। यह एक विशेष सम्मान था, क्योंकि इस पुस्तक के सभी सहयोगी लेखक जर्मन थे। इस लेख को रमन ने अंग्रेजी भाषा में लिखी थी, जिसका अनुवाद प्रकाशक ने खुद जर्मन भाषा में करवाया था। इसमें सभी प्रकार के संगीत वाद्यों की कार्यपद्धतिओं पर विस्तृत चर्चा की गई है; चर्च के घंटों से लेकर बाँसुरी, शहनाई, तुरही, पीतल के यंत्रों आदि का भी उल्लेख है। रमन ने सन् 1920 के बाद इस विषय में काम करना बंद कर दिया, क्योंकि उन्हें एक नया विषय मिल गया

था—प्रकाशिकी, जिस पर काम उन्होंने पालित प्रोफेसर बनने के बाद आरंभ किया।

कलकत्ता विश्वविद्यालय में पालित प्राध्यापक रमन

कलकत्ता विश्वविद्यालय की स्थापना सन् 1857 में ब्रिटिश सरकार ने भारत में उच्च शिक्षा उपलब्ध कराने के लिए लंदन विश्वविद्यालय के तर्ज पर किया था। उसी समय चेन्नई एवं मुंबई में भी विश्वविद्यालयों का निर्माण किया गया था। सन् 1916 में प्रसिद्ध वकील व कलकत्ता उच्च न्यायालय के न्यायाधीश सर आशुतोष मुखर्जी कलकत्ता विश्वविद्यालय के उपकुलपति नियुक्त हुए। उनकी रुचि विज्ञान में थी एवं उन्होंने महेंद्रलाल सरकार को 'भारतीय विज्ञान प्रगति संस्थान' की स्थापना करने में सहयोग दिया था। रमन के विज्ञान प्रेम से वे वाकिफ थे तथा जिस प्रकार रमन ने एक सुप्त संस्थान को विज्ञान के अंतरराष्ट्रीय नक्शे पर ला खड़ा किया था, उसके प्रशंसक थे। वे स्वयं विज्ञान शिक्षा के क्षेत्र में कुछ खास करना चाहते थे। कलकत्ता विश्वविद्यालय के उपकुलपति बनने के बाद उन्हें यह मौका मिला। सरकारी मदद के बिना ही उन्होंने कई लाख रुपए की धनराशि इकट्ठा करके कलकत्ता विश्वविद्यालय के अंतर्गत विज्ञान महाविद्यालय (साइंस कॉलेज) की स्थापना की। जिन लोगों ने अधिक आर्थिक सहायता दी थी, उनके नाम पर आचार्य पदों का प्रतिष्ठान किया गया था, जैसे सर तारकनाथ पालित चेयर एवं सर रासबिहारी घोष चेयर। आशुतोष इन पदों में केवल उच्च काबिलियतवालों को ही नियुक्त करना चाहते थे। उनकी दिली इच्छा थी कि भौतिकी विषय में पालित चेयर का पद रमन जैसे प्रतिष्ठित भौतिकविद् सँभालें, परंतु उनके मन में शंका थी कि रमन उच्च पदस्थ एवं अधिक वेतनवाली सरकारी नौकरी छोड़कर उनका प्रस्ताव स्वीकार करेंगे या नहीं। यहाँ प्रतिमाह वेतन सिर्फ 600 रुपए था, जबकि रमन को उस समय 1100 रुपए मिल रहा था। फिर भी उपकुलपति ने औपचारिक

तौर पर रमन को अपना प्रस्ताव भेजा। रमन ने तुरंत अपनी सहमति दे दी। विज्ञान के प्रति प्रेम की वजह से रमन ने सरकारी नौकरी छोड़कर आचार्य पद स्वीकार कर लिया। परंतु एक अड़चन और थी—इस पद के लिए विदेश में शिक्षित होना आवश्यक था; अतः रमन को लंदन जाने को कहा गया। रमन ने साफ इनकार कर दिया, क्योंकि वे भारतीय की प्रतिभा को मनवाने के लिए विदेशी प्रमाणपत्र की आवश्यकता का समर्थन नहीं करते थे। सर आशुतोष की मध्यस्थता के बाद उन्हें इस शर्त से छूट दी गई। इस प्रकार सन् 1917 में रमन का विज्ञान जगत् में औपचारिक आगमन हुआ। सरकारी नौकरी से सीधे प्राध्यापक पद पर नियुक्ति केवल उनकी योग्यता पर मिली थी और वह भी सिर्फ 29 वर्ष की अल्पायु में। सन् 1924 में सर आशुतोष के निधन से रमन को गहरी चोट पहुँची, क्योंकि वे अपने एक बड़े हितैषी खो चुके थे।

पालित प्रोफेसर के लिए अध्यापन कार्य का दायित्व नहीं था; उनका काम अपने विषय में शोध करना, शोधकर्ताओं का मार्गदर्शन करना और विज्ञान की प्रयोगशालाओं का निरीक्षण करना था। फिर भी रमन एम.ए. व एम.एससी. के विद्यार्थियों को नियमित रूप से पढ़ाने लगे। वे कहा करते थे, यदि किसी विषय को ठीक से समझना हो तो उस विषय पर वारता देने से काम आसान हो जाता है। रमन को व्याख्यान देना इतना पसंद था कि वे लगातार दो तीन घंटे बिना रुके बोलते थे तथा विद्यार्थी मंत्रमुग्ध होकर सुनते थे। रमन ने विज्ञान संस्थान से भी अपना नाता बनाए रखा। विश्वविद्यालय में विद्यार्थी एवं शोधकर्ताओं से निपटने के बाद वे संस्थान में प्रयोगात्मक काम करने जाया करते थे। यह आश्चर्य की बात है कि रमन खुद डॉक्टरेट की उपाधि प्राप्त नहीं थे, परंतु उनके पर्यवेक्षण में कई शोधकर्ता इस उपाधि को प्राप्त करने के लिए काम कर रहे थे। आखिरकार सन् 1921 में कलकत्ता विश्वविद्यालय ने उन्हें मानद डॉक्टरेट की उपाधि प्रदान की।

सी.वी. रमन

सन् 1921 में आयोजित ब्रिटिश विश्वविद्यालयों के सम्मेलन में भाग लेने के लिए रमन ऑक्सफोर्ड गए, जहाँ वे छह सप्ताह रहे। वे कलकत्ता व बनारस विश्वविद्यालय के प्रतिनिधि के तौर पर वहाँ गए थे। यह उनकी पहली विदेश यात्रा थी। इस दौरान उनकी मुलाकात कई प्रतिष्ठित भौतिकविदों जैसे सर जे.जे. थामसन, सर लॉरेंज. ब्रैग व लॉर्ड अर्नेस्ट रदरफोर्ड से हुई। इंग्लैंड में उन्होंने सेंट पॉल्स कैथिड्रल के मर्मरश्रावी गलियारा (ह्विस्परिंग गैलरी) पर कुछ प्रयोग किए। रमन सन् 1924 में 'ब्रिटिश विज्ञान प्रगति संघ' के आमंत्रण पर कनाडा गए। वहाँ उन्होंने टोरोंटो में आयोजित प्रकाश प्रकीर्णन विषय पर एक चर्चा का उद्घाटन किया। कनाडा से वे अमेरिका गए, जहाँ उन्होंने फ्रैंकलीन संस्थान के शताब्दी महोत्सव में भारत का प्रतिनिधित्व किया। इसके बाद रॉबर्ट मिलिकन के आमंत्रण पर उन्होंने अतिथि प्राध्यापक के तौर पर कैलिफोर्निया प्रौद्योगिक संस्थान (कैलटेक) में चार महीने बिताए। वहाँ उन्होंने विभिन्न विषयों पर व्याख्यान दिया और साथ ही गैसों के प्रतिचुंबकीय अभिलक्षण पर प्रायोगिक अध्ययन किया। रमन ने अमेरिका में कई वैज्ञानिक संस्थानों का दौरा किया। जहाँ उन्होंने कई प्रसिद्ध वैज्ञानिकों जैसे कॉम्पटन, लैंगमूर, कूलीड्ज आदि से मुलाकात की। अमेरिका से स्वदेश वापस लौटते वक्त वे यूरोप में कई जगहों पर गए। छह महीने बाद वे पुनः सोवियत संघ के दौरे पर गए, जहाँ उन्होंने लेनीनग्राद विज्ञान अकादमी के शतवर्ष पूर्ति समारोह में भाग लिया। आठ महीने बाद बर्लिन, पेरिस, जेनेवा व रोम होते हुए वे कलकत्ता लौट आए। इसके बाद कई वर्षों तक रमन विदेश नहीं गए। वे शोधकार्य में जुट गए एवं इसी दौरान 'रमन प्रभाव' की खोज की।

कलकत्ता विश्वविद्यालय में पालित प्रोफेसर बनने के बाद अनेक शोधकर्ता रमन के साथ काम करने लगे। अब वे अकेले नहीं थे, अतः काम में तेजी आई। नए क्षेत्रों में काम शुरू किए गए, जैसे स्फटिक के लचीलेपन एवं प्रकाशिकी गुणवत्ता, कलिल पदार्थों की भौतिकी, एक्स-

रे विवर्तन आदि। विज्ञान महाविद्यालय में अध्ययन का काम होता था, जबकि भारतीय विज्ञान प्रगति संस्थान में प्रायोगिक कार्य। रमन एवं उनके शिष्यों का इन दोनों जगहों में आना-जाना लगा रहता था। संस्थान अब कलकत्ता विश्वविद्यालय की एक इकाई जैसा बन गया था।

रमन जब कॉलेज में पढ़ाई कर रहे थे, तभी से वे प्रकाशिकी विषय में दिलचस्पी लेने लगे थे। दरअसल उनका पहला शोधपत्र इसी विषय पर छपा था। अब उन्होंने इस क्षेत्र में अपने शिष्यों के साथ अनेक पहलुओं पर अनुसंधान कार्य शुरू कर दिया।

प्रकाश की किरणें या प्रकाश तरंगें जब किसी माध्यम या वस्तु से होकर गुजरती हैं, तब उसमें होने वाले परिवर्तन का अध्ययन प्रकाश विज्ञान के अंतर्गत आता है। प्रकाश की किरणें वस्तुओं से टकराकर परावर्तित हो सकती हैं, वस्तु के किनारे से टकराकर विवर्तित हो सकती हैं, या वस्तु में बने दरारों (स्लिट) के बीच से गुजरकर आपसी मुठभेड़ से झालर सदृश्य छाया पैदा कर सकती हैं। विवर्तन व अणुओं द्वारा प्रकाश के बिखरने या विकीर्णन प्रक्रिया के अंतर को समझने के लिए रमन ने पदार्थ की तीनों अवस्थाओं यानी ठोस, द्रव्य तथा गैस माध्यम के जरिए प्रकाश को गुजारकर कई प्रयोग किए। वे इस निष्कर्ष पर पहुँचे कि जिस प्रकार वायुमंडल में स्थित अणुओं से प्रकाश का विकीर्णन होता है, वैसे ही पानी के अणुओं से प्रकाश की किरणें विकर्णित होती हैं।

प्रकाश का आणविक प्रकीर्णन

सन् 1919 में रमन ने प्रकाश के आणविक प्रकीर्णन पर काम शुरू किया। इसके पहले सन् 1914 में कीन और पोर्टर ने एक आश्चर्यजनक प्रयोग किया था, जिसमें श्वेत प्रकाश किरणपुंज को एक विलयन माध्यम से गुजारा गया था। इस विलयन में कुछ रसायनिक घोल डाला गया जिससे, साफ द्रव्य धीरे-धीरे धूसर होता गया, क्योंकि इसमें कुछ कण बनने लगे थे। जाहिर है कि विलयन के दूसरी ओर से प्रसारित होने

वाले प्रकाश की तीव्रता में कमी आती गई, जो एक समय बिलकुल रुक गया। परंतु थोड़े समय के बाद प्रकाश फिर से दिखने लगा, जिसका रंग बैंगनी से शुरू होकर नीला, हरा, पीला तथा अंत में श्वेत हो गया। प्रकाश की तीव्रता में कमी होने का कारण तो स्पष्ट था परंतु दोबारा इसके प्रकट होने की व्याख्या कीन व पोर्टर नहीं दे सके। इसकी व्याख्या बाद में रमन ने दी। पहले तो कणों के अवरोध से प्रकाश की तीव्रता में कमी आती है। परंतु बाद में कणों द्वारा विवर्तित प्रकाश के अंतरक्षेप से प्रकाश दिखने लगता है। इस प्रयोग ने रमन को सोचने के लिए मजबूर कर दिया कि प्रकाश का विवर्तन कणों की अपेक्षा विलयन के अणुओं से होने पर क्या होगा?

सितंबर 1921 में यूरोप से लौटते समय एस.एस. नारकुंडा नामक जहाज में उन्हें पंद्रह दिन बिताने थे। इस दौरान समुद्र की गहरी नीलिमा ने उन्हें आकृष्ट किया। वे अपने साथ एक छोटी निकेल प्रिज्म, छोटी दुरबीन, ध्रुवक विश्लेषक, विवर्तन जाली आदि सामान लेकर गए थे, अतः समुद्र के पानी के प्रकीर्णन पर काम करने लगे। जहाज में सवारी के दौरान उन्होंने अपने शोध पर दो लेख भी प्रकाशन के लिए भेजे। उन्होंने इस बात की पुष्टि की थी कि समुद्र का गहरा नीलापन उसके जल की वजह से है, न कि आसमान के प्रतिबिंब के कारण, जैसा कि रैले ने कहा था। रमन को दृढ़ विश्वास हो चला था कि वायु के अणुओं की तरह प्रकाश का प्रकीर्णन जल के अणुओं से भी होता है एवं समुद्र का नीला दिखना इसी का परिणाम है, जैसाकि आसमान के नीला दिखने का कारण वायु अणुओं द्वारा प्रकाश का प्रकीर्णन है। सूर्य की रोशनी में सभी रंगों का समावेश है, अतः वह श्वेत रंग का दिखता है। अलग-अलग रंग की प्रकाश के प्रकीर्णन की मात्रा उसकी आवृत्ति (ν) के चार घात (यानी (ν^4) के समानुपाति होती है, यानी अधिक आवृत्ति के प्रकाश का प्रकीर्णन भी अधिक होगा। नीले रंग के प्रकाश की आवृत्ति लाल रंग की आवृत्ति से काफी ज्यादा होता है, अतः नीली रोशनी का

प्रकीर्णन वायु कणों द्वारा ज्यादा होने से वह आँखों से टकराती है एवं आकाश हमें नीला दिखता है। यही कारण है कि यातायात संकेत बत्ती में गाड़ी को रोकने के लिए लाल रंग का उपयोग करते हैं, न कि नीले रंग, क्योंकि लाल रंग वायुमंडल से कम टकराने के कारण दूर से भी दिखाई देती है, जबकि नीला रंग ज्यादा टकराकर आस-पास बिखर जाता है एवं बहुत दूर से नजर नहीं आती।

यदि यह मान भी लिया जाए कि समुद्र का नीला रंग उसके जल के अणुओं द्वारा प्रकाश के प्रकीर्णन के कारण होता है। इसके लिए कोई सैद्धांतिक परिकल्पना का होना जरूरी है। उस समय उपलब्ध रैले प्रकीर्णन सिद्धांत 'सिर्फ गैसों के लिए मान्य था। गैस में निकटवर्ती अणुओं के बीच की दूरी काफी अधिक होती है' जब कि द्रव्य में वे काफी करीब होते हैं। उदाहरण के लिए वाष्प के एक ग्राम अणु का आयतन जल के उतने ही मात्रा के आयतन का 1600 गुणा होता है। परंतु इससे यह निष्कर्ष नहीं निकलता कि जल में प्रकाश का प्रकीर्णन वाष्प की तुलना में उतने गुणा ज्यादा होगा। अब प्रश्न उठता है कि गैस की तुलना से यह कितना अधिक होगा ? इसका हल रमन ने आइंसटाइन और स्मौलुचौवस्की समीकरण से दिया जो प्रकाश प्रकीर्णन पर घनत्व में स्थानीय विचलन के प्रभाव की व्याख्या करता है। परंतु इसे साबित करने के लिए रमन ने सुनियोजित ठंग से कई वर्षों तक प्रयोगात्मक अध्ययन किए, तब जाकर सफलता उनके हाथ लगी।

कलकत्ता प्रवास का स्वर्णिमकाल—'रमन प्रभाव' का आविष्कार

एक पूर्णतः समांगी (होमोजिनियस) माध्यम से प्रकाश का प्रकीर्णन नहीं होता, परंतु कोई भी माध्यम पूर्णतः समांगी नहीं होता। अतः सभी वस्तुओं से कुछ-न-कुछ प्रकाश का प्रकीर्णन अवश्य होता है। जिस प्रकार किसी वस्तु से टकराकर प्रकाश का प्रकीर्णन होता है, वैसे ही यह अणुओं के साथ से भी होता है। उस समय, यानी सन् 1922 तक प्रकाश

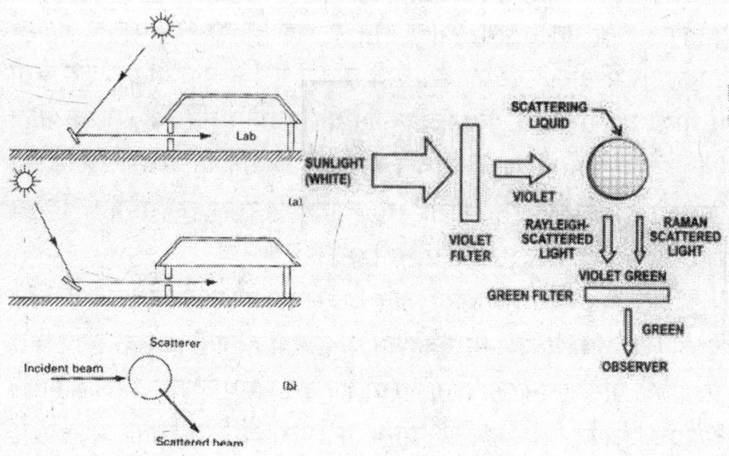

प्रकीर्णन की व्याख्या करने के लिए चिरसम्मत 'प्रकाश तरंग सिद्धांत' का ही उपयोग होता था। उसी दौरान प्रकाश के कणों जैसा व्यवहार करने के सिद्धांत यानी 'क्वांटम सिद्धांत' की सत्यता के प्रमाण मिलने लगे थे, परंतु इसपर किसी का विश्वास नहीं था। प्रकाश के कणों का अणुओं द्वारा प्रकीर्णन होने पर क्या हो सकता है, क्या कोई नए तथ्य की खोज 'क्वांटम सिद्धांत' से निकल सकती है ? इन सब प्रश्नों को उस समय रमन ने उठाया था।

यद्यपि सन् 1905 से ही आइंस्टाइन इस बात का जिक्र कर रहे थे कि विकिरण ऊर्जा कणों जैसा व्यवहार करता है, परंतु इस बात पर लोगों को यकीन सन् 1923 में 'कॉम्पटन प्रभाव' के आविष्कार के बाद ही हुआ। कॉम्पटन ने एक्स किरणों द्वारा अणुओं के प्रकीर्णन की व्याख्या 'क्वांटम सिद्धांत' से सफलतापूर्वक की थी। एक्स किरणों की आवृत्ति प्रकाश की किरणों से काफी अधिक होती है, अतः 'कॉम्पटन प्रभाव' के परिपेक्ष्य में कम आवृत्ति वाले इन विद्युत्-चुंबकीय किरणों के प्रकीर्णन होने पर क्या होगा, इसपर रमन ने शोध शुरू कर दिया।

सी.वी. रमन

प्रकाश का जब किसी वस्तु से प्रकीर्णन होता है तब उसके आवृत्ति एवं ध्रुवण में बदलाव आ सकता है। चिरसम्मत सिद्धांत आवृत्ति में बदलाव की इजाजत नहीं देता, परंतु प्रकीर्णन से ध्रुवण में बदलाव संभव है। यदि आवृत्ति में भी बदलाव होता है तो प्रयोगात्मक तरीके से उसे कैसे देखा जा सकता है ?

इसके लिए प्रकाश के तीव्र स्रोत तथा वर्णक्रममापी की जरूरत पड़ती है। उस जमाने में तीव्र प्रकाश स्रोत उपलब्ध नहीं था, और यदि था भी तो काफी महँगा जो रमन की संस्थान के बजट के बाहर था। परंतु रमन इस बात से तनिक भी विचलित नहीं हुए तथा सूर्य के प्रकाश को स्रोत के रूप में इस्तेमाल करने की ठानी। उन्होंने एक दर्पण को उपयोग में लाया, जिससे सूर्य की किरणों को प्रयोगशाला के अंदर ले जा सके। परंतु समय के साथ सूर्य की दिशा परिवर्तन होने के कारण दर्पण की स्थिति में भी उपयुक्त समायोजन करना जरूरी था, जिससे सूर्य की किरणें प्रयोगशाला के अंदर हमेशा बनी रहें। रमन ने इसके लिए यांत्रिक प्रचालित दर्पण को उपयोग मे लाया, जो उस समय उपलब्ध था।

सूर्य की किरणों में तो कई रंगों का समावेश होता है, अत: सिर्फ एक रंग की रोशनी प्राप्त करने के लिए किरणों को एक फिल्टर पट्टिका, जैसे बैंगनी रंग की पट्टिका से गुजारा गया, जिससे केवल बैंगनी रंग की रोशनी प्राप्त हुई। अब इस बैंगनी रंग की रोशनी को एक पात्र में रखे तरल पदार्थ पर पड़ने दिया गया, जिससे प्रकीर्णित होकर प्रकाश सभी दिशाओं में फैल गया। इस प्रकीर्णित प्रकाश में केवल बैंगनी रंग की किरण है अथवा दूसरे रंग की भी किरण है, इसे जानने के लिए रमन ने विभिन्न रंगों की पट्टिकाओं का इस्तेमाल किया। परंतु प्रकीर्णित प्रकाश, जिसकी तीव्रता काफी कम थी, उसे रमन अपनी आँखों से देखना चाहते थे, क्योंकि उस समय वर्णक्रमदर्शी उपलब्ध नहीं था। इसके लिए उन्होंने चारों ओर से बंद एक लकड़ी का बक्सा बनवाया, जिसमें प्रकीर्णित प्रकाश को देखने के लिए एक छेद था तथा जिसके अंदर मुश्किल से एक व्यक्ति समा सकता था।

कलकत्ता में किए गए सभी प्रयोग, जिसमें अलग अलग द्रव्यों का इस्तेमाल किया गया था, इस बात का सबूत दे रहे थे कि प्रकीर्णित प्रकाश में आपतित रंग के प्रकाश के साथ अन्य रंग की किरण भी मौजूद है, यानी आवृत्ति में परिवर्तन हो रहा था। इस प्रभाव को सबसे पहले रमन के छात्र रामनाथन ने सन् 1923 में देखा था। अब सवाल यह था कि इसकी व्याख्या कैसे करें ? क्योंकि उस समय केवल रैले प्रकीर्णन के बारे में लोगों को पता था, जिसमें प्रकीर्णित होने पर प्रकाश की किरण आपतित किरण के रंग जैसा ही रहता है। अत: इसे नया प्रकीर्णन कहने से पहले यह सुनिश्चत करना आवश्यक था कि यह कोई दूसरी प्रक्रिया की वजह से तो नहीं हो रहा था। प्रतिदीप्ति (फ्लोरोसेंस) एक ऐसी प्रक्रिया है जिसमें आपतित किरण कुछ विशेष वस्तुओं द्वारा अवशोषित होकर दूसरे रंग की किरणें पैदा करती है। अत: शुरू में इसे प्रतिदीप्ति मान लिया गया। परंतु प्रतिदीप्ति के लिए द्रव्य में कुछ अन्य वस्तुओं (गंदगी) का होना आवश्यक है, जो प्रतिदीप्ति पैदा कर सके। इस संशय को दूर करने के लिए प्रयोग में जिन द्रव्यों का इस्तेमाल हो रहा था, उन्हें कई बार शुद्ध किया गया, परंतु फिर भी प्रकीर्णित प्रकाश में आपतित रंग के प्रकाश के साथ अन्य रंग की किरणों की मौजूदगी बरकरार रही। इससे इस बात की पुष्टि तो हो गई कि द्रव्य में अशुद्धि इसकी वजह नहीं थी, परंतु ऐसा क्यों हो रहा था, इसकी उचित व्याख्या कोई भी न दे सका। इसलिए इसे 'क्षीण प्रतिदीप्ति' कहा गया।

सन् 1923 से 1925 के बीच किए गए सभी प्रकीर्णन प्रयोगों में क्षीण प्रतिदीप्ति के संकेत मिलते रहे। रमन के छात्र कृष्णन ने पचास से भी अधिक द्रव्यों में इसे देखा। यद्यपि कोई भी इसे प्रतिदीप्ति प्रक्रिया मानने को तैयार नहीं था, परंतु इसके लिए कोई दूसरा नाम भी नहीं सूझ रहा था। इस तीन वर्षों के दौरान रमन इसके बारे में गहरी चिंता करते रहे। वे सोचने लगे कि कहीं इसका संबंध प्रकाश के क्वांटम सिद्धांत से तो नहीं है ! कॉम्पटन के प्रयोग ने फोटॉन, यानी विद्युत्-चुंबकीय किरण के कणों

NEW THEORY OF RADIATION

PROF. RAMAN'S DISCOVERY

(ASSOCIATED PRESS OF INDIA.)

CALCUTTA, Feb. 29.

Prof. C. V. Raman, F. R. S., of the Calcutta University, has made a discovery which promises to be of fundamental significance to physics. It will be remembered that Prof. A. H. Compton of the Chicago University was recently awarded the Nobel Prize for his discovery of the remarkable transformation which X-rays undergo when they are scattered by atoms. Shortly after the publication of Prof. Compton's discovery, other experimenters sought to find out whether a similiar transformation occurs also when ordinary light is scattered by matter and reported definitely negative results. Prof. Raman with his research associates took up this question afresh, and his experiments have disclosed a new kind of radiation from atoms excited by light.

The new phenomenon exhibits features even more startling than those discovered by Prof. Compton with X-rays. The principal feature observed is that when matter is excited by light of one colour, the atoms contained in it emit light of two colours, one of which is different from the exciting colour and is lower down the spectrum. The astonishing thing is that the altered colour is [illegible] independent of the nature of the substance used. It changes however with the colour of the exciting radiation, and if the latter gives a sharp line in the spectrum, the second colour also appears as a second sharp line. There is in addition a diffuse radiation spread over a considerable range of the spectrum. He will deliver a lecture demonstrating these phenomena first at Bangalore on the 16th March.

First newspaper announcement of the Discovery of the Raman Effect made on 28th Feb. 1928

प्रकाश का नया सिद्धांत

प्रोफेसर रमन की खोज

(एसोशिएटेड प्रेस ऑफ इंडिया)

कलकत्ता, फरवरी २९। कलकत्ता विश्वविद्यालय के प्रो. सी.वी. रमन, एफ.आर.एस. ने भौतिकी में एक अत्यंत महत्त्वपूर्ण मौलिक अनुसंधान की है। याद कीजिए, हाल ही में सिकागो विश्वविद्यालय के प्रो. ए.एच. कॉम्पटन को परमाणु द्वारा प्रकीर्णित होने पर एक्स-किरणों में आए आसाधारण बदलाव की खोज करने के लिए नोबेल पुरस्कार से सम्मानित किया गया था। प्रो. कॉम्पटन की खोज के प्रकाशन के तुरंत बाद कई प्रयोगकर्ताओं ने यह जानने की कोशिश की क्या प्रकाश की किरणों में भी किसी पदार्थ से प्रकीर्णित होने पर वैसा ही परिवर्तन होता है, परंतु वे सभी स्पष्टत: नकारात्मक नतीजे पर पहुँचे। प्रो. रमन अपने शोधकर्ताओं के साथ इस विषय पर नये सिरे से कार्य शुरू किया तथा उनके प्रयोगों से स्पष्ट हो गया है कि प्रकाश द्वारा उत्तेजित होकर परमाणु एक नई किरण का उत्सर्जन करती है।

प्रो. कॉम्पटन की एक्स-किरणों से की गयी खोज की तुलना में यह नई खोज विशेष महत्त्व रखती है। इसमें प्रमुख तौर पर देखा गया है कि जब एकरंगी प्रकाश से कोई वस्तु उत्तेजित होती है, तब उसमें स्थित परमाणुओं से दो अलग रंगों के प्रकाश का उत्सर्जन होता है, जिसमें से एक रंग उत्तेजित करने वाले प्रकाश के रंग से भिन्न व वर्णक्रम के निचली ओर होता है। ताज्जुब की बात है कि बदला हुआ रंग वस्तु विशेष पर निर्भर नहीं करता। लेकिन उत्तेजित करने वाले प्रकाश के रंग के बदलाव से इसके रंग में परिवर्तन होता है तथा यदि वर्णक्रम में उत्तेजक प्रकाश की किरण एक तीक्ष रेखा जैसी दिखती है तो यह दूसरा रंग भी दूसरी तीक्ष रेखा के रूप में दिखाई देती है। इसके अलावा वर्णक्रम का काफी बड़ा क्षेत्र बिखरे हुए प्रकाश से भरा होता है। वे सबसे पहले बैंगलौर में 16 मार्च को प्रदर्शन के साथ इस प्रक्रिया पर व्याख्यान देंगे।

अखबार में छपी 28 फरवरी, 1928 में हुई 'रमन प्रभाव' के आविष्कार की सबसे पहली विज्ञप्ति।

के अस्तित्व की पुष्टि कर दी थी, जिसके लिए उन्हें सन् 1927 में नोबेल पुरस्कार मिला था। अब रमन और अधिक लगन से प्रयोगात्मक कार्य में जुट गए। वे इस बात का पता लगाना चाहते थे कि क्या यह भी कॉम्पटन जैसा ही प्रभाव है, जिसमें फोटॉन की ऊर्जा एक्स किरणों के बजाय दृश्य प्रकाश के संगत है ? अब इस बात का भी पता चल चुका था कि तरल पदार्थ से प्रकीर्णित किरणें ध्रुवित है, अतः प्रतिदीप्ति प्रक्रिया इसका कारण नहीं हो सकता क्योंकि प्रतिदीप्ति प्रक्रिया से निकलने वाली किरणें ध्रुवित नहीं होती। 28 फरवरी 1928 को रमन और कृष्णन ने एक प्रयोग किया जिसमें कई फिल्टरों की मदद से आपतित किरणों को एकवर्णी किरण जैसा बनाया गया तथा प्रकीर्णित किरणों को आँख द्वारा ठीक से देख पाने के लिए वर्णक्रमदर्शी का इस्तेमाल किया। उन्हें स्पष्ट तौर पर दो अलग वर्णक्रम (रंग) दिखाई दिया—एक तो आपतित किरणों का था (रैले प्रकीर्णन) तथा दूसरा उससे अलग रंग का, यानी आपतित किरणों की आवृत्ति से भिन्न। बस फिर क्या था, एक बड़ी खोज हो चुकी थी। रमन ने इसे परिवर्तित प्रकीर्णन कहा था, जो आगे चलकर 'रमन प्रभाव' के नाम से जाना गया।

विश्व को इस मूल शोध के बारे में जानकारी 'नेचर' पत्रिका के 31 मार्च, 1928 अंक में प्रकाशित रमन व के.एस. कृष्णन के लेख से मिला। बाद में रमन ने 'नेचर' पत्रिका के 21 अप्रैल, 1928 अंक में इस विषय पर विस्तार से लिखा। इसके अलावा रमन के कई शोधपत्र 'इंडियन जर्नल ऑफ फिजिक्स' में भी छपे, जिसका प्रकाशन रमन ने शुरू किया था। रमन में एक खासियत थी कि वे अपने शोधकार्य के प्रकाशन में बिलकुल देर नहीं करते थे।

आविष्कार के दो वर्ष बाद ही सन् 1930 में भौतिक विज्ञान का नोबेल पुरस्कार रमन को दिया गया, क्योंकि समाज के लिए 'रमन प्रभाव' की उपयोगिता प्रमाणित हो चुकी थी। इसकी पुष्टि इस बात से हो जाती है कि आविष्कार के डेढ़ वर्ष के अंदर ही अगस्त 1929 तक रमन प्रभाव से

संबंधित 150 से भी अधिक लेखों का प्रकाशन विश्व के कई प्रतिष्ठित पत्रिकाओं में हो गया था, जबकि नोबेल पुरस्कार के लिए नामांकन की आखिरी तारिख यानी 31 जनवरी, 1930 तक यह संख्या बढ़कर 225 हो चुकी थी। नोबेल पुरस्कार के इतिहास में शायद यह एक मिशाल है जब आविष्कार होने के इतने कम समय के अंदर किसी को यह पुरस्कार मिला हो। यह पुरस्कार तभी दिया जाता है जब मूल शोध का व्यावहारिक उपयोग साबित हो जाए।

'चिरसम्मत सिद्धांत' के आधार पर 'रमन प्रभाव' की व्याख्या

चिरसम्मत सिद्धांत के अनुसार प्रकाश को विद्युत्-चुंबकीय तरंग माना जाता है तथा अणुओं को छोटे-छोटे गोलक जो एक दूसरे से कमानी (स्प्रिंग) द्वारा जुड़े होते हैं। अणु की ऊर्जा सरल आवर्ती दोलक (हार्मोनिक ऑस्सीलेटर) के समानुपाती होता है। इस प्रकार जुड़े अणुओं में तनन (स्ट्रेचिंग) या बंकन (बेंडिग) होता है, जो प्रकाश की किरणों से प्रभावित होती है (अणु के द्विध्रुव-आघूर्ण यानी डाईपोल मोमेंट में परिवर्तन)। अणु अपनी प्रारंभिक अवस्था में आने के लिए किरणों का उत्सर्जन करती है, जिसे 'प्रकीर्णन' कहा जाता है। प्रकाश किरणों से प्रेरित होकर अणु के द्विध्रुव-आघूर्ण के दोलन में परिवर्तन होता है। यदि द्विध्रुव-आघूर्ण दोलन

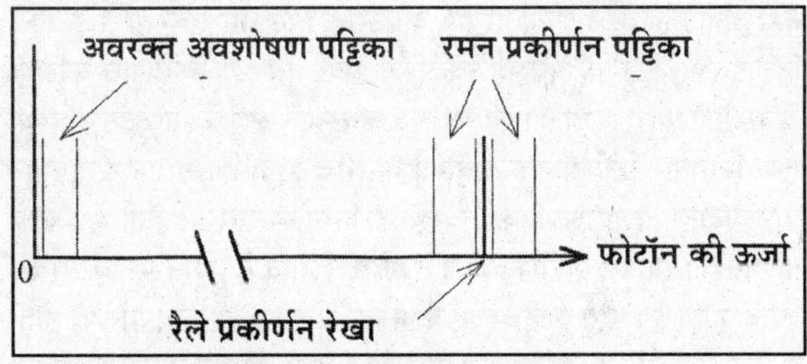

चित्र : 2.5 चिरसम्मत सिद्धांत के अनुसार अवशोषण व उत्सर्जन स्थानांतरण

की आवृत्ति अणु के कंपन आवृत्ति के बराबर होती है तो अणु उस किरण का अवशोषण करती है (अवरक्त अवशोषण)। यदि द्विध्रुव-आघूर्ण दोलन की आवृत्ति आपाती किरण की आवृत्ति के समान होती है तो उसी किरण का उत्सर्जन होता है, जिसे 'रैले प्रकीर्णन' कहते हैं। प्रकाश की किरण से अणु के द्विध्रुव-आघूर्ण में परिवर्तन ध्रुवणीयता की वजह से होता है, जो खुद आणविक कंपन से बदलता है। इस दशा में बदलाव के जोड़ व अंतर की आवृत्ति के प्रकाश का उत्सर्जन होता है। जिसे 'रमन प्रकीर्णन' कहते हैं। इनमें कम आवृत्ति की किरण को स्टोक्स प्रकीर्णन तथा अधिक आवृत्ति के किरण को 'ऐंटीस्टोक्स प्रकीर्णन' कहते हैं। चित्र 5.2 में इन स्थानांतरणों को दरशाया गया है।

'क्वांटम सिद्धांत' के आधार पर 'रमन प्रभाव' की व्याख्या

रमन प्रकीर्णन में देखा गया है कि आपाती किरण से कम आवृत्ति वाले यानी स्टोक्स प्रकीर्णन की तीव्रता अधिक आवृत्ति वाले एंटीस्टोक्स प्रकीर्णन की तीव्रता से कम होती है, जबकि चिरसम्मत सिद्धांत के अनुसार इन दोनों की तीव्रता समान होनी चाहिए। क्वांटम सिद्धांत के आधार पर इसकी व्याख्या समस्या को सुलझा देती है। इसके अनुसार अणु के कंपन की क्वांटित अवस्थाएँ होती हैं। जब आपाती फोटॉन की ऊर्जा इतनी अधिक नहीं होती कि वह अणु या परमाणु को उसके निम्नतम अवस्था से उठाकर निम्नतम इलेक्ट्रॉनिक अवस्था तक पहुँचा सके, तब अणु किसी एक कल्पित अवस्था में पहुँच जाती है। यह कल्पित अवस्थाएँ अणु के निम्नतम अवस्था तथा निम्नतम इलेक्ट्रॉनिक अवस्था के बीच होती हैं। साधारण तापमान पर ज्यादातर अणुएँ कंपनिक अवस्था के निम्नतम स्तर पर होती हैं, जिनसे फोटॉन का स्टोक्स प्रकीर्णन होता है। ऐसी स्थिति में 'रमन प्रभाव' सबसे ज्यादा होता है। परंतु कुछ अणुएँ कंपनिक अवस्था के उच्च स्तर पर होती हैं, जिनसे फोटॉन का ऐंटिस्टोक्स प्रकीर्णन होता है। फोटॉन की ऊर्जा अधिक होने पर अणु उच्च इलेक्ट्रॉनिक अवस्था में पहुँच जाता है, जिससे कुछ

पदार्थों से प्रतिदीप्ति स्पेक्ट्रम निकलता है, जो रमन स्पेक्ट्रम से काफी भिन्न होता है। आरंभ में रमन एवं उनके सहयोगियों ने खोज के दौरान नए प्रकीर्णित विकिरण को प्रतिदीप्ति ही समझा था, जिसे सुलझाने में उन्हें कई वर्ष लग गए। कुछ वस्तुओं मे अनुनादी रमन प्रकीर्णन संभव है, जिसमें अणुएँ इलेक्ट्रॉनिक अवस्था में पहुँच जाती हैं। इस प्रक्रिया का अणुप्रस्थ परिच्छेद साधारण 'रमन प्रकीर्णन' से काफी अधिक होती है।

रमन का नोबेल पुरस्कार एक वितर्क का विषय बन गया था—मुख्यतः दो कारणों की वजह से। एक तो यह कि क्या रमन ने ही सर्वप्रथम प्रकाश के प्रकीर्णन में आपाती किरण के अलावा अन्य रंग की किरणें देखी थी, और दूसरा यह पुरस्कार उन्हें अकेले क्यों मिला—अन्य सहयोगी इसके सहभागी क्यों नहीं बने। निस्संदेह रमन ही उस वर्ष नोबेल पुरस्कार पाने के योग्य थे, उनका नामांकन तो इसके एक वर्ष पहले भी हुआ था। ऐसी आलोचना से रमन बेहद दुःखी थे। उस समय प्रकाश के प्रकीर्णन पर केवल कलकत्ता में ही नहीं बल्कि कई अन्य देशों में भी शोधकार्य हो रहा था। इनमें विशेष उल्लेखनीय है सोवियत महासंघ में लेंडस्बर्ग व मैंडेलश्ताम का काम, जो क्वार्टज क्रिस्टल से प्रकाश के प्रकीर्णन पर काम कर रहे थे। लगभग रमन के आविष्कार के साथ ही इन्होंने भी प्रकीर्णित किरण में अन्य रंग के किरणों की उपस्थिति का पता लगाया था। परंतु विश्व को इस बात का पता 13 जुलाई, 1930 को चला, जब इसपर एक लेख 'नेचरवाइजेनशाफेन' पत्रिका में छपा, यानी रमन के प्रकाशन के चार महीने बाद। अतः यह जरूरी है कि शोध करने के बाद उसका प्रकाशन जल्द-से-जल्द किया जाए।

यह सच है कि रमन के कहने पर उनके छात्र के.एस. कृष्णन ने कुछ प्रयोग अवश्य किए थे, परंतु प्रयोग की परियोजना एवं प्रकाश के प्रकीर्णन की व्याख्या रमन ने ही दी थी। कृष्णन की डायरी से इस बात की पुष्टि होती है। इसके अलावा रमन के चाहते हुए भी इस पुरस्कार को

वे किसी दूसरे के साथ बाँट नहीं सकते थे, क्योंकि इसके लिए मनोनयन की जरूरत होती है, जो आविष्कारक खुद नहीं कर सकता।

कलकत्ता से प्रस्थान

जिस कलकत्ता ने रमन की प्रतिभा को पहचाना एवं उसे पनपने का मौका दिया, अब उनकी असाधारण सफलता को मानो बर्दाश्त नहीं कर पा रहा था। सफलता हमेशा ईर्ष्या को जन्म देती है एवं रमन के साथ कुछ ऐसा ही हुआ। रमन भी स्पष्टवादी व खुद्दार थे, जिससे उनके हितैशियों की संख्या में कमी आ रही थी। रमन पर यह गलत इलजाम लगाया गया कि वे केवल दक्षिण भारतीयों को प्राधन्यता देते हैं। इससे उनके अहंकार को गहरी ठेस पहुँची। यह सच है कि कुछ छात्रों को, जो सुदूर मद्रास से कलकत्ता में सिर्फ उनके साथ काम करने के उद्देश्य से आए थे, आरंभ में रमन ने उनके आवास आदि का बन्दोबस्त किया था, परंतु वे कतई बंगाली विरोधी नहीं थे। घटनाक्रम ने कुछ ऐसा मोड़ लिया कि रमन के खिलाफ साजिश करके उन्हें भारतीय विज्ञान प्रगति संस्थान के सचिव पद से हटाया गया। रमन भी ऐसी हालत में कलकत्ता छोड़ना चाहते थे। इसी दौरान भारतीय विज्ञान संस्थान, बैंगलौर के निदेशक पद के लिए योग्य व्यक्ति की तलाश हो रही थी। इसके लिए समिति लंदन से किसी को लाना चाहते थे, जैसाकि उस समय रिवाज था। जब लॉर्ड रदरफोर्ड से इस बारे में सलाह माँगी गई तो उन्होंने रमन का नाम सुझाते हुए कहा कि जब प्रार्थी आपके देश में मौजूद है तो बाहर क्यों ढूढ़ रहे हैं ? रमन जैसा योग्य व्यक्ति आपको नहीं मिलेगा। अत: रमन का इस पद के लिए चयन हो गया एवं उनके पास इसे स्वीकारने का प्रस्ताव भेजा गया। बैंगलौर में जाकर भौतिक विज्ञान के क्षेत्र में एक नए विभाग बनाने की रमन की माँग को भी मंजूरी मिल गई थी।

□

3

रमन प्रभाव को नोबेल पुरस्कार

रमन प्रभाव क्या है ? जब प्रकाश की किरण किसी तरल पदार्थ (या किसी भी ठोस या गैस वस्तु) पर पड़ती है तो उस पदार्थ के अणुओं द्वारा या तो उसका अवशोषण होता है अथवा प्रकीर्णन। प्रकीर्णन प्रक्रिया से किरण की आवृत्ति में बदलाव हो भी सकता है या नहीं। जब प्रकीर्णित किरण की आवृत्ति तथा आपतित किरण की आवृत्ति एकसमान रहती है, तो उसे 'रैले प्रकीर्णन' कहते हैं, जिसमें प्रकाश के रंग में कोई परिवर्तन नहीं होता। लेकिन जब प्रकीर्णित किरण की आवृत्ति में फर्क दिखाई देता है, यानी उसका रंग आपतित प्रकाश किरण के रंग से भिन्न हो जाता है, ऐसे प्रकीर्णन को 'रमन प्रकीर्णन' कहते हैं तथा इस प्रभाव को 'रमन प्रभाव' के नाम से जाना जाता है। यद्यपि सुनने में यह काफी सहज लगता है परंतु इसे पाने के लिए काफी समय लग गया। इस क्षेत्र में अनेक शोधकर्ता कई वर्षों से विभिन्न देशों में काम कर रहे थे लेकिन, सफलता रमन के हाथ लगी।

नोबेल पुरस्कार

नोबेल पुरस्कारों की घोषणा नवंबर महीने के द्वितीय या तृतीय सप्ताह में की जाती है। इसके चयन के लिए गठित नोबेल समिति की बैठक में काफी गोपनीयता बरती जाती है। घोषणा के लगभग एक

महीने बाद दिसंबर के मध्य में पुरस्कार वितरण समारोह होता है। रमन इस बात से वाकिफ थे कि घोषणा होने तक इंतजार करने से पुरस्कार लेने के लिए स्टॉकहोम जाने का टिकट मिलना मुश्किल होगा, अतः रमन ने जुलाई में ही अपने तथा अपनी पत्नी के लिए जहाज के दो टिकट खरीद लिये थे, जिससे दिसंबर के आरंभ में वे वहाँ पहुँच सकें। उनको पूरा यकीन था कि उस वर्ष का नोबेल पुरस्कार उन्हें ही मिलेगा। ऐसा था उनका अपने पर अटूट विश्वास! इससे पहले सन् 1929 में भी रमन को नोबेल पुरस्कार के लिए मनोनीत किया गया था, परंतु उस वर्ष एल.डी ब्रॉग्ली को यह पुरस्कार दिया गया था।

सन् 1930 के नोबेल पुरस्कार के लिए 21 वैज्ञानिकों को मनोनीत किया गया था, जिनमें एम. बॉर्न, ए. सोम्मरफिल्ड, ई. श्रॉडिंगर, डब्ल्यू. हाईजेऩबर्ग, एच.एफ. ऑसबर्न तथा मेघनाद साहा का नाम शामिल था। लेकिन चयन समिति ने इस पुरस्कार के लिए रमन को सबसे योग्य समझा, जिनका मनोनयन दस वैज्ञानिकों ने किया था—

ई. ब्लॉख (पेरिस)	डब्ल्यू.आर. वूड एवं सी.वी. रमन
नील्स बोह्र (कोपेनहेगन)	डब्ल्यू. आर. वूड अथवा वूड एवं रमन
ओ. चाल्सन (लेनिनग्राद)	रमन एवं लैंड्सबर्ग व मैंडेलस्टाम
जे. पेरे (पेरिस)	रमन अथवा रमन एवं डब्ल्यू. हाईजेऩबर्ग
एफ.आई. डी ब्रॉग्ली (पेरिस)	रमन
एच.एम. डी ब्रॉग्ली (पेरिस)	रमन
आर. फिर्फर (ब्रेसलाव)	रमन
जे. स्टार्क (ग्रॉसहेसहोए)	रमन
ई. रदरफोर्ड (कैंब्रिज)	रमन
सी.टी.आर. विल्सन (कैंब्रिज)	रमन

विल्सन और रदरफोर्ड ने अपने मनोनयन पत्र में लिखा था 'इसमें कोई संदेह नहीं है कि तरल एवं ठोस पदार्थों से होकर प्रकाश के गुजरने से आवृत्ति में जो परिवर्तन आता है उससे पदार्थ की प्राकृतिक आवृत्तियों

के बारे में अमूल्य जानकारी मिलती है, जो किसी और विधि से मिलना दुर्लभ है और भविष्य में इसका उपयोग हमारे ज्ञानवर्धन में काफी सहायक सिद्ध होगी। हम दोनों मानते हैं कि रमन एक असाधारण प्रतिभासंपन्न भौतिकविद् हैं, जो अपने ही देश में रहकर कठिन परिस्थितियों के बीच एक सफल शोधकेंद्र का निर्माण किया है, जहाँ उच्च कोटि के शोध हो रहे हैं। रमन एक ऐसे व्यक्ति हैं, जो सैद्धांतिक एवं प्रायोगिक दोनों ही क्षेत्रों में दक्ष हैं, जो उनके प्रकाशित शोधपत्रों में स्पष्ट झलकता है। हमारी राय में उनका कार्य नोबेल सम्मान पाने के उपयुक्त उच्चकोटि का है।

स्टार्क ने चयन समिति को लिखा था—आपकी माँग पर 1930 का भौतिकी क्षेत्र में नोबेल पुरस्कार के लिए मैं अपना प्रस्ताव भेज रहा हूँ, जो मेरी राय में कलकत्ता के प्रोफेसर सी.वी. रमन को उनके आविष्कार प्रकाश प्रकीर्णन से आवृत्ति में परिवर्तन के लिए दिया जाए, जिस प्रभाव को उनके नाम से जाना जाता है। इस खोज से भौतिकी के सिद्धांत की अस्थायित्वता का पता चला है 'जो एक वास्तविकता है।'

फिर्फर का मनोनयन इस प्रकार था—इस वर्ष के भौतिक विज्ञान में नोबेल पुरस्कार के लिए मैं कलकत्ता विश्वविद्यालय में भौतिकी के प्रोफेसर वेंकट रमन (एफ.आर.एस.) का नाम प्रस्ताव करता हूँ। पिछले कई दशकों से प्रोफेसर रमन अत्यंत महत्वपूर्ण शोध कार्यों में जुटे हैं, जिनमें से मैं केवल उनके कुछ कार्यों का उल्लेख करूँगा, जैसे भारतीय वाद्ययंत्रों के ध्वनिकी का विस्तृत अध्ययन तथा अणुओं द्वारा प्रकाश का विवर्तन। प्रकाश विवर्तन पर शोध करते हुए उन्होंने एक महान् खोज की, यानी उस प्रभाव की, जिसे उनके नाम से जाना जाता है (रमन प्रभाव)। इसके फलस्वरूप रमन उन वैज्ञानिकों की श्रेणी में अग्रणी हो गए, जो आधुनिक अणु भौतिकी समस्याओं पर शोध कर रहे हैं।

'रमन प्रभाव' से अणुओं के अभिलाक्षणिक आवृत्तियों के बारे में जानकारी मिलती है, यानी अणुओं के अनवेषण के लिए यह काफी उपयोगी है। इसकी विशेषता है कि इस विधि से वर्णक्रम के अवरक्त क्षेत्र में स्थित आवृत्तियों का मापन किया जा सकता है, जो स्पेक्ट्रमदर्शीय विधि से संभव नहीं है। अतः अवरक्त क्षेत्र में शोध तथा 'रमन प्रभाव' का आपस में गहरा संबंध है जिसकी वजह से नित्य नए प्रमाण मिल रहे हैं। संक्षेप में कहें तो यह मेरा दृढ़ विश्वास है कि रमन प्रभाव पिछले वर्षों का सबसे महत्वपूर्ण एवं उपयोगी आविष्कार है।'

नील्स बोह्र ने लिखा था—'यह घटना (रमन प्रभाव) जिसकी व्याख्या सैद्धांतिक क्वांटम विचार से काफी मिलती हैं, भविष्य में पदार्थों के अणुओं की अवस्थांतर के दौरान उत्सर्जित विशिष्ट वर्णक्रम के बारे में जानकारी प्राप्त करने का अत्यंत महत्त्वपूर्ण तरीका साबित होगा।'

सन् 1930 के 10 दिसंबर में स्वीडेन की राजधानी स्टॉकहोम में आयोजित एक भव्य समारोह में रमन को भौतिकी विषय में नोबेल पुरस्कार दिया गया। पुरस्कार में उन्हें एक सनद (डिप्लोमा), 23 कैरेट सोने का एक पदक (200 ग्राम वजन) तथा एक लाख तिहत्तर हजार स्वीडीश क्रोनर की धनराशि मिली थी जिसे उन्होंने अनुसंधान कार्य में लगा दिया। सन् 2001 से पुरस्कार की धनराशि बढ़ाकर एक करोड़ स्विडीश क्रोनर (लगभग दस लाख अमेरिकी डॉलर) कर दी गई है।

नोबेल पुरस्कार से सम्मानित होते समय रमन के कार्य का उल्लेख

आदरणीय महामहिम नरेश, शाही परिवार के सदस्यगण, देवियो और सज्जनो,

विज्ञान अकादमी का यह निर्णय है कि सन् 1930 का नोबेल पुरस्कार विज्ञान के क्षेत्र में सर वेंकट रमन को प्रकाश के प्रकीर्णन पर

सी.वी. रमन

किए उनके कार्य तथा उनके नाम से प्रचलित प्रभाव के लिए दिया जाए।

प्रकाश का विसरण एक प्रकाशीय घटना है, जिसके बारे में लोग वर्षों से जानते हैं। रोशनी की एक किरण का बोध तब तक नहीं होता, जब तक कि वह आँख से नहीं टकराती। परंतु जब प्रकाश का एक पुंज किसी माध्यम से होकर गुजरता है, जिसमें धूल के कुछ महीन कण भी होते हैं, तब किरणें कणों से प्रकीर्णित होकर दूसरी दिशाओं की ओर जाती हैं एवं प्रकाश का पथ अन्य दिशाओं से भी देखा जा सकता है। इसकी व्याख्या इस प्रकार दी जा सकती है—प्रकाश की किरण के वैद्युतिक क्षेत्र से धूल के कण डोलने लगते हैं, जिससे टकराकर प्रकाश सभी दिशाओं में बिखरने लगता है। इस बिखरे प्रकाश का तरंगदैर्ध्य या दोलन संख्या प्रति सेकेंड वही होती है, जो टकराने वाले प्रकाश का होता है। परंतु इस प्रभाव की मात्रा प्रकाश के तरंगदैर्ध्य के मान पर निर्भर करता है। कम तरंगदैर्ध्य वाले प्रकाश पर यह प्रभाव ज्यादा होता है, जबकि अधिक तरंगदैर्ध्य वाले प्रकाश पर कम। यही कारण है कि वर्णक्रम के नीले रंग पर यह प्रभाव लाल रंग की तुलना में ज्यादा होता है। अतः जब प्रकाश पुंज, जिसमें सभी तरंगदैर्ध्य की किरणें उपस्थित होती हैं, किसी माध्यम से होकर गुजरती है, तब पीले एवं लाल रंग की किरणें कम प्रकीर्णित होकर माध्यम के दूसरी ओर

से निकल जाती है जबकि नीले रंग की किरण ज्यादा प्रकीर्णित होकर आस-पास बिखर जाती हैं, इस प्रभाव को 'टींडल प्रभाव' के नाम से जाना जाता है।

लार्ड रैले ने 'टींडल प्रभाव' पर शोध किए थे एवं उन्होंने आसमान के नीले रंग व सूर्योदय एवं सूर्यास्त के समय रक्ताभ रंग दिखने के कारण की व्याख्या दी थी। सूर्य की किरण वायुमंडल में विद्यमान धूल की कणों या जल कणों द्वारा विसरित होने के कारण आसमान का रंग ऐसा दिखता है। बाद में सन् 1899 में रैले ने अपने वक्तव्य में सुधार करके बताया कि वायु के अणुओं से ही प्रकाश का प्रकीर्णन होता है, जिससे आसमान का रंग नीला / रक्ताभ दिखता है।

सन् 1914 में कवांस ऐसे प्रयोग करने में सफल हुए, जिससे सिद्ध हो गया कि शुद्ध तथा धूल रहित गैस से भी प्रकाश का प्रकीर्णन होता है।

विभिन्न पदार्थों जैसे ठोस, द्रव्य तथा गैसों में प्रकाश के प्रकीर्णन की प्रक्रिया पर विस्तार से अध्ययन करने पर पाया गया कि प्रकीर्णित प्रकाश पूरी तरह से 'टींडल प्रभाव' के सिद्धांत के अनुरूप नहीं है। 'टींडल प्रभाव' के सिद्धांत के अनुसार प्रकीर्णन से बगल में फैला प्रकाश ध्रुवित होना चाहिए, जबकि ऐसा नहीं पाया गया। इस विचलन ने वैज्ञानिकों का ध्यान आकर्षित किया तथा वे प्रकीर्णित प्रकाश के बारे में और अधिक जानकारी पाने के लिए शोध कार्य में जुट गए, जिसमें रमन ने सक्रिय भूमिका निभाई। वे अणुओं में पाए जानेवाले असम्मति के कारण का पता लगाने में जुट गए। प्रकाश प्रकीर्णन पर कार्य करते हुए सन् 1928 में रमन ने अप्रत्याशित एवं अत्यंत आश्चर्यजनक आविष्कार किया। उन्होंने पाया कि प्रकीर्णित प्रकाश में आरंभिक किरणों के अलावा कुछ अन्य तरंगदैर्ध्य वाले किरणें भी शामिल हैं, जो आरंभिक किरणों में नहीं थी।

इस नई किरण के बारे में स्तार से जानने के लिए आरंभिक

प्रकाश, जो शक्तिशाली पारद लैंप से उत्पन्न की जा रही थी, उसे फिल्टर करके एकक तरंगदैर्ध्य वाले किरण में परिवर्तित किया गया। इस प्रकाश को एक माध्यम से गुजारा गया तथा प्रकीर्णित प्रकाश को एक स्पेक्ट्रमलेखी द्वारा परखा गया, जिसमें हरेक तरंगदैर्ध्य या आवृत्ति की किरण अपनी एक रेखा बनाती है। रमन ने देखा कि स्पेक्ट्रम चित्र में फिल्टर किए गए आरंभिक पारद किरण रेखा के अलावा उसके दोनों ओर नई आवृत्ति की तेज रेखाएँ नजर आ रही थीं। पारद लैंप से फिल्टर करके दूसरी आवृत्ति की किरणों से भी ऐसा ही नतीजा निकला। इस प्रकार आरंभिक प्रकाश की आवृत्ति को बदलने से अलग अलग प्रकीर्णित प्रकाश नजर आए जिनमें आरंभिक आवृत्ति व नई आवृत्ति रेखाओं के बीच की दूरी हमेशा समान पाया गया।

रमन ने विभिन्न प्रकार के अनेक पदार्थों में प्रकीर्णन प्रयोग किए तथा हरेक में समान प्रभाव पाया, जिससे इसके सर्वव्यापी होने का प्रमाण मिला।

प्रकीर्णन की इस परिघटना के आविष्कारक के नाम से प्रचलित 'रमन प्रभाव' की व्याख्या उन्होंने स्वयं दी है, जो प्रकाश की आधुनिक संकल्पना पर आधारित है। इस परिकल्पना के अनुसार पदार्थ से प्रकाश का उत्सर्जन या अवशोषण किसी भी ऊर्जा की मात्रा में नहीं हो सकता, बल्कि एक विशिष्ट मान की ऊर्जा में ही होता है, जिसे 'प्रकाश क्वांटा' कहते हैं। अतः प्रकाश की ऊर्जा एक प्रकार से अणुओं जैसा व्यवहार करता है। प्रकाश क्वांटा का परिमाण किरण की आवृत्ति के अनुपाती होता है, अतः आवृत्ति के दूगने होने पर प्रकाश क्वांटा का मान भी दोगुणा हो जाएगा।

किन परिस्थितियों में एक परमाणु ऊर्जा का उत्सर्जन करता है या अवशोषण, इसे समझने के लिए बोह्र परमाणु की संरचना पर गौर करेंगे, जिसमें एक धनावेशित नाभिक के चारों ओर कई गणावेशित इलेक्ट्रॉन विभिन्न दूरियों पर गोलाकार पथ में चक्कर लगाते हैं। प्रत्येक

इलेक्ट्रॉन पथ की एक निर्दिष्ट ऊर्जा होती है, जिसका मान केंद्र में स्थित नाभिक से इस पथ की दूरी के अनुसार भिन्न होता है। केवल कुछ ही पथ स्थायी होते हैं जिनपर इलेक्ट्रॉन के घूमने से ऊर्जा का उत्सर्जन नहीं होता। परंतु जब इलेक्ट्रॉन अधिक ऊर्जा वाले कक्ष से कम ऊर्जा वाले कक्ष में आता है, यानी बाहरी कक्ष से भीतरी कक्ष में गिरता है तब प्रकाश पैदा होता है, जिसकी आवृति इन दोनों पथों की विशेषताओं पर निर्भर करता है तथा जिसकी ऊर्जा एक 'प्रकाश क्वांटा' के बराबर होता है। अत: परमाणु से उतने आवृत्तियों के प्रकाश उत्पन्न हो सकते हैं, जितने कि स्थायी कक्षों के बीच संभव है। इस प्रकार वर्णक्रम में हर आवृत्ति की अपनी विशिष्ट रेखा दिखाई देती है।

प्रकाश की किरण के किसी परमाणु से टकराने पर उसका अवशोषण संभव नहीं है, यदि उसके क्वांटा का मान उस परमाणु से निकलने वाले किसी भी संभावित प्रकाश क्वांटा के बराबर न हो। लेकिन 'रमन प्रभाव' इस नियम का उलंघन करती है। वर्णक्रम में दिखने वाली रमन रेखाएँ उस परमाणु के आवृत्तियों से मेल नहीं खातीं बल्कि वे आरंभिक प्रकाश किरण की आवृत्ति के अनुरूप बदलती है। रमन ने इस प्रभेद की व्याख्या दी है, जिसके अनुसार आरंभिक प्रकाश एवं परमाणु से निकलने वाले संभावित प्रकाश के आपसी मेल से नई 'रमन रेखाएँ' पैदा होती है। जब एक परमाणु से किसी भी परिमाण की ऊर्जा वाले प्रकाश किरण के टकराने से एक भिन्न प्रकाश क्वांटा वाले किरण का उत्सर्जन होता है, जिसमें इन दोनों क्वांटाओं का अंतर परमाणु के किसी भी आवृत्ति के बराबर होता है तो ऐसी परिस्थिति में आरंभिक प्रकाश का अवशोषण होता है। इस अवशोषण से परमाणु एक अतिरिक्त प्रकाश का उत्सर्जन करता है, जिसका मान आरंभिक आवृत्ति व परमाणु के आवृत्ति के जोड़ अथवा अंतर के बराबर होता है। यह नई आवृत्ति की किरणें आरंभिक आवृत्ति किरण के दोनों ओर फैले नजर आते हैं। आरंभिक आवृत्ति तथा उसके सबसे करीब बनी

'रमन रेखा' के बीच की दूरी परमाणु की न्यूनतम आवृत्ति यानी अवरक्त वर्णक्रम होता है। यह सारी चीजें जो परमाणु में घटती हैं, अणुओं मे भी दिखाई देती हैं।

इस प्रकार विस्थापित अवरक्त वर्णक्रम आरंभिक प्रकाश की आवृत्ति के नजदीक देखी जा सकती है। अणुओं की संरचना जानने के लिए 'रमन रेखाओं' का आविष्कार हमारे लिए अत्यंत ही उपयोगी सिद्ध हुआ है।

अब तक अवरक्त वर्णक्रम अवृत्ति के प्रकाश का अध्ययन करना दुर्गम्य था क्योंकि इसकी स्थिति फोटोग्राफी प्लेट के संवेदनशील हिस्से से काफी दूर पर होती है। अब रमन के आविष्कार से यह सब कठिनाईयाँ दूर हो गई है तथा अणुओं के नाभिक दोलन की तहकीकात करने के नए द्वार खुल गए हैं। इसके लिए आरंभिक किरण की आवृत्ति इस प्रकार चुनी जाती है, जो फोटोग्राफी प्लेट के संवेदनशील क्षेत्र में पड़ती हो। इससे 'रमन रेखा' के रूप में अवरक्त वर्णक्रम का स्थानांतरण उस जगह हो जाता है, जहाँ आसानी से यथार्थपूर्वक इनका मापन किया जा सकता है।

इसी विधि से वर्णक्रम की दूसरी छोर पर स्थित पराबैंगनी वर्णक्रम का अध्ययन भी 'रमन प्रभाव' की सहायता से की जा सकती है। अत: अणुओं के दोलन की पूरी जानकारी सही एवं सरल तरीके से प्राप्त करने का उपाय हमें मिल गया है।

स्वयं रमन एवं उनके सहयोगियों ने इस आविष्कार के बाद से अब तक गुजरे समय के दौरान अनेक पदार्थों के ठोस, द्रव्य एवं गैस अवस्थाओं से उत्सर्जित आवृत्तियों का अध्ययन किया है। इसके अलावा कई और विषयों पर अध्ययन किया गया है, जैसे भिन्न परिस्थिति में परमाणु एवं अणुओं से उत्सर्जित आवृत्ति में कोई परिवर्तन होता है या नहीं, विद्युत्-अपघटनी वियोजन में आणविक अवस्था तथा क्रिस्टलों का अवरक्त अवशोषण वर्णक्रम।

'रमन प्रभाव' से पदार्थों की रसायनिक संरचना के बारे में काफी महत्त्वपूर्ण नतीजें मिल रहे हैं तथा निकट भविष्य में पदार्थों की संरचना पर गहन अध्ययन करने के लिए एक अत्यंत ही मूल्यवान साधन 'रमन प्रभाव' हमारे हाथ लगा है।

सर वेंकट रमन जी, राजकीय विज्ञान अकादमी ने आपको, गैस के विसरण पर की गई उत्कृष्ट शोध तथा उस प्रभाव के आविष्कार के लिए, जो आपके नाम से जाना जाता है, विज्ञान में नोबेल पुरस्कार देकर सम्मानित किया है। पदार्थों की संरचना के बारे में जानने के लिए 'रमन प्रभाव' एक नई राह है तथा इसके इस्तेमाल से कई महत्त्वपूर्ण नतीजे भी मिल चुके हैं।

अब मैं आपको महामहिम नरेश के कर कमलों से यह पुरस्कार लेने के लिए आमंत्रित करता हूँ।

रमन ने तालिओं की गरगराहट के बीच राजा गुस्टाफ के हाथों पुरस्कार ग्रहण किया। उन्होंने उपस्थित लोगों के सामने अलकोहल से प्रकाश का प्रकीर्णन करके 'रमन प्रभाव' का प्रदर्शन किया। अलकोहल पर 'रमन प्रभाव' देखकर अब लोग रमन पर अलकोहल का प्रभाव देखने के लिए उत्सुक थे, परंतु उन्हें निराश होना पड़ा, क्योंकि रमन शराब नहीं पीते थे। नोबेल पुरस्कार विश्व का सर्वश्रेष्ठ पुरस्कार है तथा रमन भारत के पहले वैज्ञानिक बने, जिन्हें इस पुरस्कार से सम्मानित किया गया।

रमन का नोबेल पुरस्कार व्याख्यान

नोबेल पुरस्कार प्राप्त करने के अगले दिन यानी 11 दिसंबर, 1930 को रमन ने अपने अनुसंधान पर व्याख्यान दिया। जिसका शीर्षक था—'प्रकाश का आणविक प्रकीर्णन।' अपनी वार्ता में उन्होंने इस बात का जिक्र किया कि समुद्र के जल की नीलिमा देखकर उनके मन में जिज्ञासा जागी, जो उनके शोध की प्रेरणा बनी। इस रहस्य की खोज

के लिए उन्होंने प्रकाश के प्रकीर्णन पर विस्तारित अध्ययन किया जिससे 'रमन प्रभाव' का आविष्कार हुआ। अपनी वार्त्ता में उन्होंने अपने सहकर्मियों के योगदान का भी उल्लेख किया। उनके व्याख्यान का विवरण नीचे दिया गया है।

समुद्र के जल का रंग

विज्ञान प्रगति के इतिहास में ऐसा अकसर देखा गया है कि हमारे आस-पास की प्राकृतिक घटनाओं के अध्ययन से प्रेरित होकर ज्ञान के नए क्षेत्रों के विकास की शुरुआत हुई है। उदाहरण के तौर पर आसमान के रंग को देखकर उत्पन्न जिज्ञासा प्रकाशिकी क्षेत्र में अनेकों अनवेषण का कारण बनी है। हमारे इस वार्त्ता के विषय पर शोध आसमान के नीले रंग की व्याख्या के बाद शुरू हुई, जिसे स्वर्गीय लॉर्ड रैले ने दी थी एवं जिसकी प्रामाणिक पुष्टि बाद में हुई। समुद्र के जल के नीले रंग का रहस्य जानने की उत्सुकता भी इस क्षेत्र में शोध करने का कारण है। सन् 1921 में गरमी के मौसम में समुद्री यात्रा से यूरोप जाते समय मैंने पहली बार भूमध्यसागर के जल में अद्‌भुत नीलापन देखा। जल के नीला दिखने का कारण जल के अणुओं द्वारा सूर्य के प्रकाश का प्रकीर्णन हो सकता है। परंतु इस बात की पुष्टि करने के लिए यह जरूरी था कि तरल पदार्थ में प्रकाश के विसरण के नियमों को सुनिश्चित किया जाए। अत: कलकत्ता लौटते ही सितंबर 1921 में मैंने इस बारे में प्रयोग शुरू कर दिया। जल्द ही यह स्पष्ट हो गया कि इस विषय का महत्व केवल समुद्री जल के नीलेपन की व्याख्या से संबंधित नहीं है, बल्कि इस क्षेत्र में व्यापक अनुसंधान हो सकता है। ऐसा लगा कि प्रकाश प्रकीर्णन के अध्ययन से भौतिकी एवं रासायनिकी क्षेत्र में गहन समस्याओं का समाधान मिल सकेगा। इसी विश्वास के साथ हमने कलकत्ता में प्रकाश का अणुओं द्वारा प्रकीर्णन पर शोध प्रमुख कार्यक्रम बनाया।

अस्थिरता सिद्धांत

काम शुरू होने के कुछ ही महीने में यह स्पष्ट हो गया था कि प्रकाश का अणुओं द्वारा प्रकीर्णन एक सामान्य प्रक्रिया है, जो न केवल गैस एवं वाष्प कणों के माध्यम में होता है बल्कि द्रव्यों तथा क्रिस्टलीय व अक्रिस्टलीय ठोस पदार्थों में भी होता है। यह प्रक्रिया मूलत: माध्यम में व्याप्त आणविक क्रमभंग की वजह से होता है, जो प्रकाशिकी घनत्व में स्थानीय परिवर्तन पैदा करता है। अक्रिस्टलीय ठोस पदार्थों को छोड़कर ऐसे आणविक क्रमभंग का कारण तापीय उत्तेजना है, जिसके प्रायोगिक प्रमाण मिले हैं। द्रव्यों में स्थित अणु प्रकाश के लिए असमानुवर्ती होते हैं तथा उनमें दिशा परिवर्तन करने की क्षमता होती है, जिसकी वजह से इनमें एक अतिरिक्त व भिन्न प्रकीर्णन होता है। यह प्रकीर्णन ध्रुवीत नहीं होता, जबकि घनत्व में स्थानीय परिवर्तन की वजह से उत्पन्न प्रकीर्णन अनुप्रस्थ दिशा में संपूर्ण ध्रुवीत होता है। इस विषय में एक विस्तृत लेख कलकत्ता विश्वविद्यालय मुद्रणालय से फरवरी 1922 में प्रकाशित हुई थी, जिसमें उस समय तक प्राप्त सभी नतीजों का समावेश था।

इस लेख में जिक्र किए गए अनेक समस्याओं की छानबीन हमारे कई योग्य सहयोगियों के साथ मिलकर किया गया था। यहाँ सिर्फ कुछ ही अन्वेषणों के बारे में बताना संभव है, जिन्हें कलकत्ता में छह वर्षों के दौरान यानी सन् 1922 से 1927 के बीच किया गया। रामनाथन ने द्रव्यों में प्रकाश के आणविक प्रकीर्णन का अध्ययन दाब एवं ताप में विस्तृत परिवर्तन करके किया, जिससे अस्थिरता सिद्धांत को कुछ हद तक समर्थन मिला। उनके शोध से यह भी पता चला कि तापमान के परिवर्तन से द्रव्य एवं वाष्प द्वारा प्रकीर्णित किरणों में ध्रुवीत किरणों की तीव्रता की मात्रा में परिवर्तन होता है। कमलेश्वर राव ने द्रव्यों के मिश्रण पर शोध किया, जिससे इस बात का पता चला कि ऐसे भी

मिश्रण होते हैं जिनमें एक साथ घनत्व, संयोजन तथा आणविक दिक् विन्यास में अस्थिरता पाया जाता है। श्रीवास्तव ने स्फटिक (क्रिस्टलीय) पदार्थों पर काम किया तथा घनत्व में अस्थिरता व तापमान में वृद्धि होने से प्रकाश प्रकीर्णन पर पड़ने वाले प्रभाव का अध्ययन किया। तरल पदार्थों की सतह से प्रकाश प्रकीर्णन एवं तापमान की वजह से इसपर पड़ने वाले प्रभाव का अध्ययन रामदास ने किया, जिन्होंने तल तनाव व लचीलेपन के आपसी संबंध के बारे में जानकारी हासिल की। उन्होंने यह भी पता किया कि एक क्रांतिक तापमान पर तल लचीलेपन का परिवर्तन आयतनिक लचीलेपन में हो जाता है। सोगानी ने एक्स किरणों का द्रव्यों से प्रकीर्णन पर काम किया, जिससे इसके और प्रकाशिकी प्रकीर्णन के बीच समानताओं को समझा जा सके तथा एक्स किरणों के लिए अस्थिरता सिद्धांत के उपयोग की पुष्टि की जा सके।

अणुओं की विषम दैशिकता

जैसा कि पहले बताया गया है, द्रव्यों से प्रकीर्णित प्रकाश में ध्रुवीत प्रकाश की मात्रा अणुओं की विषमदैशिकता पर निर्भर करता है। कलकत्ता में सन् 1922 से 1927 के बीच किए गए प्रयोगों का उद्देश्य इसी गुणवत्ता से संबंधित आँकड़े इकट्ठे करने का था, जिससे विभिन्न प्रकाशिकी घटनाओं और प्रकीर्णन के बीच के संबंध को समझा जा सके। कृष्णन ने अनेक द्रव्यों की जाँच करने के बाद यह स्पष्ट कर दिया कि अणुओं की प्रकाशिकी विषम दैशिकता उसके रासायनिक संगठन पर निर्भर करता है। रामकृष्ण राव ने अनेक प्रकार के गैस एवं वाष्पों से प्रकीर्णित प्रकाश के अध्रुवीकरण का अध्ययन किया, जिसके नतीजे इस विषय के विकास के लिए काफी महत्त्वपूर्ण साबित हुए। वेंकटेश्वरन ने जलीय विलयनों के विद्युत् अपघटन से वियोजन पर प्रकाश के प्रकीर्णन प्रभाव का पता लगाया। रामचंद्र राव ने विभिन्न तापमानों पर कुछ अति दीर्घित अणुओं वाले द्रव्य तथा अति

ध्रुवीत पदार्थों से होने वाले प्रकाश प्रकीर्णन पर शोध किया तथा अणुओं के आकार व अणु संगुणन का प्रकीर्णित प्रकाश में अध्रुवीकरण की जाँच की।

तरल पदार्थों पर किए गए प्रयोगों के नतीजों की व्याख्या करने के लिए जरूरी सघन द्रव्य माध्यम में प्रकाश प्रकीर्णन के आणविक सिद्धांत के विकास करने का कार्य मैंने रामनाथन तथा कृष्णन के साथ की थी। हमने घनत्व में स्थानीय विचलन पर एक संशोधित सूत्र का अवकलन किया, जो आइन्साइन के सूत्र से भिन्न था परंतु जिसके नतीजे प्रायोगिक नतीजों के काफी करीब था। कृष्णन और मैंने शृंखलाबद्ध ठंग से इन खोजों पर कई लेख भी प्रकाशित किए, जिसमें यह दिखाया गया था कि प्रकाश प्रकीर्णन से अनुमानित अणुओं की प्रकाशिकी विषम दैशिकता का उपयोग द्रव्य के प्रकाशीय व परावैद्युतिक गुणों की जानकारी प्राप्त करने में की जा सकती है। इन अध्ययनों से यह निष्कर्ष निकला कि द्रव्यों में अवलोकित आणविक विषम दैशिकता तथा क्रिस्टलीय पदार्थों के प्रकाशिकी, वैद्युतिकी व चुंबकीय विषम दैशिकता के बीच आपसी संबंध हो सकता है।

एक नई घटना

अब तक उल्लिखित अन्वेषण मुख्य तौर पर प्रकाश के चिरसम्मत विद्युत-चुंबकीय सिद्धांत पर आधारित था, प्रकाश प्रकीर्णन को समझने के लिए जिसका उपयोग करने में प्रधानतया रैले एवं आइंस्टाइन का नाम लिया जाता है। हालाँकि प्रकीर्णन प्रक्रिया में 'कणिका सिद्धांत' के सबूत मिलने की संभावना को अनदेखा नहीं किया गया था, बल्कि इस बारे में विस्तृत चर्चा फरवरी 1922 के मेरे लेख में की गई थी, जिसका प्रकाशन कॉम्पटन के सुप्रसिद्ध एक्स किरण प्रकीर्णन खोज के कम-से-कम एक वर्ष पूर्व हुई थी। यद्यपि हमारे प्रयोगात्मक अध्ययन विशेषकर प्रकाश के विद्युत-चुंबकीय सिद्धांत का प्रतिपादन कर रहे थे,

परंतु शोध के प्रारंभिक चरण में ही ऐसी प्रक्रिया की उपस्थिति के संकेत मिलने लगे थे, जो चिरसम्मत विचार के परे था। पारदर्शी द्रव्य से प्रकाश का प्रकीर्णन काफी क्षीण होता है, दरअसल इसकी तीव्रता गंदे माध्यम से होने वाले प्रकीर्णन, जिसे 'टींडल प्रभाव' कहते हैं, से भी कम होता है। प्रयोगात्मक अध्ययन से यह पता चला था कि रैले-आइन्स्टाइन आणविक प्रकीर्णन में एक और क्षीणतम किरण भी मौजूद है, जिसकी तीव्रता चिरसम्मत प्रकीर्णित किरण की तुलना में सैंकड़ों गुणा कम है तथा जिसका तरंग-दैर्ध्य आपतित किरण के समान नहीं है। सबसे पहले रामनाथन ने अप्रैल 1923 में कलकत्ता में इस तथ्य का पता लगाया, जो इस बात की व्याख्या करने की कोशिश कर रहे थे कि क्यों कुछ तरल पदार्थों (जैसे पानी, ईथर, मीथाईल व इथाईल अलकोहल) से प्रकीर्णित हुए किरणों में अध्रुवीत किरण की मात्रा आपतित किरण के तरंग-दैर्ध्य पर निर्भर करती है। रामनाथन ने पाया कि द्रव्य के व्यापक रासायनिक शुद्धिकरण व बार-बार निर्वात में धीमे आसवन करने के बाद भी प्रकीर्णित किरण में इस नई किरण की तीव्रता पहले जैसी बरकरार रही, जो यह दरशाती है कि यह प्रक्रिया द्रव्य विशेष की अपनी पहचान है, न कि प्रतिदीप्त अशुद्धता के कारण। सन् 1924 में कृष्णन ने यही प्रक्रिया कई और द्रव्यों में देखी। मैंने यह प्रक्रिया बर्फ तथा प्रकाशीय काँच मे स्पष्ट देखा था।

कॉम्पटन प्रभाव का प्रकाशीय अनुरूप

इस पेचींदा प्रक्रिया का मूल कारण स्वाभवतः हमारे अंदर दिलचस्पी पैदा किया तथा सन् 1925 में ग्रीष्मकाल के दौरान वेंकटेश्वरन ने सूर्य के प्रकाश को रंगीन पर्दों से गुजारकर द्रव्यों से टकराने के बाद प्रकीर्णित हुए वर्णक्रम का चित्र लेकर इस पर कार्य किया; परंतु इससे कोई निर्णायक नतीजा नहीं निकला। रामकृष्ण राव ने ऐसी ही प्रक्रिया को देखने के लिए गैस व वाष्प में अध्रुवीत प्रकीर्णित किरणों की जाँच-

पड़ताल 1926–1927 के दौरान किया, परंतु उन्हें भी सफलता नही मिली। सन् 1927 के आखिर में कृष्णन ने इस समस्या पर अध्ययन शुरू किया। उनका काम जारी ही था कि अन्य एक प्रयोग में इसी तरह के प्रभाव पहली बार दिखे। उस दौरान हम काँचीय द्रव्य सदृश्य अति गाढ़े कार्बनिक द्रव्य से प्रकाश के प्रकीर्णन पर शोध कर रहे थे जिसपर काम करने की जिम्मेदारी वेंकटेश्वरन की थी। वे इस दिलचस्प नतीजे पर पहुँचे कि अति शुद्ध ग्लिसरीन से सूर्य का प्रकाश प्रकीर्णित होकर चमकीले हरे रंग का दिखता है, न कि सामान्य नीले रंग का। इस खोज में एवं रामनाथन द्वारा पानी व अलकोहल से प्राप्त नई किरण की खोज में काफी समानता थी, परंतु यह प्रकाश अत्यंत तीव्र था, जिसका अध्ययन करना काफी आसान था। हमने बिलकुल समय बर्बाद न करके इस विषय में परीक्षण जारी रखा। सूर्य की किरण के मार्ग में विभिन्न फिल्टरों को रखकर अलग अलग आपतीत संकीर्ण वर्णक्रम प्राप्त करके द्रव्य से प्रकीर्णित हुए किरण को देखने पर पता चला कि उसका रंग आपतीत किरण के रंग से भिन्न था तथा वर्णक्रम के लाल रंग की तरफ विस्थापित था। यह प्रकीर्णित किरणें अति ध्रुवीत भी था। इससे स्पष्ट संकेत मिल रहे थे कि यह प्रक्रिया 'कॉम्पटन प्रभाव' के सदृश्य है। कॉम्पटन के काम से यह तो ज्ञात हो चुका था कि प्रकीर्णन प्रक्रिया में किरण का तरंग दैर्ध्य कम हो सकता है तथा ग्लिसरीन से प्राप्त नतीजे को देखकर लगा कि जो बात हमें सन् 1923 से हैरान कर रहा था, वह वास्तव में 'कॉम्पटन प्रभाव' का ही प्रकाशीय अनुरूप है। इस विचार ने हमें दूसरे पदार्थों पर शोध करने के लिए प्रेरित किया।

इस नई खोज के गहन अध्ययन करने में जो मुख्य अड़चन आड़े आ रहा था, वह था प्रकीर्णित किरण में इसकी अत्यंत मंद तीव्रता। इस समस्या का समाधान करने के लिए एक 7 इंच अपवर्तक दूरबीन व कम केंद्रबिंदु वाले लैंस का एक साथ उपयोग करके सूर्य के किरण

को काफी संकरा बनाया गया, जिससे इसकी तीव्रता में अत्यधिक बढ़ोतरी हुई। इसके अतिरिक्त आपतीत व प्रकीर्णित किरणों के मार्ग में फिल्टरों को रखा गया, जैसाकि रामनाथन ने सन् 1923 में किया था, जिससे रूपांतरित किरण को दूसरे किरणों से पृथक् किया जा सके। उपकरणों की इस व्यवस्था से अनेकों द्रव्य में रूपांतरित किरण स्पष्ट तौर पर देखा गया तथा कुछ स्थितियों में वे ध्रुवीत थे। कृष्णन, जो उन दिनों इन खोजों में मेरी सहायता कर रहे थे, कई कार्बनिक पदार्थ के वाष्पों में भी इस प्रभाव को देखा तथा उनके रूपांतरित किरण में ध्रुवीकरण अवस्था का निर्धारण किया। संपीडित गैस जैसे कार्बन मोनोऑक्साईड व नाइट्रस ऑक्साईड तथा क्रिस्टलीय बर्फ एवं प्रकाशीय काँचों से भी प्रकीर्णित हुए किरण में रूपांतरित किरण दिखाई दिए। इन सब निरीक्षणों को देखकर निस्संदेह कहा जा सकता था कि यह प्रकाशीय प्रकीर्णन कॉम्पटन प्रभाव के समरूप है।

इस नए प्रभाव की वर्णक्रमीय विशेषताएँ

सन् 1925 में इस नए प्रभाव के वर्णक्रमीय जाँच करते समय कोई सुनिश्चित परिणाम न मिलने के कारण इस विषय पर शोध करना बंद कर दिया गया था जो अब 7 इंच अपवर्तक दूरबीन के इस्तेमाल से प्राप्त शक्तिशाली चमकीला आपाती किरण की वजह से करना संभव लग रहा था। मुझे आपतीत किरणों और कार्बनिक द्रव्यों के बीच जाईश कोबाल्ट-ग्लास का फिल्टर रखने पर प्रकीर्णित किरणों के वर्णक्रम में नीले-पीले रंग के क्षेत्र में एक पट्टा (बैंड) दिखाई दिया, जो फिल्टर से पारगत हुए आपाती जामुनी-बैंगनी रंग से अलग था तथा इन दोनों रंगों के बीच इन्हें अलग करता हुआ एक काला क्षेत्र था। वर्णक्रम के इन दोनों रंगों के क्षेत्र काफी तीक्ष्ण नजर आए, जब पारगत किरण को और भी संकुचित करने के लिए आपतीत किरण के पथ पर एक अतिरिक्त फिल्टर लगाया गया। इस निरीक्षण से यह स्पष्ट था कि सूर्य

की किरणों के वजाए मर्करी आर्क लैंप के साथ एक चौड़े द्वारक (ऐपर्चर) वाला संग्राही लैंस व कोबाल्ट-ग्लास फिल्टर का इस्तेमाल करना अधिक उपयुक्त होगा, जो एकवर्णी प्रकाश देता है। इन उपकरणों की सहायता से विभिन्न द्रव्यों तथा ठोस पदार्थों से प्रकीर्णित हुए किरणों का वर्णक्रम आँखों से देखा गया तथा एक आश्चर्यजनक तथ्य का पता चला कि इस वर्णक्रम में विसरित पृष्ठभूमि के साथ कई रेखाएँ या पट्टा मौजूद थे जोकि मर्करी आर्क लैंप के प्रकाश में नहीं थे।

क्वार्टज मर्करी लैंप इतना शक्तिशाली एवं सुविधा जनक एकवर्णी तीव्र प्रकाश स्त्रोत था कि तरल व ठोस पदार्थों के लिए प्रकीर्णित किरणों के वर्णक्रम का चित्रण करने में कोई विशेष कठिनाई नहीं हुई। आरंभ में वर्णक्रम का चित्र लेने के लिए एक बहुत ही छोटा क्वार्टज स्पेक्ट्रोग्राम का इस्तेमाल किया गया था, जिसे हिल्जऱ कंपनी ने बनाया था। इसी प्रकार का एक बड़े आकार के स्पेक्ट्रोग्राम से कृष्णन ने द्रव्य व क्रिस्टलीय पदार्थों के प्रकीर्णित वर्णक्रम का चित्र लिया, जो काफी संतोषजनक था तथा जिससे पहली बार बैंगनी रंग की ओर विस्थापित हुए वर्णक्रमी रेखाओं की उपस्थिति का पता चला। स्पष्टतः गैस व वाष्पों के अध्ययन के लिए काफी प्रयोगात्मक कठिनाइयाँ आड़े आ रही थीं, जो उच्च दाब पर प्रयोग करने से कुछ हद तक कम होती थी। एक उन्नत चौड़े द्वारक (*/1.8), वाले उपकरण के उपयोग से रामदास ने पहली बार वायुमंडलीय दाब पर गैसीय पदार्थ (ईथर वाष्प) के वर्णक्रम का चित्रण किया।

इस जाँची परखी प्रक्रिया की व्याख्या करने के लिए इसके समरूप 'कॉम्पटन प्रभाव' का सहारा लिया गया था। कॉम्पटन के कार्य को आम स्वीकृति मिल चुकी थी जिसके अनुसार किरणों का प्रकीर्णन एक ऐसी प्रक्रिया है, जिसमें अविनाशिता सिद्धांत का पालन होता है। इसके अनुसार कण और क्वांटा में टकराव के बाद यदि प्रकीर्णित हुए कण

की ऊर्जा में वृद्धि होती है तब क्वांटा की ऊर्जा में उतनी ही कमी आती है, जिससे प्रकीर्णन के बाद किरण की आवृत्ति कम हो जाती है। ऊष्मागतिक सिद्धांत के आधार पर इसके विपरीत प्रक्रिया का होना भी संभव है। इन सब विचारों पर गौर करके प्रायोगिक नतीजों की व्याख्या की जा सकती है, तथा विस्थापित वर्णक्रम रेखाओं व अणुओं के अवरक्त आवृत्तियों के बीच प्राप्त समानता से यह जाहिर है कि इस नई विधि ने पदार्थ की संरचना पर प्रयोगात्मक अनुसंधान करने के असीमित राह खोल दिए हैं।

प्रभाव का विवरण

यहाँ यह स्पष्ट कर देना उचित होगा कि यद्यपि कॉम्पटन का अविनाशिता सिद्धांत प्रयोग से प्राप्त प्रभाव की व्याख्या करने में सहायक है, परंतु यह खुद प्रभाव के बारे में जानकारी देने के लिए पर्याप्त नहीं है। आणविक वर्णक्रम के अध्ययन से यह ज्ञात है कि गैस के अणु में चार प्रकार की ऊर्जा—जैसे स्थानांतरीय गति, घूर्णन, कंपन और इलेक्ट्रॉनीय उत्तेजना की वजह से होती है। इनमें से प्रथम ऊर्जा को छोड़कर शेष तीनों प्रकार की ऊर्जा को पूर्णांक क्वांटम संख्याओं द्वारा दरशा सकते हैं। अतः अणु की कुल ऊर्जा का मान अनेकों संभावित ऊर्जाओं में से एक हो सकता है। यदि हम यह मान लें कि अणु और प्रकाश क्वांटा के बीच संघात से ऊर्जा का विनिमय होता है तथा संघात के बाद अणु की ऊर्जा आपाती क्वांटा की ऊर्जा से कम हो जाती है तो यह निष्कर्ष निकलता है कि प्रकीर्णित किरण के वर्णक्रम में अनेकों नई रेखाएँ उपस्थित होनी चाहिए, जिससे यह वर्णक्रम उत्सर्जन या अवशोषण वर्णक्रम (स्पेक्ट्रम) की तुलना में बहुत जटिल दिखेगी, परंतु प्रयोग से एक विशिष्ट तथ्य सामने आया कि जटिल बहुपरमाणुक अणुओं के प्रकीर्णित वर्णक्रम भी अति सरल थे, जबकि इनके उत्सर्जन या अवशोषण वर्णक्रम अत्यंत पेचीदे। वर्णक्रम की यही सादगी प्रकाश

प्रकीर्णन के अध्ययन को विशेष महत्त्व एवं मान्यता देती है। अब यह स्पष्ट है कि जो प्रभाव प्रयोग द्वारा देखा गया था, उसकी भविष्यवाणी अविनाशिता सिद्धांत से करना संभव नहीं था।

नील्स बोह्र के 'क्वांटम एवं चिरसम्मत सिद्धांत' के बीच 'संगति नियम' (कॅरोस्पंडेन्स प्रिंसीपल) के आधार पर इस प्रक्रिया की परख की जा सकती है। चिरसम्मत सिद्धांत के अनुसार, यदि एक अणु, जो गतिशील, घूर्णित या कंपित अवस्था में है, प्रकाश का प्रकीर्णन करता है तो प्रकीर्णित किरण में कुछ ऐसे प्रकाश हो सकते हैं जिनकी आवृत्ति आपतीत किरण से भिन्न हो। आश्चर्य की बात तो यह है कि चिरसम्मत सिद्धांत की इस व्याख्या से प्रयोग से प्राप्त नतीजे काफी मिलते हैं। इससे पता चलता है कि क्यों आवृत्ति में विस्थापन तीन वर्गों—यथा स्थानांतरीय गति, घूर्णन तथा कंपन की वजह से विभिन्न परिमाणों में होता है। यह वरण नियम (सिलेक्शन रूल) की व्याख्या करता है। उदाहरण के लिए क्यों अणु के कंपन से पैदा होने वाली आवृत्तियों में से केवल प्रमुख आवृत्ति ही प्रकिर्णित किरण में पाई जाती है, न कि उसके अधिस्वरक व संयोजक आवृत्तिआँ, जो उत्सर्जन या अवशोषण वर्णक्रम में प्रमुख तौर पर। चिरसम्मत सिद्धांत के उपयोग से और भी बहुत कुछ जाना जा सकता है, जैसे परिवर्तित आवृत्ति के प्रकाश की तीव्रता व ध्रुवण का अनुमानित परिमाण। हालाँकि प्रयोग से प्राप्त तथ्यों का गुणात्मक विवरण देने के लिए चिरसम्मत सिद्धांत में फेरबदल करने की आवश्यकता है और इसके लिए क्वांटम सिद्धांत के उपयोग की जरूरत पड़ेगी। प्रायोगिक नतीजे को समझने के लिए क्रैमर्स और हाईजनबर्ग के कार्य तथा क्वांटम यांत्रिकी में हुए हाल के विकास, जो बोह्र संगति नियम पर आधारित है, आशाजनक प्रतीत हो रहे हैं। परंतु जब तक अणुओं की संरचना के बारे में हम और अधिक नहीं जान जाते एवं पर्याप्त मात्रा में प्रायोगिक तथ्य इकट्ठा नहीं कर लेते, यह

मान लेना जल्दबाजी होगी कि क्वांटम यांत्रिकी इसकी पूरी व्याख्या कर पाएगी।

प्रभाव का महत्त्व

इस प्रक्रिया की व्यापकता, प्रायोगिक सुविधा एवं प्राप्त वर्णक्रम की सादगी के लिए यह प्रभाव भौतिकी एवं रसायनीकी क्षेत्र के अनेकों समस्याओं का समाधान करने में उपयोगी होगी। वास्तव में इस प्रभाव का प्रमुख महत्व इसी गुण की वजह से है। प्रयोग से प्राप्त वर्णक्रम में आवृत्ति फर्क, स्पेक्ट्रमी रेखाओं की चौड़ाई, प्रकीर्णित किरण की तीव्रता व ध्रुवण अवस्था से प्रकीर्णी पदार्थ की परम संरचना के बारे में हमें जानकारी मिलती है। प्रयोग करके पता चला है कि वर्णक्रम की ये विशेषताएँ भौतिक अवस्थाओं—जैसे तापमान व समुच्चयन, भौतिक-रसायनिक अवस्थाओं—जैसे मिश्रण, विलयन, आणविक संगुणन व बहुलीकरण, तथा विशेषकर रासायनिक रचना से काफी प्रभावित होती हैं। स्पेक्ट्रम विज्ञान का यह नया क्षेत्र पदार्थ की संरचना संबंधित समस्याओं के अध्ययन का असीमित अवसर प्रदान करेगा।

कुछ अंतिम बातें

सरलतम अणुओं में भी की गई इस प्रभाव के मात्रात्मक अध्ययन से मौलिक अनुसंधान क्षेत्र में विशाल प्रगति होने की आशा है। इस क्षेत्र में मैकलेलान द्वारा द्रवित गैस पर किए गए शोध एवं आर.डब्ल्यू. वुड तथा रासेट्टी के निरिक्षण अग्रगामी हैं, जिनकी काफी सराहना हुई है। सरल रासायनिक रचनावाले क्रिस्टलीय पदार्थ में इस प्रभाव का मात्रात्मक अध्ययन निश्चित रूप से काफी महत्त्वपूर्ण है। क्रिस्टलीय पदार्थ हीरा पर शोधकार्य विशेष उल्लेखनीय हैं, जो रामस्वामी, रॉबर्टसन, फॉक्स और विस्तार से भगवंतम ने किया था। इस पदार्थ से काफी आश्चर्यजनक नतीजे मिले हैं, जो क्रिस्टलीय अवस्था के लक्षण की

पूरी जानकारी प्राप्त करने में मुख्य भूमिका निभाएँगे। मैं कृष्णमूर्ति के उस कार्य का भी जिक्र करना चाहूँगा जिसमें उन्होंने प्रकीर्णित वर्णक्रम रेखाओं की तीव्रता और रासायनिको आबंध (केमिकल बांड) के बीच संबंध का पता लगाया है तथा रासायनिक संयोग में समध्रुवी से विषमध्रुवी स्थानांतरण का अध्ययन किया है। कृष्णमूर्ति की यह खोज कि क्रिस्टलों के अणु चुंबकत्व से वर्णक्रम की विस्थापित रेखाओं की तीव्रता प्रभावित होती है, शोध के इस नए क्षेत्र में असाधारण उपलब्धि थी। □

4

बैंगलौर में रमन

कलकत्ता के कड़वा अनुभव के बाद रमन बैंगलौर चले गए, परंतु वहाँ भी वे चैन से न रह सके। अप्रैल 1933 में उन्होंने भारतीय विज्ञान संस्थान के निदेशक का पदभार ग्रहण किया, जिसकी नियुक्ति चयन समिति ने की थी। इस पद पर पाँच वर्ष रहने के बाद परिस्थिति का शिकार होकर उन्हें इस्तीफा देना पड़ा, परंतु सेवानिवृत्त होने तक वे विज्ञान विभाग के अध्यक्ष बने रहे। यह रमन की खासियत थी कि इतने विपत्तियों के बावजूद वे अपना शोधकार्य करते रहे।

भारतीय विज्ञान संस्थान में रमन

बैंगलौर में स्थित भारतीय विज्ञान संस्थान देश के उत्तम शैक्षणिक संस्थानों में से एक है। इसकी स्थापना सर जमशेदजी टाटा ने सन् 1911 में की थी। उन दिनों संस्थान के निदेशक अंग्रेज ही हुआ करते थे, जिन्हें अप्रत्याशित 3000 रुपए प्रति माह वेतन मिलता था तथा रहने के लिए एक बड़ा बंगला। संस्थान का प्रचालन एक परिषद् करती थी, जिसमें रमन भी एक सदस्य थे। रमन से पहले यहाँ के निदेशक डॉ. मार्टिन फॉर्सटर थे, जो सन् 1933 में सेवानिवृत्त होने वाले थे। उनके उत्तराधिकारी के चयन के लिए गठित समिति इग्लैंड में योग्य व्यक्ति की तलाश कर रहे थे, जैसाकि उस समय नियम था। इस दौरन लॉर्ड रदरफोर्ड से भी

चित्र : 4.1 भारतीय विज्ञान संस्थान के निदेशक रमन

सलाह ली गई, जिन्होंने रमन के नाम का प्रस्ताव दिया। अंततः इस गौरवशाली पद के लिए रमन को चुना गया, जिन्होंने इसे स्वीकार कर लिया। इस प्रकार रमन 'भारतीय विज्ञान संस्थान' के प्रथम भारतीय निदेशक बने।

आरंभ में यहाँ केवल तीन क्षेत्रों में काम शुरू हुआ था—सामान्य व अनुप्रयुक्त रासायनिकी, कार्बनिक रासायनिकी तथा विद्युत् प्रद्यौगिकी। बाद में जीव रासायनिकी विभाग में भी काम आरंभ हो गया था।

रमन के आने से पहले संस्थान में ज्यादातर काम व्यावहारिक क्षेत्र में होता था तथा मौलिक अनुसंधान की तरफ कम ध्यान दिया जाता था। इस नामी संस्थान की पहचान तब तक उसके काम से नहीं हुई थी। रमन को यहाँ का वातावरण काफी सुस्त लगा। रमन ने आते ही नए भौतिक विज्ञान विभाग का गठन किया, जिसकी अनुमति उन्हें पहले ही मिल चुकी थी। तथा कुछ विभागों में प्रशासनिक दृष्टिकोण से फेरबदल किया। उन्होंने कार्यशाला को उपयोगी बनाने के लिए उसमें भी कुछ परिवर्तन किए, जिससे अनुसंधान के लिए आवश्यक उपकरणों का निर्माण वहीं किया जा सके। रमन का मानना था कि उत्कृष्ट काम करने के लिए संस्थान में श्रेष्ठ व योग्य लोगों की जरूरत है। वे वहाँ चल रहे

काम से बेहद नाखुश थे। अतः वे जर्मनी के कुछ प्रसिद्ध वैज्ञानिकों से संपर्क किया, जो उस समय हिटलर के डर से देश छोड़कर भाग रहे थे। यद्यपि रमन के विचार में संस्थान की उन्नति के लिए यह सब करना अत्यंत आवश्यक था, परंतु संस्थान के पुराने लोगों को यह सब अच्छा नहीं लगा। कुछ वरिष्ठ प्राध्यापक, जो खुद निदेशक बनने की तमन्ना रखते थे, रमन के कार्य करने के ढंग की आलोचना करने में जुट गए तथा उनके विरुद्ध षड्यंत्र रचने लगे। ऐसा कहा गया कि उनमें प्रशासनिक दक्षता नहीं है। उनपर केवल भौतिकी विषय को बढ़ावा देने का आरोप लगाया गया। रमन भी कहाँ हार माननेवाले थे! वे खुद को औरों से श्रेष्ठ समझते थे तथा उन्हें इस बात का गर्व था। अतः आपसी समझौता होने के बजाए तनाव बढ़ता गया। उन्होंने सैद्धांतिक भौतिकी के प्रसिद्ध जर्मन वैज्ञानिक मैक्स बॉर्न को बैंगलौर बुलाया तथा उनके कार्य से प्रभावित होकर उन्हें प्राध्यापक पद पर नियुक्त करने की सिफारिश की। इस विषय को लेकर बात इतनी बढ़ गई कि लोग मैक्स बॉर्न के बारे में उल्टा-सीधा बोलने लगे। बॉर्न अत्यंत व्यथित होकर वापस लौट गए। रमन के कार्य-प्रणाली की समीक्षा करने के लिए जनवरी सन् 1936 में एक समिति गठित की गई, जिसने रमन के खिलाफ रिपोर्ट दी। रमन एकदम अकेले पड़ गए थे। उनके पास निदेशक पद से इस्तीफा देने के सिवा और कोई दूसरा उपाय नहीं बचा था। अंततः सन् 1938 में वे निदेशक पद से हट गए, लेकिन प्राध्यापक पद पर बने रहे। उनके वेतन में भी कटौती हुई। यद्यपि रमन में मनोबल की कमी नहीं थी, लेकिन फिर भी कलकत्ता तथा बैंगलौर के हादसों ने उन्हें झकझोर कर रख दिया। परंतु उनके वैज्ञानिक गतिविधियों में कोई कमी नहीं आई। उनका शोध-कार्य जारी रहा।

होमी भाभा इग्लैंड से लौटने के बाद सन् 1940 में 'भारतीय विज्ञान संस्थान' के भौतिकी विभाग में रीडर के पद पर काम करने का निर्णय लिया क्योंकि वहाँ रमन विभाग के अध्यक्ष थे, यद्यपि उन्हें देश के कई

जगहों से नियुक्ति के लिए आमंत्रित किया गया था। बैंगलौर में भाभा छह वर्ष रहे जहाँ उन्होंने ब्रह्मांड किरणों पर प्रायोगिक कार्य किया। बैंगलौर का यह प्रवास न केवल उनके जीवन में बल्कि देश में भी एक नया मोड़ लाया। यहीं आकर भाभा ने देश की प्रगति के लिए बड़े सपने देखे जिन्हें साकार करने के लिए बैंगलौर का संस्थान छोटा लगने लगा। दोराब टाटा न्यास से प्राप्त वित्तीय सहायता से उन्होंने टाटा मूलभूत अनुसंधान संस्थान (टी.आई.एफ.आर.) का निर्माण किया जिसका मुख्य उद्देश्य विज्ञान में मूलभूत शोध करना था। बाद में उन्होंने भारत में नाभिकीय ऊर्जा कार्यक्रम की शुरुआत की। विक्रम साराभाई भी कुछ वर्ष रमन के साथ भारतीय विज्ञान संस्थान के भौतिकी विभाग में कार्य किया था। साराभाई विज्ञान में उच्च शिक्षा प्राप्त करने के लिए इग्लैंड गए थे परंतु द्वितीय विश्व युद्ध छिड़ जाने से उन्हें देश वापस आना पड़ा। तब विक्रम के पिता अम्बालाल के अनुरोध पर रमन ने विक्रम को संस्थान में नियुक्ति दी एवं वे भाभा के साथ ब्रह्मांड किरणों पर काम करने लगे। युद्ध की समाप्ति के बाद साराभाई इंग्लैंड जाकर पीएच.डी. की उपाधि हासिल कर वापस देश लौट आए। यहाँ आकर उन्होंने अहमदाबाद में सन् 1947 में 'भौतिक अनुसंधान प्रयोगशाला' (पी.आर.एल.) की स्थापना की, जिसके प्रथम निदेशक रमन के छात्र के.आर. रामनाथन बने। भारत में अंतरिक्ष विज्ञान व प्रद्योगीकी के संस्थापक साराभाई थे। शायद देश में रहकर कुछ करने की प्रेरणा भाभा एवं साराभाई को रमन से ही मिली थी। रमन का दृढ़ विश्वास था कि देश में विज्ञान की जड़ मजबूत करने व इसकी प्रगति के लिए प्रसिद्ध विदेशी वैज्ञानिकों को भारत में लाना जरूरी है। परंतु उनके इस प्रयास को सफल नहीं होने दिया गया। यहाँ इस बात का जिक्र करना उचित होगा कि इसी विचारधारा को अमेरिका ने अपनाया, जो आज विश्व में महाशक्ति के रूप में उभरा है।

संस्थान में रमन के दफ्तर का दरवाजा हमेशा खुला रहता था,

जिससे कोई भी आकर उनसे मिल सके। पाँच वर्षों तक वे निदेशक रहे, जब उन्हें प्रशासन का काम भी देखना पड़ता था। वे तड़के छह बजे शिष्यों से उनके काम के बारे में विचार-विमर्श करने संस्थान चले जाते थे। नौ बजे वापस घर आकर, जो संस्थान के परिसर में ही था, नहा धोकर नाश्ता करके पुनः दस बजे दफ्तर पहुँच जाते थे, जहाँ शाम तक प्रशासनिक कार्यों में व्यस्त रहते थे। उसके बाद फिर अपने शोधकार्य में या शिष्यों के साथ उनकी समस्याओं पर चर्चा करने के लिए रात नौ बजे तक संस्थान में ही रहते थे। यह था उनका दैनंदिन कार्यक्रम!

बैंगलौर में भौतिकी

जब रमन बैंगलौर आए, तब विश्व के कई देशों में नाभिकीय भौतिकी पर जोरों से काम चल रहा था। रमन भी इस विषय पर अनुसंधान कार्य शुरू करना चाहते थे, परंतु उसके लिए आवश्यक

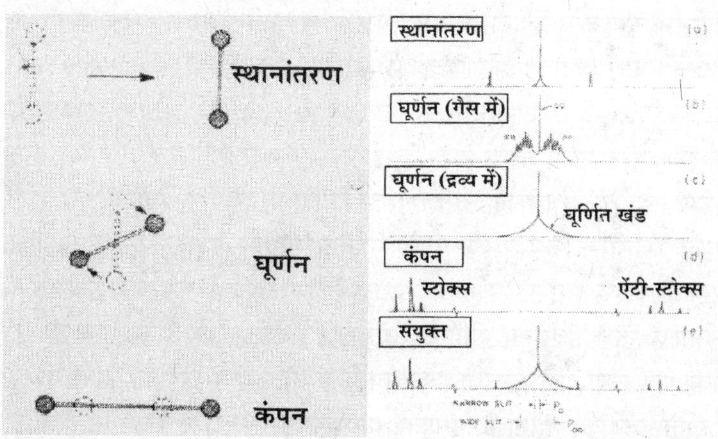

चित्र : 4.2 किसी पदार्थ के अणुओं में तीन प्रकार की गति हो सकती है— स्थानांतरण, घूर्णन व कंपन।

चित्र : 4.3 किसी माध्यम से प्रकीर्णित प्रकाश का वर्णक्रम। आवृत्ति में परिवर्तन अणुओं के स्थानांतरण, घूर्णन व कंपन की वजह से होता है

धनराशि व समर्थन के अभाव के कारण वे ऐसा नहीं कर सके। बैंगलौर में भी रमन अपने पुराने विषय प्रकाश भौतिकी व ध्वनिकी पर शोधकार्य जारी रखा, जिससे कुछ रोचक तथ्य सामने आए। यहाँ मुख्यत: उनके शिष्यों ने दो प्रमुख विषयों पर काम किया—एक, विभिन्न क्रिस्टलों द्वारा रमन प्रकीर्णन तथा दूसरा, विशेषतया द्रव्यों द्वारा होने वाले ब्रिलुवाँ व उस जैसा प्रकीर्णन।

किसी माध्यम द्वारा प्रकाश के प्रकीर्णन होने पर उसकी आवृत्ति में बदलाव माध्यम के अणुओं में लगातार होने वाली हलचल से पैदा होता है। यह हलचल या गति तीन तरह की होती है, जो अणुओं के स्थानांतरण, घूर्णन व कंपन की वजह से होता है। रमन ने जिस प्रकीर्णन को पहली बार देखा था, वह अणुओं के कंपन से पैदा होती है। इसमें आपाती किरण के दोनों ओर कुछ दूरी पर खिसकी हुई रेखाओं के रूप में दिखाई देती है। रमन की खोज के तुरंत बाद घूर्णन से पैदा होने वाले प्रकीर्णन की भी खोज हुई। इसके स्पेक्ट्रम में आपाती किरण (क्यू-क्यू) के काफी करीब दोनों ओर रेखाओं की पट्टी (पी-पी तथा आर-आर) दिखाई देती है। अणुओं के स्थानांतरण होते रहने के कारण माध्यम के घनत्व में उताव-चढ़ाव होता है, जिससे केवल रैले प्रकीर्णन का होना संभव है, यानी प्रकाश की आवृत्ति में कोई बदलाव नहीं होना चाहिए। परंतु ब्रिलुवाँ ने दिखाया कि घनत्व में उतार-चढ़ाव के फलस्वरूप प्रकीर्णित प्रकाश की आवृत्ति में बदलाव आता है, जो डॉप्लर प्रभाव की वजह से होता है। इसे 'ब्रिलुवाँ प्रकीर्णन' कहते हैं। सारांश यह है कि जब प्रकाश किरणें किसी द्रव्य से प्रकीर्णित होती हैं तो उसके अणुओं के कंपन से 'रमन प्रकीर्णन' होता है, अणुओं के घूर्णन से भी रमन प्रकीर्णन होता है, अणुओं के स्थानांतरण से 'ब्रिलुवाँ प्रकीर्णन' होता है तथा शेष 'रैले प्रकीर्णन' होता है। इन सब विभिन्न प्रकीर्णनों के बारे में विस्तारित जानकारी रमन ने दी।

'रमन प्रभाव' की खोज के बाद से ही वे हीरे की ओर आकृष्ट हुए थे। उनके पास विभिन्न प्रकार के हीरों का संकलन था, जो उन्होंने अपने पैसों से खरीदा था। हीरों की गुणवत्ता का अध्ययन करना एक गंभीर विषय बन गया। एक समय ऐसा था जब रमन के हर छात्र हीरे के किसी-न-किसी पहलू पर काम कर रहे थे। हीरे की संरचना के बारे में जानकारी हासिल करना भी एक मुख्य कार्य था। यद्यपि ज्यादातर हीरे अवरक्त (तरंगदैर्ध्य = 8 माईक्रॉन) किरणों का अवशोषण करते हैं, परंतु कुछ हीरे ऐसे भी होते हैं जो इनका अवशोषण नहीं करते।

अब हम रमन के एक और उत्कृष्ट कार्य के बारे में चर्चा करेंगे। रमन को तो तरंगों से प्यार था ही, अतः जब प्रकाश व ध्वनि दोनों तरंगें साथ हों तो क्या कहना! हम रमन और नगेंद्र नाथ के शोध पर बात कर रहे हैं, जो काफी उच्चकोटि का था। यह पराध्वनिक तरंग (अल्ट्रासोनिक) द्वारा प्रकाश के प्रकीर्णन से संबंधित है। इसे समझने के लिए ब्रिलुवाँ प्रकीर्णन के बारे में जानना होगा। किसी भी पदार्थ में चाहे वह ठोस या द्रव्य हो, उसके अणु हमेशा गतिशील रहते हैं। यद्यपि इनकी गति काफी जटिल होती है, परंतु उन्हें गणितीय विधि द्वारा समझा जा सकता है। उनकी इस अस्त-व्यस्त गति को कई भिन्न आवृत्तिवाले सुव्यवस्थित तरंगों के जोड़ के बराबर माना जा सकता है। यह तरंगें अणुओं के हिलने-डुलने से पैदा होती हैं। माध्यम का तापमान बढ़ने से इन तरंगों की संख्या में भी वृद्धि होती है। इनमें वे तरंग जिनका तरंगदैर्ध्य सबसे लंबा होता है, ध्वनि तरंग जैसे होते हैं। ध्वनि तरंग क्या होता है? यह दरअसल संपीडन (कम्प्रेशन) व विरलन (रेरीफेक्शन) क्षेत्र का बना होता है। जब ध्वनि तरंग किसी माध्यम से होकर गुजरती है तो यह उस माध्यम के घनत्व में आवर्ती (पीरिऑडिक) परिवर्तन करता है, यानी कुछ जगहों पर घनत्व ज्यादा हो जाता है एवं कुछ जगहों पर कम। ब्रिलुवाँ के अनुसार, घनत्व का यह आवर्ती परिवर्तन वैसा ही है, जैसे मानो माध्यम में एक विवर्तन ग्रेटिंग रखा

है। अतः माध्यम पर पड़ने से आपाती प्रकाश किरणों का विवर्तन होता है। प्रयोगात्मक अध्ययन करने के लिए तरल माध्यम में एक ध्वनि तरंग जनित्र लगाया गया, जिससे तरल पदार्थ के घनत्व में आवर्ती–परिवर्तन पैदा हो सके। इस द्रव्य में प्रकाश की किरणें फेंकी गई, जो विवर्तित होकर विवर्तन बैंड बनाया, यानी ऐसा चित्र जहाँ प्रकाश की तीव्रता कहीं ज्यादा थी तो कहीं कम। आवृत्ति में बदलाव किए बिना ध्वनि तरंग के आयाम (ऐम्प्लिज्यूड) में परिवर्तन करने पर पाया गया कि बैंड की कई रेखाएँ ज्यादा तीव्र हो गई तो कईयों की तीव्रता में कमी आई। यह ब्रिलुवाँ सिद्धांत के प्रतिकूल था, क्योंकि इसके अनुसार सभी रेखाओं की तीव्रता में समान परिवर्तन होना चाहिए था। रमन ने इसका कारण भिन्न क्षेत्रों में प्रकाश की गति में परिवर्तन होने की वजह बताया। जब प्रकाश की किरणें संपीडित क्षेत्र से गुजरती है तो उसकी गति कम हो जाती है एवं विरलित क्षेत्र से

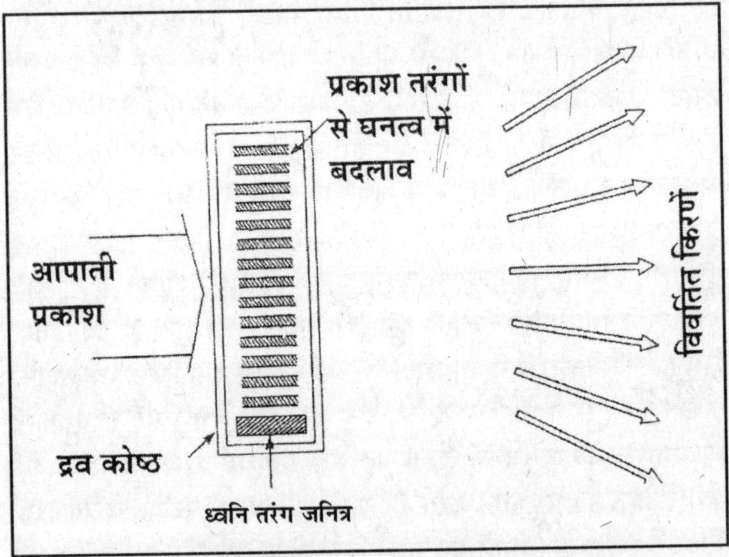

चित्र : 4.4 ध्वनि तरंगों से द्रव के घनत्व में आवधिक बदलाव होने पर प्रकाश के विवर्तन पट्टिका का बनना

गुजरते समय गति में वृद्धि होती है। परंतु प्रकाश की गति तो स्थिर होती है, लेकिन यह केवल निर्वात (वेक्यूम) में लागू होती है। जब प्रकाश किसी माध्यम से गुजरती है तो इसका वेग उस माध्यम के अपबर्तनांक (रिफ्रैक्टिव इन्डेक्स) मान के अनुसार बदलता है। यह मान माध्यम के घनत्व पर भी निर्भरशील होता है। अत: जब ध्वनि तरंगों की वजह से माध्यम के अलग अलग क्षेत्रों के घनत्व में बदलाव होता है तो उसका असर माध्यम से गुजरते हुए प्रकाश में देखा जा सकता। प्रकाश की प्रावस्था (फेज.) सभी जगह समान नहीं होती, बल्कि बारी-बारी से घटती बढ़ती रहती है। इस प्रकार लहरदार (कॅरोगेटेड) तरंग-अग्र (वेवफ्रंट) पैदा होती है। इसी विचार का गणितीय व्याख्या रमन और नाथ ने मिलकर दिया, जिसे 'रमन-नाथ' सिद्धांत के नाम से जाना जाता है।

रमन-नाथ सिद्धांत का व्यावहारिक उपयोग क्या है? प्रकाश द्वारा सूचना प्रसारण में इसका व्यापक उपयोग होता है। तंतु-प्रकाशिकी संचार (फाइबर ऑप्टिक कम्यूनिकेशन) के क्षेत्र में इसका योगदान अतुलनीय है। लेजर के आविष्कार के बाद तो इसकी उपयोगिता और भी बढ़ गई है।

रेगिस्तान में दूर स्थित कोई वस्तु जैसे पेड़ की परछाँई ऐसी दिखती है मानो उस जगह पानी है, परंतु वास्तव में ऐसा नहीं होता। इसे मरीचिका (मिराज) कहते हैं। सूर्य की किरणों से जब धरती अत्यधिक गर्म हो जाती है तो धरती के करीब वायु का तापमान बढ़ जाता है, जबकि ऊपर के वायु का तापमान अपेक्षाकृत कम रहता है। इसके फलस्वरूप जमीन से ऊ पर की ओर वायु के घनत्व में कमी होने के बजाए वृद्धि होती रहती है। ऐसी स्थिति में दूर की वस्तु से प्रकाश सीधे आने के साथ-साथ वक्र पथ द्वारा भी आता है, जिससे देखनेवाले को परावर्तन की भ्रांति होती है। इस विवरण में कुछ गलतियाँ हैं, क्योंकि किरणें नीचे की ओर आकर ऊपर नहीं उठ सकती। रमन ने इसका स्पष्टीकरण 'तरंग सिद्धांत' के आधार पर दिया एवं इसे साबित करने के लिए उन्होंने कुछ प्रयोग भी किए।

अकादमी की स्थापना

वर्ष 1934 व 1935 भारतीय इतिहास के लिए अत्यंत महत्त्वपूर्ण माना जा सकता है, क्योंकि इन दिनों रमन ने भारतीय विज्ञान अकादमी की स्थापना की तथा विज्ञान पत्रिका 'प्रोसिडिंग्स ऑफ दि इंडियन अएकेडेमी ऑफ साइंस का प्रकाशन शुरू किया। यद्यपि विज्ञान संबंधी विषयों पर चर्चा करने के लिए इससे पहले भारतीय विज्ञान कांग्रेस की स्थापना सन् 1914 में हुई थी, परंतु इसका अधिवेशन वर्ष में केवल एक बार ही होता है। अत: यह सोचा गया कि देश में एक राष्ट्रीय विज्ञान अकादमी का निर्माण हो, जहाँ विभिन्न वैज्ञानिक संस्थानों में हो रहे शोधकार्यों की समीक्षा व चर्चा करने के लिए नियमित रूप से बैठक बुलाई जा सके। इसके लिए एक समिति गठित की गई, परंतु सदस्यों के एकमत न होने के कारण दो अकादमी बनाने का निर्णय लिया गया। बैंगलौर में रमन की अध्यक्षता में भारतीय विज्ञान अकादमी (इंडियन अकेडेमी ऑफ साइंसेस) की स्थापना 27 अप्रैल, 1934 में हुई, जबकि कलकत्ता में 'राष्ट्रीय विज्ञान संस्था' (नेशनल इंस्टि्टयूट ऑफ साइंसेस) की स्थापना 3 जनवरी, 1935 में हुई। सन् 1946 में 'राष्ट्रीय विज्ञान संस्था' का कार्यालय दिल्ली में स्थानांतरित हो गया एवं सन् 1970 से इसे 'भारतीय राष्ट्रीय विज्ञान अकादमी' (इंडियन नेशनल साइंस एकेडेमी) कहा जाने लगा।

चित्र : 4.5 अध्ययन में मग्न रमन

भारतीय विज्ञान अकादमी का मुख्य उद्‌देश्य शोधकार्य के नतीजों की चर्चा करने

के लिए बैठकों का आयोजन करना, खास विषयों पर संगोष्ठी करवाना एवं शोधपत्रों का प्रकाशन करना था। रमन जब तक जीवित रहे, अकादमी के अध्यक्ष पद पर चुने जाते रहे। अकादमी की बैठकों के आयोजन अकसर कॉलेज या विश्वविद्यालयों में किए जाते थे, जिससे विद्यार्थियों को विज्ञान के क्षेत्र में हो रही गतिविधियों की जानकारी मिल सके। इससे काफी युवा विज्ञान के प्रति आसक्त हुए। अकादमी से 'विज्ञान' पत्रिका का प्रकाशन नियत समय पर होता था। रमन का मानना था कि ऊँचे दर्जे की पत्रिका का नियमित प्रकाशित होना अत्यंत आवश्यक है। समय के साथ पत्रिका में प्रकाशन के लिए काफी शोधपत्र आने लगे। अतः सन् 1977 से पत्रिका का प्रकाशन विषयवस्तु के आधार पर छह भागों में बाँट दिया गया—रसायनिक विज्ञान, भूमंडलीय विज्ञान, गणितीय विज्ञान, जैविक विज्ञान, वनस्पति विज्ञान तथा अभियंत्रिकी विज्ञान। भौतिक विज्ञान संबंधी शोधपत्रों का प्रकाशन 'प्रमाण' नामक पत्रिका में छपता है। अब तो अकादमी का काम काफी बढ़ गया है। गौरवमय रमन प्रोफेसर पद की स्थापना की गई है, जिसे प्रतिष्ठित वैज्ञानिकों को सेवानिवृत्त होने के बाद शोधकार्य जारी रखने के उद्देश्य से प्रदान की जाती है। कई उत्कृष्ट युवा वैज्ञानिकों को कुछ वर्षों के लिए अकादमी का सदस्य मनोनित किया जाता है। भारतीय विज्ञान अकादमी का सदस्य मनोनीत होना आज भारतीय वैज्ञानिकों के लिए गर्व की बात है।

साठ वर्ष की उम्र के बाद नवंबर 1948 में रमन 'भारतीय विज्ञान संस्थान' से सेवानिवृत्त हुए, परंतु विज्ञान से नहीं। उनका शोधकार्य जीवन के अंतिम क्षण तक जारी रहा।

□

5

रमन स्पेक्ट्रमिकी एवं इसकी उपयोगिता

जब प्रकाश की किरणें किसी पारदर्शी माध्यम, ठोस, द्रव अथवा गैस से होकर गुजरती है तब उसका कुछ अंश प्रकिर्णित होकर चारों ओर बिखर जाता है। इस प्रक्रिया को 'टींडल प्रभाव' के नाम से जाना जाता है। यदि आपाती किरण एकवर्णी है तो रैले सिद्धांत के अनुसार प्रकीर्णित किरण भी उसी रंग का होता है। परंतु रमन ने यह साबित किया कि एकवर्णी किरण से होने वाले प्रकीर्णित किरण में आपाती किरण (रैले प्रकीर्णन) के साथ कुछ अन्य रंग की किरणें भी होती हैं (रमन प्रकीर्णन)। रमन प्रकीर्णन की मात्रा रैले प्रकीर्णन का केवल एक करोड़वाँ भाग (10–7) ही होता है। इस क्षीण प्रकाश की प्रायोगिक पहचान पहली बार रमन ने किया, जिसके लिए उन्हें नोबेल पुरस्कार दिया गया।

दरअसल प्रकाश किरण या फोटॉन जब किसी पदार्थ के अणु या परमाणु से टकराता है तो उनके बीच निम्न प्रक्रियाएँ होती हैं जिसे चित्र 5.1 में ऊर्जा अवस्था के आधार पर दिखाया गया है—

(i) अवशोषण/प्रतिदीप्ति प्रक्रिया—आपाती फोटॉन की ऊर्जा इतनी हो कि वह अणु को इलेक्ट्रॉनिक न्यूनतम अवस्था की स्थिति से

उठाकर उत्तेजित अवस्था में ले जा सके। कभी-कभी इस उत्तेजित अवस्था से वापस न्यूनतम अवस्था में आते समय फोटॉन का उत्सर्जन होता है, जिसकी ऊर्जा आपाती फोटॉन की ऊर्जा से कम होती है। इसे 'प्रतिदीप्ति प्रक्रिया' कहते हैं। अवशोषण प्रक्रिया में फोटॉन नहीं निकलता, बल्कि वह ताप में परिवर्तित हो जाता है।

(ii) प्रकीर्णन प्रक्रिया—यदि आपाती फोटॉन की ऊर्जा इतनी अधिक नहीं है कि वह अणु को न्यूनतम अवस्था से उठाकर उत्तेजित अवस्था तक ले जा सके तो फोटॉन का प्रकीर्णन होता है। इसमें फोटॉन आरंभिक कंपन अवस्था से कल्पित अवस्था में स्थानांतरित होकर उसी कंपन अवस्था में वापस आ जाता है। इसमें प्रकिर्णित फोटॉन की ऊर्जा आपाती फोटॉन की ऊर्जा के समान रहती है। इसे 'रैले प्रकीर्णन' कहते हैं।

'रमन प्रकीर्णन'—इसमें आपाती फोटॉन आरंभिक कंपन अवस्था से कल्पित अवस्था में स्थानांतरित होकर दूसरे कंपन अवस्था में वापस

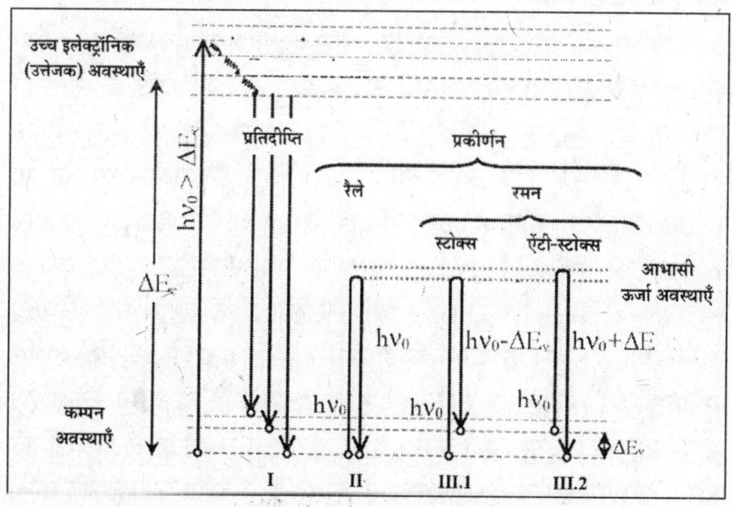

चित्र 5.1 : फोटॉन का पदार्थ के साथ पारस्परिक क्रिया से होनेवाली मुख्य प्रक्रियाएँ

आता है। इस दूसरे कंपन अवस्था की ऊर्जा आरंभिक कंपन अवस्था से अधिक या कम हो सकती है। यदि प्रकिर्णित फोटॉन की ऊर्जा आपाती फोटॉन की ऊर्जा से कम होती है तो इसे स्टोक्स कहते हैं (iii-1) जबकि प्रकिर्णित फोटॉन की ऊर्जा आपाती फोटॉन की ऊर्जा से अधिक होने पर उसे ऐंटीस्टोक्स कहते हैं (iii-2)। इस प्रकीर्णन प्रक्रिया में फोटॉन की ऊर्जा में कमी या वृद्धि अणु के आरंभिक व अंतिम कंपन अवस्था की ऊर्जा का अंतर होता है, जो उस अणु की विशेषता होती है।

प्रकीर्णन प्रक्रिया में ज्यादातर प्रत्यास्थ प्रकीर्णन (एलास्टिक स्कैट्रिंग) होता है, जिसमें फोटॉन की ऊर्जा में कोई कमी नहीं आती। परंतु कुछ फोटॉनों का अप्रत्यास्थ प्रकीर्णन (इनएलास्टिक स्कैट्रिंग) होता है, जिससे फोटॉन की ऊर्जा (तरंगदैर्ध्य) में बदलाव आ जाता है, यानी उसकी ऊर्जा में कमी व वृद्धि होती है।

रमन स्पेक्ट्रमिकी

किसी भी वस्तु की कंपनिक अवस्था, घूर्णन अवस्था एवं अन्य कम आवृत्ति वाले दोलन अवस्था के अध्ययन के लिए रमन स्पेक्ट्रमिकी तकनीक का इस्तेमाल करते हैं। इसके लिए प्रकाश स्रोत की आवश्यकता पड़ती है, जिसके क्षेत्र का फैलाव दृश्य प्रकाश, करीबी-अवरक्त या करीबी-पराबैंगनी होता है। अणु से अप्रत्यास्थ प्रकीर्णन करने पर फोटॉन के तरंगदैर्ध्य में हुए परिवर्तन या विस्थापन उस अणु की रसायनिकी व संरचनात्मक जानकारी देती है। विभिन्न पदार्थों के लिए यह परिवर्तन भिन्न होती है। अणु के कंपनिक अवस्था (वाईब्रेशनल स्टेट) के आधार पर प्रकीर्णित फोटॉन के तरंगदैर्ध्य में वृद्धि होती है या कमी आती है। प्रकीर्णित फोटॉन की ऊर्जा कम (यानी तरंगदैर्ध्य में वृद्धि, लाल वर्ण की ओर) होने पर इसे 'स्टोक्स प्रकीर्णन' कहते हैं, जबकि फोटॉन की ऊर्जा बढ़ने पर (यानी तरंगदैर्ध्य में कमी, नीले वर्ण की ओर) इसे 'एंटिस्टोक्स' प्रकीर्णन कहते हैं जैसा कि पहले बताया गया है। अतः स्टोक्स एवं

'एंटिस्टोक्स' में हुए रमन विस्थापन की माप अणु की कंपन-ऊर्जा की जानकारी देता है।

'रमन प्रभाव' के आविष्कार के तुरंत बाद विश्व के अनेक देशों में इस विषय पर गहराई से शोध कार्य होने लगा। पदार्थों की संरचना के बारे में जानकारी पाने के लिए यह अत्यंत उपयोगी विधि थी, जिससे विभिन्न पदार्थों की पहचान की जा सकती थी। इसमें केवल दो चीजों की जरूरत होती है—एकवर्णी प्रकाश स्रोत तथा स्पेक्ट्रममापी। जाँच की जा रही वस्तु को एकवर्णी प्रकाश व स्पेक्ट्रममापी के बीच रखा जाता है। वस्तु से प्रकीर्णित हुए किरणों में ज्यादातर रैले किरणें होती हैं, अत: 'रमन किरणों' को देखने के लिए इसे अलग करना जरूरी होता है। पहले रैले किरणों को अलग करने के लिए होलोग्राम विवर्तन ग्रेटिंग का इस्तेमाल किया जाता था, जबकि अब खाँच (नॉच) फिल्टर का उपयोग होता है। 'रमण किरणों' को स्पेक्ट्रममापी से देखा जाता है, जो विभिन्न वस्तुओं के लिए अलग अलग होती है। इस पूरे औजार को 'रमन स्पेक्ट्रमिकी' कहते हैं।

वर्णक्रम को दो अलग विधियों से एकत्रित किया जाता है, एक ग्रेटिंग विधि से तथा दूसरा फूरिये रूपांतरण द्वारा। ग्रेटिंग विधि में एकवर्णित्र (मोनोक्रोमेटर) का इस्तेमाल करते हैं, जो प्रकाश स्रोत के केवल एक चुने हुए संकीर्ण बैंड को ही गुजरने देता है, जिसे संसूचक, जैसे प्रकाश-इलेक्ट्रॉन संवर्धक (फोटोमल्टीप्लायर ट्यूब) में देखा जाता है। एकवर्णित्र में दरअसल दो रेखाछिद्र (स्लिट) होते हैं, एक आगमन के लिए तथा दूसरा निर्गमन के लिए। इनके छिद्र की चौड़ाई घटाई या बढ़ाई जा सकती है। किस आवृत्ति की किरण निर्गमन रेखाछिद्र से गुजर कर संसूचक में रिकॉर्ड होगा, यह रेखाछिद्रों के बीच के कोण पर निर्भर करता है। अत: विभिन्न कोणों पर रेखाछिद्र को रखकर अलग-अलग आवृत्ति की किरणों की तीव्रता मापी जाती है। दूगुनी एवं तीगुनी एकवर्णित्र

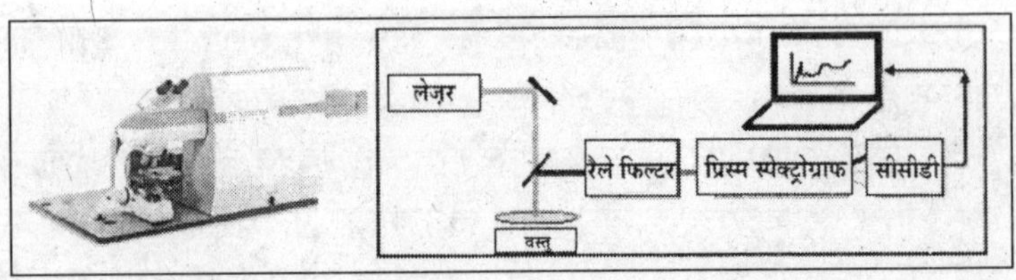

के इस्तेमाल से अवांछित रोशनी को संसूचक में प्रवेश करने से रोका जा सकता है।

फूरिए रूपांतरण में माइकलसन व्यतिकरणमापी (इंटरफेरोमीटर) मुख्य उपकरण होता है। इसमें दो दर्पण एक-दूसरे के लंबवत् दिशा में रखे होते हैं जिनमें एक स्थिर तथा दूसरा गतिशील होता है। इसके अलावा एक किरणपुंज विपाटक (स्प्लिटर) तथा एक प्रतिकारी (कंपेन्सेटिंग) प्लेट होता है। जब प्रकाश की किरणें विपाटक से टकराती हैं तो यह दो हिस्सों में बँटकर दर्पणों से परावर्तित होकर पुनः विपाटक पर आ जाती है, जहाँ इन दोनों किरणों का व्यतिकरण होता है। एक

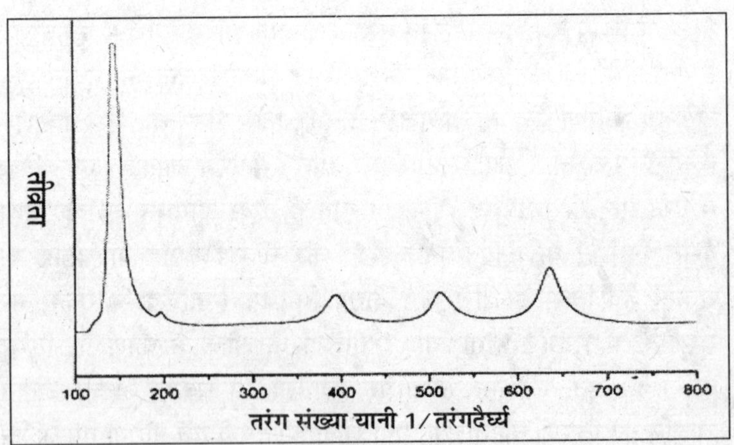

किरण विपाटक से गुजरकर दर्पण से टकराती है तो दूसरी किरण प्रतिकारी प्लेट से गुजरकर लंबवत दिशा में रखे दर्पण से टकराती है। दोनों प्लेटें एक ही वस्तु से बनी होती है, ताकि आपाती किरण से विभक्त हुए दोनों किरणों में इनकी वजह से एक जैसा प्रभाव हो। इन दोनों किरणों द्वारा तय की गई दूरी के अंतर के अनुसार रचनात्मक या विनाशात्मक व्यतिकरण होता है। इस अंतर का नियंत्रण गतिशील दर्पण की स्थिति बदल कर किया जाता है। सारी प्रक्रिया को दूर रखे संसूचक से देखा जा सकता है जिसमें प्रकाश किरणों की तीव्रता दर्पणों की विभिन्न दूरियों के अनुसार बदलती रहती है। इस प्रकार पैदा हुए पैटर्ण को व्यतिकरण चित्र कहते हैं। यदि प्रकाश स्रोत एकवर्णी है तो 'व्यतिकरण चित्र' कहते हैं। यदि प्रकाश स्रोत एकवर्णी है तो व्यतिकरण चित्र का पैटर्ण साइन तरंग जैसा होगा। इस पैटर्ण का फूरिये रूपांतरण करने पर प्रकाश स्रोत के तरंगदैर्ध्य (या आवृत्ति) के मान का पता चलता है।

तकनीकी विकास के साथ, विशेषतय सन् 1960 में लेजर के आविष्कार के बाद फूरिए रूपांतरण रमन स्पेक्ट्रमिकी का उपयोग पदार्थों के विश्लेषणात्मक अध्ययन के क्षेत्र में काफी फैल गया। लेजर वस्तुतः उच्च तीव्रता का एकवर्णी प्रकाश स्रोत है, जो अत्यंत संबद्ध (कोहरेंट) होता है। उन्नत व्यतिकरणमापी के बनने से कई नई तकनीकों का विकास हुआ। अब संसूचक के तौर पर आवेश-युग्मित कैमरा (सीसीडी) के इस्तेमाल से इस तकनीक की दक्षता काफी बढ़ गई है, जिससे काफी कम तीव्रता वाले 'रमन प्रकीर्णन' को भी देखा जा सकता है।

वर्तमान रमन स्पेक्ट्रममापी में मुख्यतः चार उपकरण होते हैं, प्रकाशीय सूक्ष्मदर्शी (ऑप्टिकल माइक्रोस्कोप), उत्तेजन लेजर, एकवर्णित्र तथा एक दक्ष संसूचक जैसे सीसीडी। रमन स्पेक्ट्रम को ठीक से देखने के लिए सही तरंगदैर्ध्य वाले लेजर का चयन करना अति आवश्यक है।

आधुनिक रमन उपकरण में विभिन्न तरंगदैर्ध्य वाले लेजरों से प्रयोग करके उपयुक्त तरंगदैर्ध्य वाले लेजर का चयन किया जाता है, जिससे

रमन किरणों को स्पष्ट देखा जा सके। उदाहरण के लिए ऐसी कई वस्तुएँ होती हैं, जिनसे प्रतिदीप्ति विकिरण निकलती हैं। इन वस्तुओं पर हरे रंग की किरणें (तरंगदैर्ध्य 532 नैनोमीटर) पड़ने पर प्रतिदीप्ति विकिरण की मात्रा इतनी ज्यादा निकलती है कि रमन किरणें दिखाई नहीं देती। ऐसी स्थिति में कम ऊर्जावाले जैसे लाल (तरंगदैर्ध्य 633 नैनोमीटर) या अवरक्त (तरंगदैर्ध्य 785 नैनोमीटर) किरणों का इस्तेमाल करना उचित होगा, जिससे प्रतिदिप्ति विकिरण पैदा ही न हों और 'रमन विकिरण' अच्छी तरह देखी जा सके। यहाँ इस बात का जिक्र करना उचित होगा कि प्रतिदीप्ति प्रक्रिया का अणुप्रस्थ परिच्छेद (क्रॉस-सेक्शन) का मान 10^{-16}/सेमी^{2} प्रति अणु है, जबकि रमन प्रकीर्णन के लिए यह मान सिर्फ $10^{-31}-10^{-26}$ /सेमी^{2} है।

रमन विकिरण की तीव्रता व रमन विस्थापन (तरंग संख्या यानी 1/तरंगदैर्ध्य) के आलेख (प्लॉट) को 'रमन स्पेक्ट्रम' कहते हैं, जैसाकि चित्र 5.3 में दिखाया गया है। 'रमन स्पेक्ट्रमिकी' से प्राप्त पदार्थ के वर्णक्रम में कई शिखर (पीक) दिखाई देती है, जो उस पदार्थ में उपस्थित अलग अलग घटकों को दर्शाती है। विभिन्न शिखरों की तीव्रताओं का अनुपात घटकों की मात्रा बतलाती है। शिखर चौड़ाई से उस पदार्थ की उत्कृष्टता के बारे में पता चलता है।

रमन स्पेक्ट्रमिकी के कई प्रगत किस्मों का विकास हुआ है, जैसे अनुनाद (रेजोनेंस) रमन, सतह संवृद्धि (सर्फेस एनाहंस्ड, एसईआरएस) रमन, ध्रुवित (पोलराईज्ड) रमन, उद्दीपित (स्टीमुलेटेड) रमन तथा पारगमन (ट्रांसमिशन) रमन। इसमें दक्षता 1014-1015 गुणा बढ़ जाती है जिससे अब केवल एक अणु से प्रकीर्णित रमण किरण को भी देखा जा सकता है। 'रमन प्रभाव' के आविष्कार के अस्सी वर्ष बाद आज भी तकनीकी विकास के साथ नए क्षेत्रों में इसके उपयोग की खोज जारी है।

रमन स्पेक्ट्रोमापी का उपयोग

रमन स्पेक्ट्रमिकी तकनीक का क्षेत्र अब काफी विस्तृत हो गया है। यद्यपि मूलतः विधि वही है, जो रमन ने अपने प्रयोग में इस्तेमाल किया था, परंतु आवश्यकतानुसार इसके कई परिवर्तित रूप उभरकर आए हैं। समय के साथ इसके नए उपयोगों की सूची में भी वृद्धि हो रही है। 'रमन स्पेक्ट्रमिकी' का उपयोग न केवल शैक्षणिक क्षेत्र में सीमित है बल्कि उद्योग जगत् में भी यह काफी उपयोगी है। 'रमन स्पेक्ट्रमिकी' का उपयोग अब विज्ञान के हर क्षेत्र में हो रहा है। इससे खनिज पदार्थों की विशेषताएँ, कार्बनिक पदार्थों की उपस्थिति की जानकारी तथा जैविक तत्त्वों, जैसे प्रोटीन, डीएनए, अमीनो एसिड व वनस्पति वर्णक की पहचान आसानी से की जा सकती है। चिकित्सा विज्ञान के क्षेत्र में इस विधि से जीन के विश्लेषण से लेकर रोगाणुओं की पहचान की जाती है। अंतरिक्ष में उपस्थित पदार्थों का पता लगाने के लिए 'रमन स्पेक्ट्रोस्कोप' विशेष तौर पर उपयोगी है। पॉलिमरों के अध्ययन में इसका काफी उपयोग हो रहा है। इसके अलावा पर्यावरण (भौमजल में अन्य तत्त्वों की उपस्थिति), रंग उद्योग (बहुलकीकरण प्रक्रिया पर निगरानी), विधि रसायन (अवैध नशीली पदार्थों की पहचान), पेट्रोलियम (खनिज तेल में संघटकों की पहचान) आदि अनेक क्षेत्र हैं, जिनमें 'रमन स्पेक्ट्रमिकी' का इस्तेमाल हो रहा है।

रमन की खोज आज एक महत्वपूर्ण तकनीक बन चुकी है, जिसका उपयोग कई क्षेत्रों में व्यावहारिक तौर पर हो रहा है। कम वजनवाले इतने छोटे परीक्षण यंत्र (स्कैनर) बनाए गए हैं, जिन्हें हाथ से पकड़कर आसानी से इस्तेमाल किया जाता है। नशीली दवाएँ, ड्रग, विस्फोटक पदार्थ, खतरनाक रसायन व गैस आदि की शनाक्त करने के लिए हवाई अड्डों में इनका उपयोग किया जा रहा है। फॉरेंसिक कामों के लिए पुलिस विभाग इनका प्रयोग करते हैं।

इन परीक्षण यंत्रों से परीक्षित वस्तुओं की आण्विक संरचना का पता लगाया जाता है। हरेक अणु का रमन स्पेक्ट्रम (जो उसकी आणविक संरचना का प्रतिबिंब होता है) विशिष्ट होता है, यानी 'रमन वर्णक्रम' को उस पदार्थ का 'फिंगरप्रिंट' कहा जा सकता है। जिस प्रकार लोगों की पहचान उनके फिंगरप्रिंट से की जाती है (सबके फिंगरप्रिंट अलग होते हैं), उसी प्रकार किसी भी वस्तु के प्रकीर्णित विकिरण से उसकी पहचान की जाती है।

आमतौर पर किसी पदार्थ की रसायनिक संरचना के बारे में जानकारी पाने के लिए उसकी रासायनिक व भौतिकी परीक्षण करना पड़ता है, जिसे करने में काफी समय लग जाता है। यही नहीं, इसके लिए उस पदार्थ में से काटकर छोटा सैंपल लेना पड़ता है, जो परीक्षण के दौरान नष्ट हो जाता है, यानी यह 'भंजक विधि' है। परंतु यही जानकारी रमन विधि (स्कैन) द्वारा केवल चंद सेकेंडों में प्राप्त की जा सकती है, जिसमें न तो पदार्थ को काटने की जरूरत पड़ती है और न ही परीक्षण के दौरान यह नष्ट होता है।

अंतरिक्ष, दूरसंचार जैसे विज्ञान के कई नए क्षेत्रों में 'रमन प्रभाव' के भावी उपयोगों की खोज की जा रही है। इसके लिए क्षीण रमन प्रक्रिया को देखने के लिए नए तरीकों का प्रयोग किया जा रहा है। अब तो लेश मात्रा में उपस्थित वस्तुओं (सौ करोड़ में एक) की पहचान 'रमन स्कैनर' से की जा सकती है। ऐसे स्कैनरों से जीवाणु (बैक्टीरिया), प्रदूषक (पलूटंट) तथा कम मात्रा में छुपाए गए विसफोटक पदार्थों की शिनाख्त की जाती है।

स्टैनफोर्ड विश्वविद्यालय में 'रमन स्कैनर' का प्रयोग कैंसर के निदान के लिए किया जा रहा है। शायद वर्तमान में चिकित्सा के लिए किए जानेवाले 'रक्त परीक्षण' की जगह भविष्य में 'रमन स्कैनर' द्वारा ही रक्त में ग्लूकोज, कोलेस्टेरल, यूरिक एसिड तथा अन्य वस्तुओं की मात्रा का पता लगाया जा सकेगा। एलर्जी से पीड़ित व्यक्ति वायु में

उपिस्थित पराग (पोलेन) की जानकारी तुरंत पा सकेंगे जिससे वे उचित सावधानी बरत सकेंगे।

औषध विज्ञान—'रमन स्पेक्ट्रमिकी' से औषधि बनाने की विधि के विभिन्न स्तरों पर विश्लेषण करके उस पर निगरानी एवं नियंत्रण रखा जाता है।

कार्बन तथा हीरा उद्योग—कार्बन उद्योग में रमन स्पेक्ट्रमिकी एक आवश्यक औजार है, जिससे हीरे जैसी विलेपन (कोटिंग) की गुणता नियंत्रण, कार्बन नैनोट्यूब के अभिलक्षण आदि की जाँच की जाती है।

पदार्थ विज्ञान—विभिन्न किस्म के पदार्थ जैसे संघटक (कम्पोजिट), पौलिमर्स, उत्प्रेरक (कैटालिस्ट) आदि के विश्लेषण में 'रमन स्पेक्ट्रमिकी' का उपयोग होता है।

'रमन प्रभाव' के व्यापक उपयोग

'रमन पद्धति' से किसी भी वस्तु की गुणात्मक या मात्रात्मक विश्लेषण बड़े ही आसानी से एवं कम समय में की जा सकती है। उसके रासायनिक संघटन व संरचना की विस्तृत जानकारी पाने के लिए यह विधि अत्यंत कार्यकारी है। वस्तु के स्थानगत या प्रसारित क्षेत्र की जानकारी सूक्ष्मदर्शी की सहायता से माइक्रो रमन द्वारा की जाती है। आजकल तो अति सूक्ष्म वस्तुओं को देखने के लिए 'नैनो रमन' का भी इस्तेमाल हो रहा है। 'रमन स्पेक्ट्रमिकी' अविनाशी परिक्षण है, जिससे पदार्थ नष्ट नहीं होता। परीक्षण के लिए वस्तु की बहुत ही कम मात्रा (मिलीग्राम या उससे भी कम) की आवश्यकता होती है।

रमन स्पेक्ट्रमिकी अर्ध-चुंबकीय पदार्थों के प्रचक्रण तरंग (मैगनॉन या स्पिन वेव) की भी पहचान कर सकता है। 'रमन स्पेक्ट्रमिकी' से पदार्थ की विभिन्न गुणवत्ताओं का पता चलता है, जैसे—

उसकी रासायनिक संरचना व संघटकों की पहचान,

आण्विक संरचना अभिलक्षण,

प्रावस्था संक्रमण (फेज.ट्रांजिशन) का मॉनीटरन,
माइक्रॉन एवं उससे भी छोटे आकार के घटकों की पहचान,
अभिक्रियाओं की निगरानी आदि।

आमतौर पर मापन सामान्य तापमान पर ही किया जाता है, परंतु जरूरत पड़ने पर पदार्थ को गरम या ठंडा करके भी मापन किया जा सकता है। भूविज्ञान के क्षेत्र में अध्ययन करने के लिए पदार्थ को उच्च दाब पर रखकर भी प्रयोग किए जाते है।

रमन स्पेक्ट्रमिकी का उपयोग शोध एवं उद्योग जगत् के विभिन्न क्षेत्रों में हो रहा है, जैसे—

औषध उद्योग,
शैल रसायन, औद्योगीक रसायन, पॉलीमर,
पदार्थ विज्ञान,
अर्धचालक, प्रकाश वोल्टीय पदार्थ,
जीव विज्ञान,
भूविज्ञान व खनिज विज्ञान, मणि विज्ञान,
नाभिकीय विज्ञान
विधि रसायन विज्ञान (फॉरेन्सिकस)
चित्रकला, पुरातत्त्व, रंग, रंगद्रव्य (पिग्मेंट) आदि।

□

6

रमन अनुसंधान संस्थान में आगमन

सन् 1948 में 'भारतीय विज्ञान संस्थान' से सेवानिवृत्त होने के बाद रमन अपने जीवन के एक नए पड़ाव में प्रवेश किया, जब उन्होंने बैंगलौर में स्थित 'रमन अनुसंधान संस्थान' के निदेशक का पद सँभाला। इस संस्थान का निर्माण कार्य उन्होंने भारतीय वैज्ञानिक संस्थान में अपने कार्यकाल के दौरान ही शुरू कर दिया था, जिससे बिना वक्त गँवाए सेवानिवृत्त होने के बाद वे अपना शोध जारी रख सकें। उनका निवास स्थान संस्थान के परिसर के अंदर ही था। उन्होंने परिसर को सुंदर बगीचों व नीलगिरी (यूकेलिप्टस) के पेड़ों से सजाया था। वे कहा करते थे—'हिंदू धर्म के अनुसार वृद्धावस्था में मनुष्य को वन में जाकर रहना चाहिए' (वनप्रस्थ) परंतु मैंने वन को ही अपने पास बुला लिया है। यहाँ आकर उन्होंने राहत की साँस ली, क्योंकि अब वे सारा वक्त अपने मनपसंद कार्य में लगा सकते थे। उनके शोध का क्षेत्र काफी व्यापक था, जैसे भौतिक विज्ञान, भूविज्ञान, जीव विज्ञान, शरीर विज्ञान आदि। इन सबमें, विशेषतया ध्वनि तथा रंगों से उनका ज्यादा लगाव था। यद्यपि उनके छात्रों की संख्या में भारी कमी आई थी, परंतु वे अपने काम से संतुष्ट थे। भारत सरकार ने उन्हें 'राष्ट्रीय प्रोफेसर' के खिताब से नवाजा था, जिससे उन्हें जो तनख्वाह मिलती थी, उससे

चित्र : 6.1 रमन अनुसंधान संस्थान

उनका गुजारा तो हो जाता था, परंतु संस्थान को चलाने में आवश्यक रकम की कमी हमेशा रहती थी। इसे पूरा करने के लिए उन्होंने जो रासायनिक उद्योग की स्थापना की थी, उससे प्राप्त धनराशि को भी संस्थान के दैनंदिन कामों में खर्च कर देते थे। इससे समझा जा सकता है विज्ञान से उनका लगाव व प्यार! चित्र–6.1 में 'रमन अनुसंधान संस्थान' की इमारत दरशाई गई है।

रमन का यह अंतिम काल तीन भिन्न चरणों में देखा जा सकता है। आरंभ में संस्थान में प्रवेश करने के बाद वे उसे बनाने व संगठित करने में लगे रहे, क्योंकि उस समय संस्थान पूरी तरह तैयार नहीं हुआ था। तब वे काफी उत्साहित एवं प्रफुल्ल रहते थे। उन्होंने संस्थान के निर्माण में अपने खुद का धन भी लगाया था। उसके बाद परिस्थिति ने ऐसा मोड़ लिया कि वे लंबे अरसे तक निराशा के अंधकार में डूब गए एवं एकांतवासी हो गए। यहाँ तक कि संस्थान के फाटक के सामने

उन्होंने एक सूचनापट्ट लगवा दिया था, जिसमें लिखा था—'यहाँ आगंतुकों का प्रवेश वर्जित है' आखिरकार अंत में रमन बीती बातों को भूलाकर पुनः प्रसन्नचित दिखने लगे एवं लोगों से मिलने लगे। यद्यपि उनकी मनोदशा में कई उतार-चढ़ाव आए, परंतु अनुसंधान से उन्होंने कभी नाता नहीं तोड़ा। अपनी ढलती उम्र के दिन उन्होंने प्रकृति की सुंदरता की जाँच-पड़ताल करने में व्यतीत किए। ऐसी क्या वजह थी जिससे इतने बड़े जाने माने वैज्ञानिक को यह सब सहना पड़ा?

एकांतवास

सन् 1947 में आजादी के बाद नेहरूजी के कुशल व विलक्षण नेतृत्व में देश प्रगति के रास्ते पर चल पड़ा। सदीयों के पिछड़ेपन से छुटकारा पाने के लिए विज्ञान पर ही सबका भरोसा था। 'विज्ञान से प्रगति' एक मूल मंत्र बन गया। नेहरूजी का दृढ़ विश्वास था कि विज्ञान व प्रौद्योगिकी के माध्यम से ही आधुनिक समाज का निर्माण संभव है। यही धारणा रमन की भी थी, जिन्होंने कहा था—'देश की अर्थव्यवस्था को सुधारने का केवल एक ही उपाय है, और वह है विज्ञान, सिर्फ विज्ञान'। शांतिस्वरूप भटनागर की अध्यक्षता में वैज्ञानिक व प्रौद्योगिकी परिषद् (सी.एस.आइ.आर.) की स्थापना हुई तथा भाभा ने देश में परमाणु ऊर्जा कार्यक्रम की शुरुआत की। इन दोनों को नेहरू का पूर्ण समर्थन प्राप्त था। विज्ञान से संबंधित गतिविधियों में द्रुत प्रगति लाने के लिए उपकरणों का आयात किया जाने लगा जो शायद रमन को नहीं भाया। वे शुरू से ही आत्मनिर्भरता पर विश्वास करते थे। उनका मत था कि बढ़िया विज्ञान की खोज दिमाग से की जाती है, न कि कीमती उपकरणों से। इसके पक्ष में वे अकसर कलकत्ता में किए गए अपने नोबेल पुरस्कार वाले प्रयोग का जिक्र किया करते थे, जो एक अति साधारण उपकरण से किया गया था एवं जिसे बनाने में 300 रुपए से भी कम खर्च हुआ था। जिस तरह देश की प्रगति की

दुहाई देकर विज्ञान का इस्तेमाल हो रहा था, उसमें न तो वैज्ञानिक उत्कृष्टता और न ही इसका उद्‌देश्य पनप सकता था। चापलूसी का बोलबाला था, जिससे उच्च पद पर ऐसे लोग नियुक्त होने लगे, जिसके लिए वे योग्य नहीं थे। अपने चारों ओर यह सब होते देख रमन व्यथित थे। वे इन सबके विरोध में अपने विचार खुलकर व्यक्त करने लगे। ऐसा नहीं था कि केवल रमन ही सरकार की इस नीति की आलोचना कर रहे थे, परंतु वे कूटनीतिज्ञ नहीं थे। उनका मानना था कि कोई संवेदनशील व भावुक व्यक्ति यदि अपनी अनुभूति को मुक्त कंठ से व्यक्त नहीं करता तो वह भी विवश एवं लाचार हो जाता है। यदि वह अपनी नापसंदवाली बात को अपने पसंद की बात बतलाने के लिए अपने आपको मजबूर करता है तथा जिसमें उसका विश्वास नहीं है, उसमें विश्वास व्यक्त करता है तो ऐसा करने के लिए उसे एक बड़ी रकम चुकानी पड़ती है, जो उसकी स्वाभाविक व सृजनात्मक क्षमता को खत्म कर देती है। यानी उनका मानना था कि अंतरात्मा की आवाज ही सर्वोपरि है।

इसके अलावा कुछ लोग उनकी वैज्ञानिक क्षमता पर भी अंकुश लगा रहे थे। ऐसे दूषित वातावरण से रमन अपने आपको दूर रखने लगे। उनके कई छात्र शोध कार्य छोड़कर या तो सरकारी नौकरी करने लगे अथवा विदेश चले गए। रमन को भारत सरकार ने देश का उपराष्ट्रपति बनने का आग्रह किया था, जिसे उन्होंने स्वीकार नहीं किया। वे चाहते तो सेवानिवृत्त होने के बाद कोई उच्च सरकारी पद पा सकते थे, परंतु उनकी खुद्‌दारी आड़े आई। एक बार उन्होंने दुःखी मन से कहा था—मेरा जीवन अत्यंत विफल रहा है। मैंने सोचा था कि देश में सच्चा वैज्ञानिक वातावरण बनाऊँगा। परंतु अब तो विज्ञान में उन्हीं लोगों की भीड़ है जो पश्चिमी देशों की नकल करते हैं।'

अनुसंधान कार्य

रमन के अनुसंधान का क्षेत्र समय के साथ बदलता रहा। अपने संस्थान में आकर उन्होंने प्राकृतिक वस्तुओं जैसे खनिज पदार्थ, कीमती

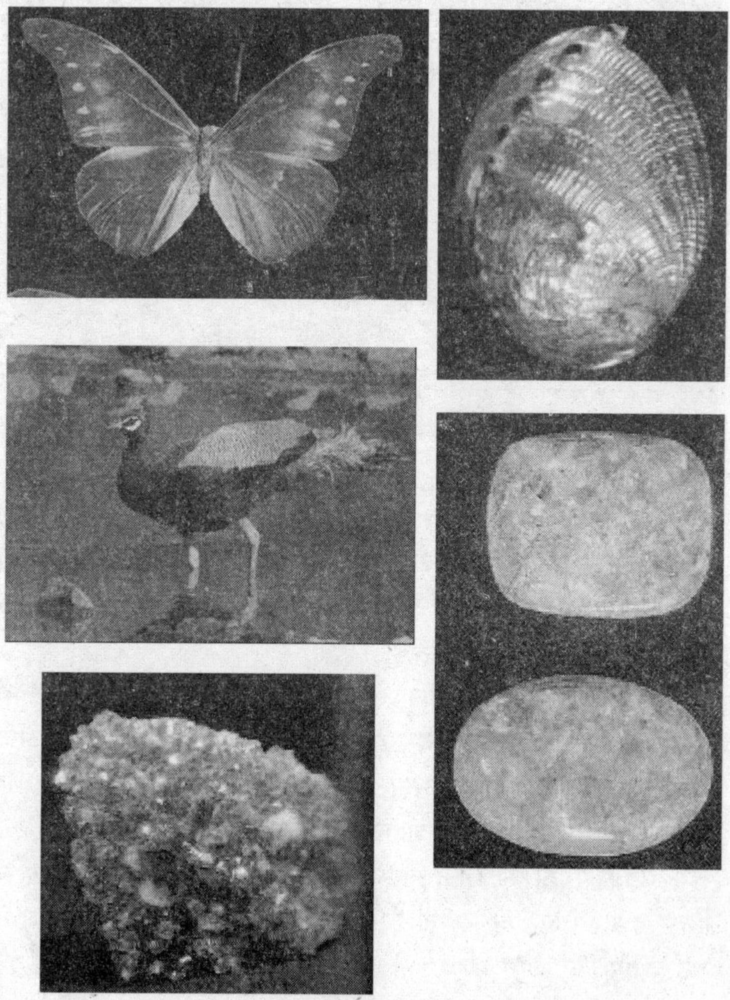

चित्र : 6.2 'रमन अनुसंधान संस्थान' के संग्रहालय में रखी कुछ वस्तुएँ।

चित्र : 6.3 रमन अनुसंधान संस्थान के परिसर में बच्चों के साथ रमन

पत्थर, पक्षियों, तितलियों व फूलों के रंगों पर अध्ययन शुरू किया। उन्होंने आँखों से रंगों का बोध कैसे होता है, इस पर गहन अध्ययन किया। उनके संग्रहालय में रखी कुछ वस्तुओं की तसवीर चित्र में दिखाई गई है—

'रमन अनुसंधान संस्थान' के बगीचों में विभिन्न प्रकार के गुलाब व अन्य फूलों का समावेश था। फूलों के रंगों का अध्ययन, उससे अर्क निकालकर उन पर शोध करना, पंखुड़ियों से प्रकाश के प्रकीर्णन पर प्रयोग करना आदि उनका शौक बन गया।

आकाशगंगा को बिना किसी बाहरी उपकरण के केवल आँखों से किस प्रकार देखा जा सकता है, इसकी व्याख्या देने के लिए रमन ने कई प्रयोग किए और मनुष्य की दृष्टि के कई विचित्र लक्षणों को सामने लाए। प्रकृति के कण-कण में व्याप्त सौंदर्य को निहारना उन्हें बहुत

अच्छा लगता था। जीवन की संध्या में बच्चों के प्रति उनका प्यार उमड़ पड़ा था। वे स्कूल के बच्चों को अपने संस्थान में आमंत्रित करते और उन्हें फूलों पर हो रहे शोधकार्य दिखाते। बच्चे भी उनके साथ घुल-मिलकर खुश रहते थे। रमन संस्थान बच्चों की किलकारियों से गूँजने लगा (चित्र-6.3)।

रमन ने सन् 1935 में भारतीय विज्ञान अकादमी की स्थापना की थी। अकादमी के वार्षिक सम्मेलन में वे अपने नए खोजों के बारे में घोषणा किया करते थे। जैसे सन् 1967 के सम्मेलन में उन्होंने पृथ्वी के घूमने से इसके चारों ओर फैली गैसों पर पड़ने वाले प्रभाव की चर्चा की थी। उसके अगले वर्ष आँखों से किसी वस्तु को देखने की

चित्र : 6.4 सन् 1954 में रमन को मिला 'भारत रत्न' सम्मान

व्याख्या दी थी। उनके सेवानिवृत होने के बाद भी देश-विदेश के कई संस्थान उन्हें विभिन्न उपाधियों एवं पुरस्कारों से सम्मानित करते रहे। भारत सरकार ने देश के सर्वोच्च नागरिक सम्मान 'भारतरत्न' से उन्हें सन् 1954 में अलंकृत किया था (चित्र—6.4)। उसी वर्ष से इन पुरस्कारों की शुरुआत हुई थी।

रमन को संगीत से बेहद लगाव था। वे कहा करते थे—मुझे अनेक वर्षों तक जीना है, क्योंकि अभी मैंने वे सब संगीत नहीं सुने हैं, जिन्हें मैं सुनना चाहता हूँ। वे अकसर वाद्ययंत्र की दुकानों में जाया करते थे एवं विभिन्न प्रकार के वाद्ययंत्रों को खरीदकर अपने संग्रहालय में रखते थे। अमेरिका के कैटगुट ध्वनिकी समिति (ऐकाउस्टिकल सोसाइटी) ने उन्हें सम्माननीय सदस्य मनोनित किया था।

रमन ने जब भौतिकी क्षेत्र में अनुसंधान कार्य की शुरुआत की थी, उस समय आधुनिक विज्ञान अपने शैशव में था। परंतु समय के साथ विकसित होकर इसकी कई शाखाएँ बन गईं। शोधकर्ता प्रयोगात्मक अध्ययन करने के लिए आधुनिक उपकरणों का इस्तेमाल करने लगे, जो साठ साल पहले उपलब्ध नहीं थे। इसके फलस्वरूप वैज्ञानिक केवल एक सीमित क्षेत्र में ही अपना शोध करके उसके विशेषज्ञ बन गए। परंतु रमन के शोध का क्षेत्र काफी व्यापक था एवं अंत तक उनकी रुचि वैसी ही बनी रही। उन्होंने कभी एक संकीर्ण दायरे में अपने को सीमित नहीं रखा। बैंगलौर में एक बार चिकित्सकों के सम्मेलन का उद्घाटन करते समय रमन ने मजाक में कहा था—'जिस प्रकार एक सामान्य चिकित्सक सभी प्रकार के रोगों का इलाज करता है' उसी प्रकार मैं विज्ञान का सामान्य चिकित्सक हूँ, कोई विशेषज्ञ नहीं। 'उन्हें हर प्रकार की वैज्ञानिक समस्याओं में दिलचस्पी थी' चाहे वह आसान हो या कठिन। उनका हल ढूँढ़ने से ही उन्हें संतुष्टि मिलती थी। सन् 1969 में उन्होंने भूकंप व पृथ्वी की आकार के बीच संबंध पर शोध

करने की बात कही थी। रमन उन व्यक्तियों में नहीं थे, जो अपनी पिछली उपलब्धियों से संतुष्ट रहे। वे हमेशा नए तथा चुनौतीपूर्ण समस्याओं का अध्ययन करना चाहते थे। उनका कहना था कि समुद्र के पानी का रंग उन्हें ज्यादा आकर्षित करता है, न कि उसमें रहनेवाली मछली को। वे विज्ञान के क्षेत्र में हो रहे तात्कालिक शोधों के बारे में पूरी जानकारी रखते थे। अपने जीवन के अंत तक वे अनेक वैज्ञानिकों से संपर्क बनाए रखा। विभिन्न वैज्ञानिक संस्थाओं से प्रकाशित होने वाले शोधग्रंथों एवं पुस्तकों को वे पढ़ा करते थे। इस प्रकार जीवन के अंत तक उनकी ज्ञान पिपासा बरकरार रही। उनका दृढ़ विश्वास था कि शोधकर्ता तभी अपने कार्य में सफल हो सकता है यदि वह उस कार्य को करने की प्रेरणा अपने आप से प्राप्त करे। केवल पैसे खर्च करने से सफलता हासिल नहीं होता।

रमन बड़े अच्छे वक्ता थे। उनके व्याख्यान में वास्तविक तथ्यों के साथ दिल्लगी का भी समावेश रहता था, जिसे श्रोता मंत्रमुद्ध होकर

चित्र : 6.5 भाषण देते हुए रमन

सुनते थे। 2 अक्तूबर सन् 1959 को गांधी जयंती के दिन रमन ने आम जनता के लिए लोकप्रिय विज्ञान वार्त्ता देना प्रारंभ किया जिसे वे हर वर्ष देते रहे। उनकी पहली वार्त्ता का विषय था—'प्रकाश, रंग और दृष्टि।' सन् 1970 में उन्होंने कान के परदे तथा सुनने की क्रिया के बारे में नए सिद्धांतों की जानकारी दी थी। यही उनका अंतिम व्याख्यान भी था।

अपने 82 वर्षगाँठ के कुछ समय पूर्व रमन ने अकादमी के सदस्यों के लिए सितंबर 1970 में एक सप्ताहव्यापी संगोष्ठी का आयोजन किया था, जिसमें उन्होंने विशेषकर युवा वैज्ञानिकों को उनके शोधकार्यों का

चित्र : 6.6 रमन वृद्धावस्था में

ब्योरा प्रस्तुत करने के लिए आमंत्रित किया था। विज्ञान के विभिन्न क्षेत्रों में हो रही प्रगति के बारे में चर्चा हुई थी जैसे नाभिकीय भौतिकी, रेडियो खगोलिकी, मौसम विज्ञान, भूकंप विज्ञान, क्रिस्टल विज्ञान, आनुवंशिकी, कृषि विज्ञान तथा तंत्रिकार्यिकी (न्यूरोफिजियोलॉजी)। यद्यपि उस समय रमन अस्वस्थ्य थे, फिर भी उन्होंने सभी सत्रों में उपस्थित रहकर युवा वैज्ञानिकों का उत्साह बढ़ाया।

अंतिम घड़ी

अपने जन्मदिन से कुछ दिन पहले रमन अपनी प्रयोगशला में काम करते हुए अचानक गिर पड़े। उनकी शारीरिक अवस्था इतनी खराब हो गई थी कि उन्हें अस्पताल ले जाना पड़ा। उन्होंने चिकित्सकों से कहा—'यदि मैं संपूर्ण रूप से स्वस्थ्य नहीं हो सकता तो मैं जीना नहीं चाहता।' कुछ दिन इलाज के बाद हालत में थोड़ी सुधार होते ही उनके इच्छानुसार संस्थान में अपने निवास स्थान पर वापस लौट आए। परंतु बिस्तर से उठकर संस्थान के परिसर पर टहलना उनके लिए संभव नहीं था। उन्होंने अपनी मजबूरी जाहिर करते हुए कहा कि 'यदि मुझे पहले से पता होता कि मेरी मृत्यु यहीं होनी है तो मैं इस कमरे की खिड़कियों को थोड़ा नीचे बनवाता' यानी वे बिस्तर में लेटे ही संस्थान के परिसर में बने सुंदर बगीचों को निहार पाते। जल्द ही उनके बिस्तर का स्तर ऊँचा किया गया, जिससे वे लेटे-लेटे ही बगीचों को देख सकें। रमन की अंतिम तसवीर चित्र-6.6 में दिखाई गई है।

रमन को शायद इस बात का अहसास हो गया था कि अंतिम घड़ी निकट आ रही है। अब तक वे अकेले ही संस्थान एवं अकादमी का कार्यभार सँभाल रहे थे। उन्होंने बिस्तर में लेटे-लेटे ही अपने घनिष्ठ मित्रों से भविष्य की योजना के बारे में परामर्श किया। 'रमन अनुसंधान संस्थान' के बारे में उन्होंने अपने विचार लिखवाए—

'सन् 1948 में इस संस्थान का निर्माण मैंने मूल शोध करने के

चित्र : 6.7 वह पौधा जिस जगह रमन को दफनाया गया था

उद्देश्य से किया था तथा मैं चाहता था कि देश में एक उच्चकोटि की विज्ञान संस्था हो। उसी को ध्यान में रखते हुए इसकी प्रशासन व्यवस्था मैंने अपने ढंग से की है। यहाँ का वातावरण दूसरे संस्थानों से भिन्न है, जो मूल शोध करने की प्रेरणा देता है। विज्ञान का अनुशीलन मेरे लिए सुरुचिपूर्ण एवं आनंददायक अनुभव रहा है। यह संस्थान मेरा आश्रयस्थल रहा है, जहाँ मैंने अपनी रुचि के मुताबिक अनुसंधान किए हैं। जाहिर है, मेरे जाने के बाद इसकी कार्य-पद्धति में बदलाव आएगा। मैं चाहता हूँ कि यह संस्थान ज्ञान अर्जन करने का एक विशाल केंद्र बनकर उभरे जिससे देश-विदेश के वैज्ञानिक इसकी ओर आकर्षित हों। इसके सुंदर बाग-बगीचे, बड़ा पुस्तकालय, विशाल संग्रहालय आदि उच्च शिक्षा प्राप्त करने का केंद्र बनने में मददगार होंगे। विज्ञान तभी पनप सकता है जब इस क्षेत्र में काम करने की आंतरिक इच्छा हो। बाहरी दबाव से इसका विकास संभव नहीं है। विज्ञान में मौलिक अनुसंधान किसी के निर्देश व जोर-जबरदस्ती से नहीं हो सकता। यही कारण है कि मैंने तय किया था

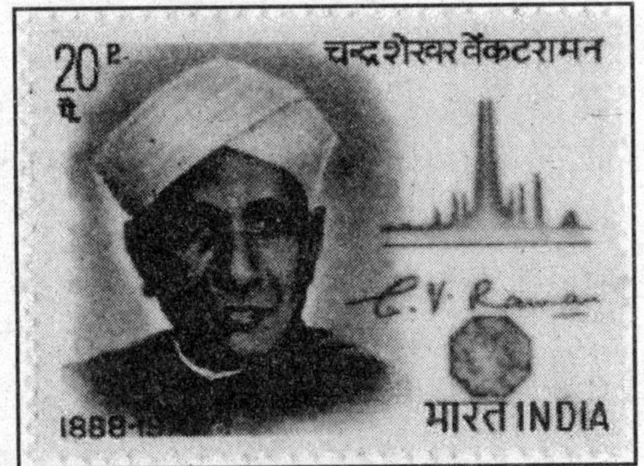

चित्र : 6.8 रमन के सम्मान में डाक टिकट

जहाँ तक संभव हो, सरकार से किसी प्रकार की सहायता न लेने का। मैंने अपनी सारी संपत्ति संस्थान के नाम कर दी है। परंतु शायद यह पर्याप्त नहीं होगा, जिससे संस्थान को ज्ञान उपार्जन करने के केंद्र के रूप में विकसित किया जा सके। अतः सरकारी सहायता न लेने की शर्त पर कायम न रहकर अब संस्थान को उन सभी से मदद लेनी होगी, जो बिना किसी शर्त के सहायता करने को तैयार हैं।'

इसके दो दिन बाद, 21 नवंबर, 1970 के तड़के 82 वर्ष की उम्र में शांतचित्त रमन ने सदा के लिए अपनी आँखें बंद कर लीं। उनके इच्छानुसार संस्थान में ही उन्हें दफनाया गया, जिसके चारों ओर फैले परिवेश में वे हमेशा खुश रहते थे तथा इससे उनका गहरा लगाव भी था। कोई धार्मिक अनुष्ठान भी नहीं किया गया। उस जगह एक पौधा लगाया गया था (चित्र-6.7)। वहाँ कोई स्मारक चिन्ह नहीं है। भारत सरकार ने उनके सम्मान में एक डाक टिकट जारी किया (चित्र-6.8)।

रमन ने विज्ञान के क्षेत्र में बिना किसी प्रलोभन के प्रवेश किया था, क्योंकि उन्हें इससे काफी लगाव था। रमन को प्रकृति की हर चीजों के बारे में जानने की अभिलाषा थी। आरंभ में ध्वनि एवं प्रकाश के रहस्य से आकर्षित होकर वे विज्ञान की दुनिया में तेजी से अग्रसर हुए। अपनी उपलब्धियों एवं आत्मसम्मान के बल पर उन्होंने विज्ञान जगत् में भारत का नाम रोशन किया। वैज्ञानिक संस्थानों का निर्माण,

वैज्ञानिक पत्रिकाओं का प्रकाशन तथा युवा-वर्ग को विज्ञान में कार्य करने के लिए प्रेरित करके उन्होंने देश में आधुनिक विज्ञान परंपरा की स्थापना की। उनका कहना था—'प्राचीन काल में हमारा देश विद्वता, दर्शनशास्त्र व विज्ञान के क्षेत्र में अपनी महिमा दिखा चुका है, परंतु अब ज्ञान प्राप्त करने के लिए निराशाजनक ढंग से हम पश्चिमी देशों पर निर्भरशील हैं। भारत को पिछलग्गू न बनकर विज्ञान के क्षेत्र में अग्रज होना चाहिए। हमें अपनी समस्याओं को चिन्हित कर उनका हल खुद ढूँढ़ना चाहिए।' एक बार छात्रों को संबोधित करते हुए रमन ने कहा था—'मेरे सामने बैठे हुए युवक एवं युवतियों को मैं कहना चाहता हूँ कि वे कभी आशा व हौंसला न छोड़े। सफलता तभी मिल सकती है, जब काम को साहस एवं निष्ठापूर्वक किया जाए। मैं यह बात दृढ़ता के साथ कह सकता हूँ कि भारतीयों का दिमाग उत्कृष्टता की दृष्टिकोण से किसी से कम नहीं है। शायद हममें हिम्मत की कमी है, उस प्रेरक बल की कमी है, जो हमें कोई भी काम करने के लिए उत्साह देती है। मेरे विचार से हम हीनभावना से ग्रस्त हैं। आज देश में फैली इस पराजयवादी भावना को नष्ट करने की जरूरत है। हमें जीत के जोश को जाग्रत् करना होगा, जो हमें हमारी असली पहचान वापस दिला सके।'

रमन मानते थे कि किसी भी देश का भविष्य उसके ज्ञान भंडार तथा युवा पीढ़ी पर निर्भर करता है। उनका कहना था—'देश का सबसे बड़ा उद्योग ज्ञान का उत्पादन व उसका वितरण है। किसी भी संस्थान व मनुष्य का सर्वश्रेष्ठ कार्य वहाँ की युवा पीढ़ी को तंदुरुस्त व मजबूत करना है, जिन पर देश का भविष्य निर्भर है।' रमन महिलाओं के कार्य के समर्थक थे। एक बार उन्होंने कहा था—'मैं समझता हूँ कि यदि भारतीय महिलाएँ विज्ञान के क्षेत्र में आएँ तथा इसके विकास व प्रगति में दिलचस्पी लें तो वे वह हासिल कर पाएँगी, जो पुरुषों ने भी हासिल

नहीं किया है। नारी में एक स्वाभाविक गुण है और वह है निष्ठा। विज्ञान में सफल होने के लिए इस गुण का होना अत्यंत आवश्यक है। अत: यह सोचना गलत होगा कि विज्ञान में काम करने का हक केवल पुरुषों को ही है।'

रमन में जिस प्रकार बालसुलभ कौतूहल था, उसी प्रकार प्रकृति के गूढ़ रहस्यों को समझने की अनुभूति भी थी। वे सच्चे अर्थों में एक महान् वैज्ञानिक थे। उनके शोध कार्य को आरंभ से छह दशकों में विभाजित किया जा सकता है, जैसे ध्वनि विज्ञान (1910-1920), प्रकाश प्रकीर्णन (1920-1930), द्रवों व मणियों में प्रकाश का बिखराव (1930-1940), हीरे व मोती पर अध्ययन (1940-1950), खनिज पदार्थ (1950-1960) तथा रंग बोध (1960-1970)। अपने जीवनकाल में उन्होंने 475 से भी अधिक शोधपत्र लिखे तथा छह शोधग्रंथों की रचना की।

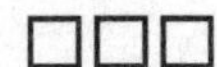